EMBROIDERED MACHINE NETS
LIMERICK AND WORLDWIDE

This book is a comprehensive study of the various methods by which plain machine nets were embroidered from the late 18th century onwards to create what was described in 1882 as 'a very effective and fashionable imitation of lace'.

Limerick in south-west Ireland was one of the most successful areas of commercial production, working for special commissions and for export. For nearly a century, nets were decorated with run, tambour and cut cambric embroidery.

Chapter 1 traces the history of Limerick lace from its origination by Charles Walker in 1829 to its extinction as a major trend-setter in 1914, and beyond that to its atrophied survival.

Chapters 2, 3 and 4 are practical accounts of the materials, methods of working and stitches used in the three techniques. They are based on antique laces in the author's collection and on records contemporary with their manufacture. The pages on which the various examples occur are listed in detail in the Contents.

During the period of Limerick's flourishing, machine nets were embroidered throughout the world, by hand and by machine. In ch.5. various geographical areas are studied, and some distinctive styles emerge, particularly in agricultural communities where decorative stitches take the form of mini-motifs rather than Limerick-style casket fillings. However the passion for domestic embroideries burgeoning in the second half of the nineteenth century has tended towards a loss of national identity.

The designs for working are intended to suggest novel ideas and encourage a blending of the three main techniques, and the mixing of thread colours and textures.

In ch.2 there is a brief overlap with the preliminary section of *Limerick Run Laces*, published in 1992. However, additional information is given, different examples have been chosen and many new stitches described.

The names given to the stitches are either descriptive or based on the embroidery stitches which they resemble. In general, the local names of the stitches, which vary with the site of production, have been omitted.

One technical term may need clarification:
Guipure = laces where the open areas are crossed by bars. These are strands of thread, usually strengthened by a covering of stitches. In the absence of a meshwork, the bars hold the motifs in place. The term can be applied to all techniques, whether bobbin, needle, cut cambric, crochet, or Schiffli chemical laces. Thus a lace referred to simply as *Guipure* leaves the question of its identity unanswered.

ACKNOWLEDGEMENTS

Photographic and other acknowledgements for the illustrations are given with the individual captions.

The following abbreviations are used for longer names:

DAR = The Museum of the Daughters of the American Revolution, Washington DC.

Metropolitan = The Metropolitan Museum of Art, New York.

Nottingham = The Museum of Costume and Textiles, Nottingham.

V and A = The Board of Trustees of the Victoria and Albert Museum, London.

R.B. = Ronald Brown.

The following details of acknowledgements are from The Museum of the Daughters of the American Revolution:

p.5. Object no. 510
Gift of Mrs Daniel H. Wetzel.

p.6. Object no. 54.10
Gift of Mrs Sydney Hale Minor.

p.166. Object no. 53.56.1
Gift of Mrs Ellen Sydney Kent, in memory of Florence Kent Harris, State Regent of Cuba, 1935-1946

p.207 (a) Object no. 2616.1
Gift of Mrs Harold B. Barney.

(b) Object no. 3819
Gift of Cecile S. Belford honoring Mrs E.T. Collins.

(c) Object no. 61.146.2
Gift of Mrs Clair Thompson.

p.208 (a) Object no. 2423.1
Gift of Mrs Abbie S. Sargent.

(b) Object no. 4660
Gift of Mrs Eugene L. Webber.

(c) Object no. 75.60.

In addition, my grateful thanks are due to:

Linda Ballard of the Ulster Folk and Transport Museum, Belfast; Jennifer de Simas of the Museum of the Litchfield Historical Society, Connecticut; Mairead Dunlevy of the Hunt Museum, Limerick; Jeremy Farrell of the Museum of Costume and Textiles, Nottingham; Sister Francis of St Clare's Convent, Kenmare; Marianne Gachter-Weber of the Textilmuseum, St Gallen; Elizabeth McCrum of the Ulster Museum, Belfast; and Monica Strassle of the Textilbibliothek, St Gallen, for so kindly making available for study the embroidered machine nets, and designs, in their collections;

to Marjolein Amphlett, Gillian and Michael Hartz, and Anne Swift for their very helpful translations;

and to Lysiane Brulet, Vibeke Ervø, Elise Hoste de Vynck and Olwyn Scott for directing my attention to areas of production I might otherwise have missed.

Except where otherwise credited, photographs and line drawings are by the author.

EMBROIDERED MACHINE NETS

****

LIMERICK & WORLDWIDE

PAT EARNSHAW

GORSE PUBLICATIONS · GUILDFORD

British Library Cataloguing-in-Publication Data

A catalogue record for this book is available from the British Library

ISBN 0 9513891 5 7 Hardback
ISBN 0 9513891 7 3 Paperback

Printed by Staples Printers Rochester Ltd, Rochester, Kent ME2 4LT

Typeset by Alastair Thompson, Paris, France
in Helvetica, Avant Garde and Garamond 3.

for Gorse Publications, PO Box 214,
Shamley Green, Guildford GU5 0SW,
England

Hardback edition of 500
This copy number: 231

Pat Earnshaw.

By the same author:

- The Identification of Lace
- A Dictionary of Lace
- Bobbin and Needle Laces, Identification and Care
- Lace in Fashion from the Sixteenth to the Twentieth Centuries
- Lace Machines and Machine Laces
- Needle-made Laces
- Youghal and other Irish Laces
- Youghal Lace, the Craft and the Cream
- Threads of Lace from Source to Sink
- Embroidery Skills: Needlelace
- Outlines and Stitches, a Guide to Design, with special reference to Halas Needle Laces
- Limerick Run Laces, an Introduction

CONTENTS

1. THE EMBROIDERED LACES OF LIMERICK

Names and Definitions

Limerick laces are defined as hand embroideries on machine net. Run, tambour, or cut cambric techniques are used.

The abbreviated form 'embroidered nets' should be avoided, since this also refers to the much older technique of filet in which a fabric of hand-knotted net is decorated by running or darning stitch.

The French word *tulle* could be substituted for 'machine net', but it has its own ambiguity. In English it is narrowed to 'silk machine net', while in French it may be broadened to include any meshwork, hand- or machine-made - or alternatively restricted, with questionable accuracy, to filet, since this net was in the past manufactured in the town of Tulle.

The naming of Irish laces by a geographical location was a late and reluctant introduction. Merchants favoured names which, regardless of truth, sold products. The Irish cut cambric technique for years referred to as 'Limerick' or 'Irish' appliqué and guipure, was named 'Carrickmacross' only in 1872. It was not unknown for Limerick run laces to be marketed as 'Nottingham' lace, and the chain-stitch embroideries of Coggeshall in Essex to appear in the shops as 'Brussels' or 'Limerick'. Lace names should thus always be viewed with an open mind, and stitches and techniques of manufacture are perhaps a more reliabe guide to geographical origin. We shall see.

Run
(Courtesy: Nottingham)

Tambour
(Courtesy: Nottingham)

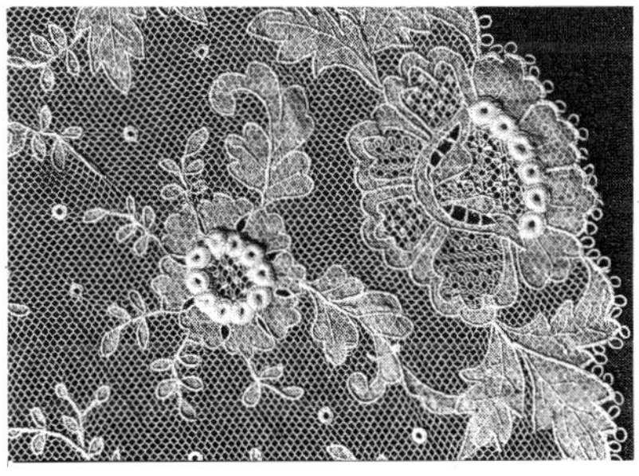

Appliqué

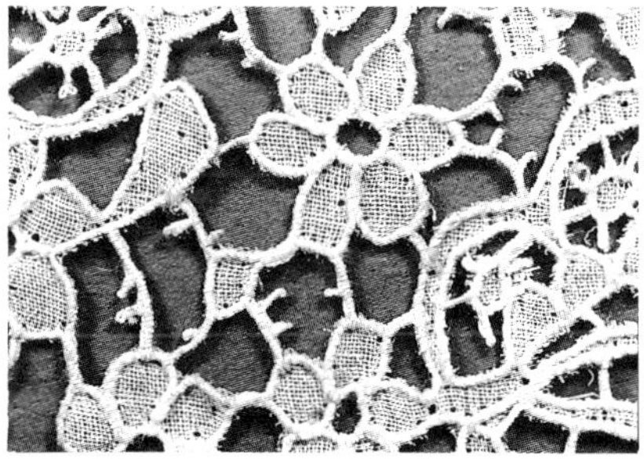

Guipure

Limerick techniques

Run laces are made by passing the sewing needle and thread in and out of the meshes of the net. Many of the decorative stitches bear a superficial resemblance to those of pulled threadwork, or more rarely of bobbin laces.

Tambouring refers to chain-stitch worked with a small hook which wraps around a length of thread drawing it upwards through the net. In this way a series of linked loops progresses rapidly across the surface leaving a trail of shapes which, nourished by the imagination of the embroiderer, can be almost limitless in variety. The work is considerably quicker than run embroidery and less of a strain on the eyes.

The two techniques, run and tambour, often appeared in the same piece of lace[1]. Run stitches might fill the decorative areas of a mainly tambour lace, or vice versa.

But, according to *Needlecraft,* c1902, 'this fashion did not appear to take'[2].

Cut cambric consists of design motifs, formed of a near-translucent woven fabric, which in the *appliqué* technique are stitched to machine net and, in the *guipure* technique, linked to each other by short buttonholed bars, i.e. there is no net.

Decorative fillings of the appliqué laces consist of running stitches, those of the guipure laces of needlelace stitches or needleweaving.

Appliqué and guipure techniques often occured together in the same lace.

Origin and History

Limerick laces began in 1829, but skilled needlewomen had long existed in Ireland, and the later techniques of Limerick were already anticipated by the mid-18th century. The Irish were particularly famous for their 'sewed' or 'worked' muslin. In 1755 the Dublin Society offered a prize of £5 for 'the best Dresden work'[3].

This type of embroidery, originating in Saxony, was a delicate pulling together of the warps and wefts of a flimsy woven muslin in a sumptuous variety of ways. Shadow-, chain- and run-stitches also occured. The meticulous counting of threads which this technique required was easily carried over to the counting of meshes in Limerick embroideries of machine net.

Other Irish 18th-century forerunners were formal run-stitch patterns rimming the borders of large handkerchiefs of muslin or filmy lawn, a technique inspired perhaps by the popular brocaded muslins imported from India[4].

Cambric appliqué on net was foreshadowed by a domestic occupation: 'To appliqué with net, muslin or cambric was a favourite work in England during the latter part of the previous century (18th) and the first years of the present, and the work so made was largely used in place of lace, the foreign laces of that period being subject to so heavy a duty as to render them only within the reach of the wealthy'[5].

Forerunners of tambour work in Ireland are less obvious, though another prize, also offered by the Dublin Society in 1755, indicates that 'cheyne [presumably chain] lace' was manufactured[6]. This

Above: Dresden work

Below: Run on muslin

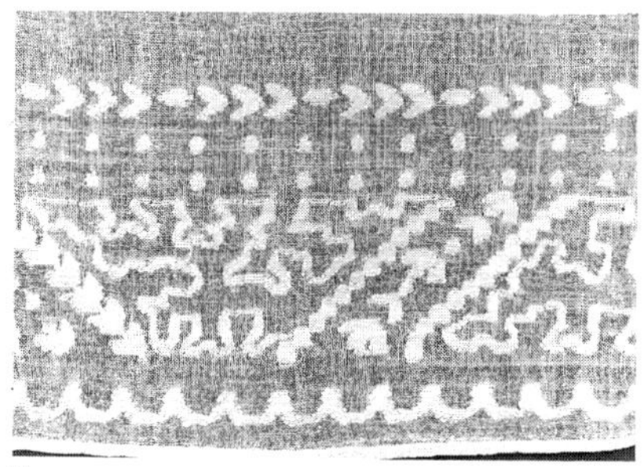

Muslin-on-muslin appliqué

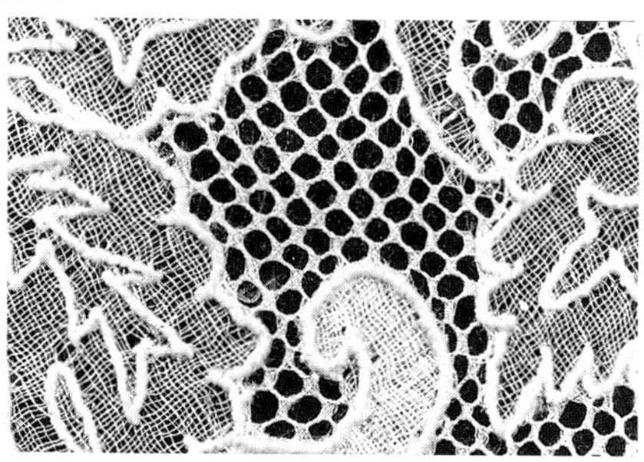

would be not an embroidery, but a true lace independent of any backing fabric, with the stitches attached only to each other, similar to that being produced in Spain in the early 17th century[7].

Chain-stitch embroidery on silk gauze or fine cotton muslin reached Europe from the Far East during the 1750s, the loosely-woven texture of these cloths providing a light openwork appearance. By the 1760s tambouring was a popular time-passer for the ladies of the French Court quick to realise its potential for elegant poses of the head and hand. The original sewing needle was replaced by a fine hook which greatly speeded the process. By 1787, muslin-tambouring was a commercial venture in Scotland, and shortly afterwards at Coggeshall in Essex[8]. Like other Scottish embroidery techniques, the industry spread to Ulster.

Above: A lady tambouring. Note the round frame. Angelica Kauffman, c1780 (Courtesy: Christie's)

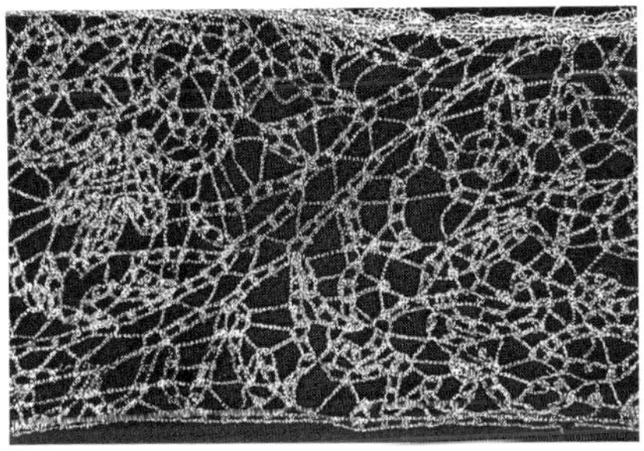

Cheyne lace

Below left: Tambour on muslin

For the Irish people, the first two decades of the 19th century were a time of great poverty. The Home Rule Bill of 1800 had drawn over 86% of the landed gentry back to England, taking their rents and patronage with them[9]. One of those who remained, and sought to relieve the misery of the country folk, was Lady O'Brien who played a major role in preparing the soil for the implantation of true Limerick laces. In 1822 she established a school at her demesne at Dromoland (pronounced Dro-mo'-land), a short distance to the north-west of Limerick, and there revived satin stitch embroidery, sometimes known as 'Irish work', teaching in all some 400 children, who were soon to form a core of eager, skilled, attentive and experienced needlewomen[10].

Thus the ground was ready for the seed, and in 1829 the sower arrived, in the person of the 'highly intelligent and enterprising' Charles Walker. Though a native of Oxford, he had married a lady who was mistress of an extensive lace manufactory in Essex[11]. He thus came into contact with the local tambour industry, now extended from muslin to net[12], which had reputedly been introduced into Coggeshall by a French emigré, Drago, during the Napoleonic Wars.

Walker himself, around 1823, set up a lace school at Ongar, taking six children from the surrounding area as apprentices at a premium of £4 each, with the intention of keeping them on as workers until the age of 21[13]. His name appears in *Pigot's Directory* for that year.

Plain machine nets had by this time been in existence for several decades. Weft- and warp-knitted nets (patent nets) in white or black silk had been on the market since before 1800, and cotton two-twist bobbinets were invented in 1808. Compared with totally hand-made laces such fabrics were inexpensive, and even when decorated cost only one-sixth to even one-tenth as much.

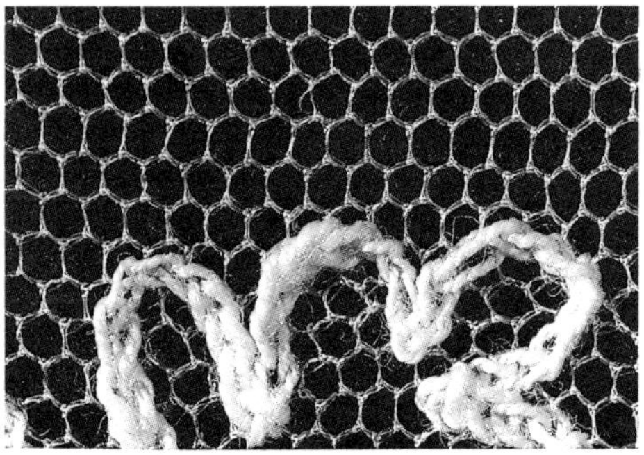

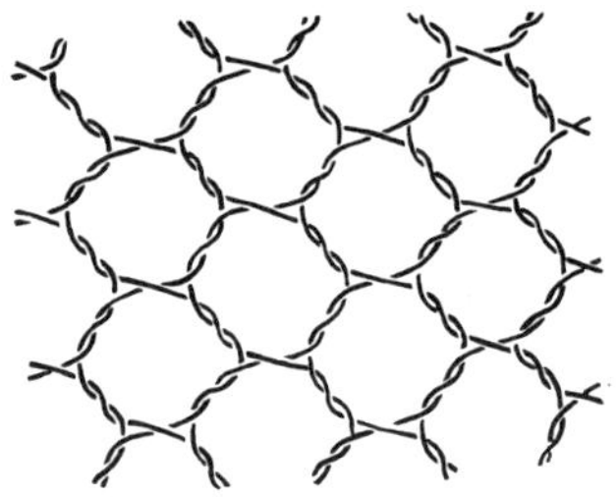

Top: weft-knitted net (Photo: RB)
Middle: Warp-knitted net
Below: Two-twist bobbinet

Charles Walker's school appears not to have been a success. Nevertheless, his name appears in Pigot's Directory for London, Middlesex and Essex in 1826-7, as the occupier of a retail warehouse in Cheapide, East London.

It was the custom for provincial manufacturers to have distributing centres in the City and John Heathcoat, the inventor of bobbinet, also had offices there, though his main business was located at Tiverton in Devon.

It was possibly through this connection that Walker became acquainted with the decoration of machine nets by run embroidery, Nottingham being at that time the centre of a huge industry in the 'running' of veils, aprons, collars and other accessories, employing thousands of outworkers.

These skills of run and tambour embroidery Walker took with him across the Irish Sea in the hands of some two dozen Englishwomen[14]. His choice of Ireland for the setting up of his manufactories was probably influenced by the cheapness of Irish labour[15], but his selection of the city of Limerick undoubtedly rested on the established success of its delicate white embroideries and the dexterity of its workers.

He first made a tour through the country to acquaint himself thoroughly 'with the capacities, habits and conditions of the people'. Then, after careful consideration, bought a large building at Mount Kennet, near the water front, at a cost of £20,000[16]. This scheduled factory combined the essential requisites of ample room, good light and thorough ventilation[17]. The serious production of Limerick laces was about to begin.

Walker's factory soon acquired a reputation for being well run, and he himself was so confident of the quality of his products that he is said to have offered 'a large wager that he would select 100 Irish girls from among his workers who would produce any given piece of lace, to be wrought in a state superior to any similar work to be made by the like number of girls to be found in France, Flanders, Saxony or Germany'[18]. His boast indicates the existence at this early date of similar manufactures in four other European countries. Later in the 19th century the tally could be extended to Spain, Italy, Denmark and Switzerland though, through a paucity of documentation, and the apparent restriction of the manufacture to a domestic level and internal trade, historical evidence concerning them tends to be vague.

During the fourteen years between the setting-up of Walker's factory in 1829 and his death in November 1843, the total number of 'young persons of the softer sex' engaged in embroidering nets rose to 2,000[19]. At the same time the manufacture of Limerick laces spread to the convents and cottages of the nearby counties of Kerry, Cork and Clare, with St Joseph's School attached to the Convent of Mercy, Kinsale, being a particularly important centre. The name 'Limerick' was retained for these disseminated productions.

Walker sold his business in 1841 but the purchaser becoming bankrupt, no money was forthcoming and he suffered the sad fate of so many other hard-working innovators, dying in poverty[20].

Following Walker's death, the factory was taken over by his friend Lambert[21]. Another associate, Forrest, set up his own business. But even during

Walker's lifetime there had been problems of social unrest. A historian writing of 1834 reported: 'The lace trade which has recently been created in Limerick and gives employment to about 1,300 young people, was in jeopardy, by reason of a projected combination and dispute among the work people'[22]. The incipient sluggishness in the economy was shortly counteracted by the shocking famines of 1846-7 which drove women to seek work in the hope of saving their families from starvation. A voluntary organisation, The Ladies' Industrial Society for Ireland, was established in Dublin in 1847 to market the increased output. Unfortunately it was a hybrid of benevolence and ignorance, untutored in commerce, and it failed within a few years[23].

Nevertheless Limerick laces made a good showing at the Crystal Palace Exhibition of 1851[24], while at Cork in 1852 they formed 'one of the most important and attractive features of the entire Exhibition'[25]. In 1855 Forrest & Sons won a second class medal at the prestigious Paris Exhibition - which also featured the Empress Eugenie's dress of Alençon needle lace, and gold embroideries by Worth[26]. Fashionable patterns featured pine cones, copied from the Kashmir shawls and their British imitations[27].

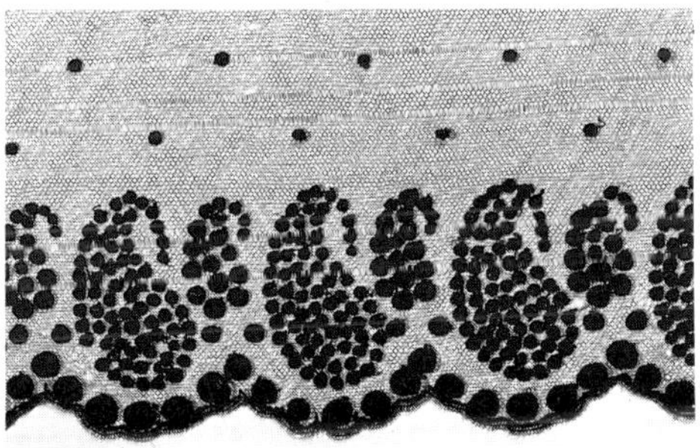

Black silk cones on point net (Courtesy: D.A.R.)

Tambour work apparently did not share in this explosion of regard: 'Tambour work, once so fashionable, is now so entirely disused, that we need only observe it was done with a small needle, resembling a very fine crochet hook, and the pattern worked on muslin or net in chain-stitch'[28]. This decline was undoubtedly linked to a caprice of fashion which swung towards broderie anglaise, favouring in general an aggressive ostentation which displaced the subtle delicacy of embroidered nets, damaging these industries not only in Ireland, but in all the centres in France, Germany and Scotland, grossly reducing the number of workers and causing wages to be cut by half[29].

At the same time, attempts to meet the sudden popularity of Limerick run resulted in goods being too hastily produced, and insufficient attention given to quality and design. Sales diminished, and boom was followed by slump. The temptation to the workers was obvious - if they could earn more by quickly producing 'twelve bad collars for 5d [2p] rather than one good one for 5s [25p]' it was natural for cottagers, given their near-destitute circumstances, to do so, although as Ben Lindsey, founder of the Irish Lace Depot in Dublin pointed out, '12 inferior specimens of work can more readily alienate 12 patrons than one good specimen can do to regain them'[30]. Thus the later 1850s and the 1860s were a period of decline.

Poverty of design was scathingly attacked by Susannah Meredith, redoubtable historian of *The Lacemakers*. Speaking of the setting up of Government Schools of Design in the early 1860s, she refers to their futility when 'the instructions afforded convey no information of the introduction of the artistic element'. There was no system of inspection, and the pupils were left to their own devices[31].

Some further idea of the then prevalent attitude towards design was given by the state of the sewed muslin industry (a blanket term covering satin stitch embroidery and pulled threadwork) which also, in 1865, was suffering a severe depression. In desperation the workers implored Queen Victoria to accord them her royal patronage and so restore the embroidery to public favour. She replied sternly 'that the commodity should be better cultivated', and directed that the best instructions, and the newest patterns, should be sent for to France for this purpose[32].

Many of the best lacemakers emigrated to America[33], or took up less skilled but more lucrative jobs in the jam factories[34]. In 1859, Fraser's *Handbook for Travellers in Ireland* said of Limerick: 'Its manufactories are very limited; they chiefly consist of cotton spinning, blonde and lace'[35]. By 1869, fewer than 500 Limerick workers were still engaged in embroidered nets[36]. But Limerick was not the only lace to have fallen into disrepute. Nearly all the hand-made laces had been hit by the application to the Pusher and Leavers machines of the Jacquard automatic patterning device following, around 1840, the successful patterning of the warp-knitting machines a decade earlier[37].

In addition, the recurrent Court mournings, particularly prolonged after the death of Prince Albert in 1861, favoured black laces, and while the colour of the lace made not one iota of difference to the machines, for human workers the dark colour strained the eyes and slowed the work. Examples of elaborately beautiful run and tambour laces in black silk do exist, but they are rare.

Black tambour embroidery on point net (Courtesy: D.A.R.)

Another adverse factor was the American Civil War of 1861-5 which caused not only a famine of raw cotton and thus a shortage of thread, but closed an extremely profitable export market.

Further, in 1865, the Cornely embroidery machine was invented[38] and at once began to imitate tambour work with frightening verissimilitude.

Susannah Meredith, an unashamed crochet-phile, wrote in 1865: 'Limerick lace, a species of work which had almost run its course before crochet had begun . . never attained any high degree of cultivation; and none of it is now as good as similar lace produced elsewhere'[39]. Cole threw a typically cogent light on this topic. Referring to Limerick work at the Convent of Mercy, Kinsale, he wrote:

> 'Thirty years ago [i.e. c1860] this convent did much in the locality towards organizing the peasant and fish-women's cottage work, and for a time success was secured for it. However a rage ensued for making crochet - a class of thread work more easily made than embroidery on net. "Wanton-eyed" women standing at their doors and chattering with anyone who would stop and talk, greatly favoured crochet-making. Crochet was readily bought up, but as quickly earned a bad name - not merely on account of its artistic deficiencies, but also because of its socially demoralizing effects . . '[40].

By 1879 the trade in Limerick laces had almost died out, and that which remained was so degenerate as to be 'extremely unattractive', of a coarse and inferior kind, made 'principally for an export trade'[41], that is outside Britain, intended mostly for the American market regarded, at that time, as incurably provincial. Bowles remarked in 1889: 'The material latterly used is much coarser and more suitable for furniture than dress'[42].

The slump continued into the 1880s, fostered by the disastrous coincidence of inadequate training and supervision with a decadent taste in fashion 'when women are content to smother their costumes in mile after mile of common cheap laces which would have filled their grandmothers with horror. Quantity and not quality appear to be the desideratum of a large majority, while many women cannot distinguish machine-made from hand-made lace'. 'Probably fewer than 300 Limerick laceworkers are now active, with not a single young girl in the city learning to make lace as a means of earning her livelihood'[43].

But undercurrents of the Arts and Crafts Movement, begun in 1875, were slowly rising to the surface. In 1883, Mrs Robert Vere O'Brien, related by heredity and marriage to both Lady de Vere of Curragh and to Lady O'Brien of Dromoland, set up a lace and embroidery school in Limerick. She improved the methods of work, the quality of the materials, the designs and the marketing process. She introduced her own original designs, rich with Irish flowers of the hedgerow such as dog roses, fucshias, ivy leaves, daisies, bryony and woodbine, all of near-naturalistic form and enlivened by the popular coiling tendrils. The clear open form and simplicity of outline of her drawings contrasted advantageously with the heavily over-detailed laces cascading off the machines as if manufacturers quailed in horror that some gap might appear in the closely cluttered fabric. Mrs O'Brien also adapted for Limerick work designs from the early 19th-century Alençon needle laces of France, such as a flounce made for the Empress Josephine, first wife of Napoleon I. At the same time the Limerick School of Art instituted prizes to stimulate apprentice designers[44].

Mrs O'Brien was ably assisted by the female penitents of the Convent of the Good Shepherd, where a group of Belgian nuns had set up a practical industry in the mid-19th century. Some 20% of the inmates were taught lacemaking, including Valenciennes and Brussels bobbin laces, but they could turn their hand also to embroidery on net. The appearance of strongly Brussels-based designs among

A lace fan worked under Mrs Vere O'Brien's supervision, after a design from the School of Art, Cork (Courtesy: V and A)

Limerick laces may perhaps be attributed to the influence of these sisters, though perhaps alternatively to that piracy of designs from the Continent so firmly recommended by Queen Victoria. From her Limerick school, Mrs O'Brien sent teachers to Tralee and Cork, thus extending the industry[45].

However, it was the dynamic Head of the Cork School of Art, James Brenan, enlisting the help of firebrand Alan Summerby Cole of the South Kensington Museum, who entered most vigorously on a program of reform. Having unhesitatingly declared Limerick laces to be 'irretrievably moribund', Cole then set about galvanising them into a spectacular resurrection. His electrifying series of lectures, begun in 1884, pointed out the failures of the production system, the defeatism of dependence on charitable marketing, and the essential need to keep a keen and observant eye on the vicissitudes of fashion.

Brenan wrote in 1883, 'The hand-made article can never expect to keep its place in the market, unless it can prove its superiority to the machine-made work. If the worker is merely a lace-producing machine, with no desire to express beauty of form in the material, then the machine is bound to conquer, for it will certainly outrival her in accuracy and neatness'[46].

Ben Lindsey of the Irish Lace Depot, contrasting the continued success of Belgian laces with the failure of the Irish, pointed out that 'the Belgian Government have lace-making and drawing taught in the public schools'[47]. As early as 1851 there were 369 lace schools in Belgium, with a total of over 17,000 pupils[48]. They were, at the same time, critisised for their sweat-shop production. Bowles went further, with a clarion call to the women of Ireland to 'stand shoulder to shoulder', to lay their wants before the House of Commons, to get 'the special training which it is the duty of the State to furnish'. In addition, he agreed with Cole that inefficient marketing was an impedimentary feature: 'There is, in Paris alone, a sufficiently large market to take in all the hand-made lace in Ireland'[49].

By the late 1880s, a major upturn was on the way. The reorganised Schools of Art, often attached to Convents where the laces were made, produced specialist designers such as Michael Hayes (Limerick), Emily Anderson (Cork), Alice Jacobs (Metropolitan, Dublin) and, a little later, Cecilia Keyes of the Convent of Mercy, Kinsale. Following a three-year scholarship to the School of Art in South Kensington, Cecilia returned to Kinsale where, oblivious of changing trends, she drew plants of an idealised naturalism, and continued to design Limerick laces until her death in 1936[50]. Amazingly beautiful designs in Chinese white survived until 1992 at St Clare's Convent, Kenmare, for both Limerick and cambric appliqué laces. Limerick lace was made, additionally, at the Dominican Convent, Cabra, and at the Convent of Mercy, Golden Bridge, Dublin[51].

In 1887 the Dublin Horse Show set aside a space for an Art Industrial Exhibition, complete with prizes for the best work, and this practise continued until 1923[52], when it was replaced by a National Art Competition featuring paintings and drawings. In

1967, as interest in making lace began to awaken from its long torpor, the Royal Dublin Society Crafts Competition was reinstated, and continues to this day to display Irish laces[53].

Design by Cecilia Keyes, Convent of Mercy, Kinsale, March 1912 (Courtesy: Nat. Mus. Ireland)

The Irish Lace Depot in Grafton Street, Dublin, was bought in 1887 by the Countess of Aberdeen, following the death of Ben Lindsey. The Countess further set up, after the Edinburgh International Exhibition of 1886, the Irish Industries Association. In 1893 she master-minded one of the Irish villages created at the Chicago World Fair, to publicise Irish work[54].

In 1910, at the fourth Exhibition of the Arts and Crafts Society of Ireland and Guild of Irish Art-workers, held in Dublin, a Limerick fan leaf, flounce, scarf and bridal dress were displayed, as well as a Carrickmacross collar, all designed and worked by students of the Dublin, Cork and Cabra Schools of Art and the Convent of St Louis in Carrickmacross[55].

Ravishing art nouveau designs now emerged, and demand once more began to exceed supply, but this time the market was an extremely prosperous one and a high quality of materials and workmanship could be sustained.

An art nouveau stole with grapes, ivy leaves and shamrocks

In 1904, both Limerick and Carrickmacross featured in *The Times* list of the six most popular 'real' laces. The writer added 'though when it is possible to obtain old lace, that is always worn in preference to any modern creation'. For ball and evening gowns and evening blouses the preferred laces were either Limerick or old French Alençon and Argentan - two needle laces at that time being resurrected in antique style by the firm of Lefebure in Bayeux[56].

In 1906, Liberty's catalogue of *Choice Examples of Inexpensive Lace* showed splendid fichus, scarves and wraps of run and tambour laces, though without indicating their place of origin. Limerick laces were among those specially created for the Vice-Regal ball of 1907, at which the wearing of Irish lace was obligatory. Between 1898 and 1909 the Paris designer Paquin created gowns expressing Irish themes and trimmed with Irish laces, such as the luxurious *Habit d'Irlande* created for the Irish Grand

A design for Limerick lace painted by Emily Anderson. Note the outline pattern on the right, and the filled-in design to the left. c1888. (Courtesy: St Clare's Convent, Kenmare)

Steeplechase of 1904. Limerick lace formed the ballooning sleeves, the elegant bodice, the double flounce, and the long ties spreading over the protruberant back of the ankle-length skirt[57]. In 1908, Robinson and Cleaver of Belfast produced a catalogue of *Irish Hand-made Lace*, 'the beauty of which needs no words of recommendation'. Some fine examples of Limerick run and tambour, and 'Carrick-me-cross' fichus, shoulder scarves, plastrons, berthes, handkerchiefs and table centres were illustrated.

At the *Daily Mail* exhibition of the same year, Limerick laces from Kinsale were shown by Messrs Hayward of Old Bond Street. Cannock & Co, by now the largest manufacturers of Limerick lace, showed tambour, needlerun, and Carrickmacross guipure and appliqué in the form of handkerchiefs, scarfs, collars, berthe, yokes, blouses, bridge coats and bridal veils. Similar entries came from the County Cork Industrial Association, while from the Countess of Dunraven came designs of Limerick run lace, old tambour Limerick and samples of 'silk run lace copied from foreign laces'[58].

These references to 'Carrickmacross' and 'silk run lace' are both of interest. Just as Limerick laces were made not only in Limerick but in several neighbouring counties - and even, from 1850, as far away as Newry, Co. Down[59] - so the making of Carrickmacross appliqué and guipure was not restricted to Co. Monaghan, but extended to any centre capable of satisfying the market demand[60].

The Monaghan industry, begun in 1820, is said to have been inspired by old Italian work. Though not openwork, and in no way a lace, onlaid appliqué exploited the idea of stitching or glueing one layer of fabric to another, then cutting away all parts of the surface layer superfluous to the design. Another source of inspiration may have been the muslin-on-muslin appliqué popular in England in the second half of the 18th century, where the meshwork around the two-layered motifs was formed by the counting and pulling together of warps and wefts in a regular manner.

Limerick's cut-cambric productions may have been imitations of Monaghan work, but they may equally well have arisen independently when Lady de Vere of Curragh Chase (pronounced *Cu'rra*), ten miles to the south and slightly west of Limerick city, while studying a piece of Brussels bobbin lace appliqué in her possession, saw how its design areas might be represented by a machine-woven fabric and the surrounding openwork by machine net, thus enabling similar designs to be created with a saving of months of work and a considerable amount of money[61].

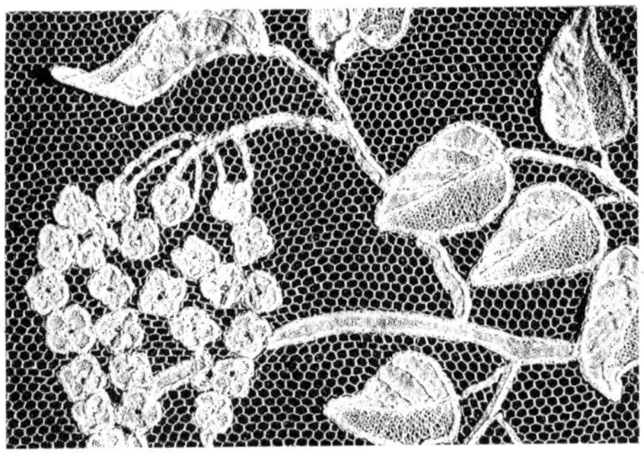

Brussels bobbin appliqué on droschel

Boyle puts the origin of Limerick cut-cambrics slightly later, at the time of the famines of 1846-8 when, 'in the desperate pursuit of buyers, the Limerick centres at various times copied appliqué and guipure work'[62]. The definition of Carrickmacross in Caulfield and Saward, 1887, appears to give priority to Limerick: 'The appliqué is worked on net like that made at Limerick, but the designs are better and more elaborate'[63].

In the *Illustrated Exhibitor* of 1852 the technique was not attributed to any particular area: 'Muslin appliqué is another style, and is at present very much in vogue. It consists of muslin or cambric laid on net, the design being in the thicker material, and the ground in net. Ornamental stitches are often introduced in this kind of work'.[64]

The obscure term 'Curragh lace' is obviously associated with Lady de Vere, but it is unclear whether the fabric applied was cambric or a second layer of net. The term appears in various references, but without definition.

By the time of the Dublin Exhibition of 1853, the better Limerick workers were able to make 'some of the finest guipure produced in Ireland'. Ben Lindsey, in 1886, acknowledged that 'Guipure has been made in Limerick quite equal if not superior to any of the Carrickmacross schools. Yet for some cause or other the production of this class has almost if not altogether ceased'[65]. A possible explanation of its fall from favour may have been its cost. The price of a 12-inch deep flounce of Carrickmacross guipure, quoted by *The Times* of December 3, 1904, was twice that of Carrickmacross appliqué, and four times that of Limerick (run or tambour not specified).

As for the 'silk run lace copied from foreign laces', mentioned in the *Daily Mail* catalogue, the original foreign laces were almost certainly the Spanish blondes, in which massively lustrous ivory or black flowers were set in a filmy meshwork of similarly coloured thread. Laces of this type were made extensively in Catalonia and Valencia between 1775 and 1875[66], and in Normandy for the Spanish market during the later 19th century. In both areas less expensive imitations, made by run embroidery on silk net, were produced. A run lace, made in Kells, reputedly about 1830, is shown in the accompanying photograph. Though the embroidery is in silk, the base is a cotton bobbinet[67], and the general impression is of the lighter *blonde de Caen* from northern France. Silk bobbinets would have been available at that time, though expensive. According to Cole, Limerick run laces sought to imitate blonde, though the similarity is by no means obvious. Certainly black blonde run in heavy embroidery silk on silk net, was made in Limerick in the late 19th century.

An intriguing adjunct to this was provided by an entry in *The Illustrated Exhibitor* for 1851: 'The trial of black lace embroidered with colours has been less happy: this may be an innovation destined to obtain

This Kells lace, from Co. Meath, won a silver medal in 1834. (Courtesy: Nat. Mus. Ireland)

some success in the colonies . . The darning thread is a thick loosely-spun silk, the pattern almost like satin on the nearly transparent ground'[68]. The geographical origin of this specimen is obscure, and though Limerick occasionally produced black silk run on white net, and black net appliquéd with white cambric, there is no evidence of the use of more garish colours.

Some indication has already been given of the high regard in which Limerick laces were held during the first decade of the 20th century (p.8). *The Times* of Dec 3, 1904, reported: 'Carrickmacross is increasing in favour almost every day' and, on Dec 6: 'The efforts which have been made in recent years by the Government and by private persons [notably the Countess of Aberdeen] to improve the condition of Irish industries has had the effect of very materially increasing the manufacture of lace in that country for which there has been, and still is, a great demand'. The article continued, 'It is doubtful whether at any period since the manufacture of lace became general, it has been in greater favour than it is at the present day, when to possess a good collection of lace with many ladies is an even greater ambition than to possess fur or jewellery'. During that year Dickens and Jones sold a total of 1.5 million yards of lace - 'every month enough is sold to link London to Rome', while in the USA buyers at the Irish lace counter of a leading department store were 'thick as bees about a bung-hole'[69].

The extraordinary popularity in the following years of 'waists', 'yoke-vests', 'neckwear', motoring veils, flounces and over-blouse tunics reaching sometimes to the knees, favoured the embroidered machine nets of Limerick. Flounces in particular were likely to appear at the most unlikely social occasions and in the most unexpected bodily places. Indeed: 'This summer a dress without lace garniture of some kind is hardly worthy to be called a dress at all'[70].

Above: The end of a stole, black run on white net (Photo: RB)

Left: a flouncing in 'imitation hand embroidery', highly popular for the races of Longchamps and Deauville. Note the familiar floral bouquets. Oct. 1912

Right: More lace at the races. Embroidered machine net flounces at Auteuil, March 1911

However a glance through the pages of the American *Lace and Embroidery Review* for 1911 to 1913 reveals that the situation was not quite simple nor entirely happy. As *The Times* had said: 'This is an age of artistic imitation and there is very little that is scarce or valuable that has not been skilfully counterfeited'.

Ruthless and unrelenting competition came not only from the three embroidery machines - the Handmachine, Schiffli and Cornely; and the two patterning lace machines - the Pusher and Leavers; but from foreign manufacturers shamelessly copying Irish designs and techniques.

The excellent quality of the Plauen embroideries on net (sometimes called Saxony laces), and of the St Gall guipures - both emanating from the steam-driven Schiffli machine forever extending its length and its sophistication, as well as the delicate bobbin finings of diaphanous silk or cotton created by the Leavers, provided such elegant substitutes that they were often hard to distinguish from - and might even appear more attractive than - 'real' lace. Further, since the genuine article was both slower to make and likely to cost three times as much, and its eclectic cachet almost certain to be unappreciated by the ignorami who, vastly outnumbering the cognoscenti, would fail to recognise themselves inferior in face of the proud display of a status symbol which they did not understand, its purchase was rendered to some extent pointless.

Plauen lace showing an imitation run design

A Harrods catalogue of the period lists lace robes in 'imitation' Limerick, needlerun black lace scarves and fichus, and guipure and Saxony (i.e. Schiffli) lace collars, while its *Album de Luxe of Fashion*, 1913, describes an Empire Tea Gown of satin, 'daintily trimmed with Fine Tambour and Tinsel lace'.

St Gall lace imitating Carrickmacross, new in 1912

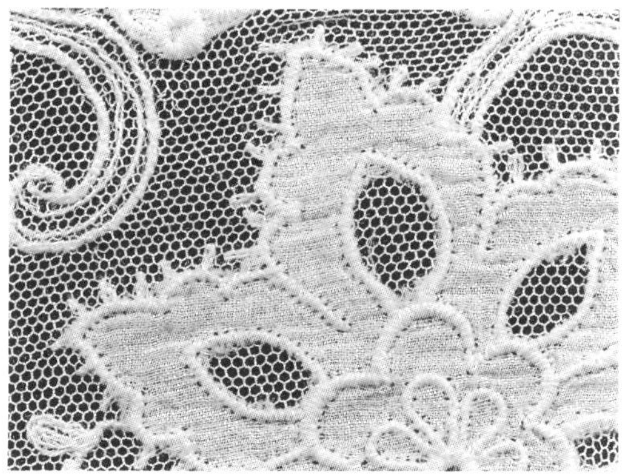

Cut fabric attached to net by Schiffli blatt stitch

Deliberate attempts to copy by hand or machine all the forms of embroidered bobbinet associated with the name Limerick were natural enough while the laces remained popular and there was a market to be invaded. In 1911 the firm of Colin C. Duncan of New York City was marketing 'at a reasonable price' 'artificial Carrickmacross' which up to then had been 'almost impossible to make by machine'[71]. Although the cutting away of the applied fabric would still need to be done by hand, labour was cheap. There would be no couched outline thread, but since both the Cornely chain-stitch and the Schiffli zigzag (blatt) stitch gave a very adequate attachment of fabric to net, and designs could be quite closely copied, the facsimile - bearing in mind the vast quantity of lace to be assessed on every socialite - could still be safe from discovery.

The imitation tambouring produced by the Cornely had flourished since the mid-1860s. It was so identical to the hand technique that even with close scrutiny it was often difficult to be totally certain of the origin. The machines tended to use linked-up patterns to avoid constantly cutting and restarting the thread, their laces lacked the knots of thread-joinings on the reverse side occuring in the hand-made form, and they were usually worked in thicker materials to survive the strong tensions generated by mechanical action - but these were subtleties that might go undetected. As a result, hand-tamboured nets never truly flourished on a large commercial scale after their early decline, and the run embroidery was generally regarded as more valuable, presumably because its machine imitations were less attractively convincing. The run work also cost more since meshes had to be counted, and it therefore took longer to make. The introduction to *Needlecraft's* instructions for making Limerick lace, c1902 says that tambour lace was 'at one time the most popular style of Limerick lace. Although much beautiful work is now done in this style, the run lace is the favourite'. Some excellently designed and executed pieces of tambour work continued to emerge from Mrs O'Brien's school in Kinsale during the 1890s.

A Cornely embroidery imitating hand tambour work (Photo: RB)

As for the imitation of run laces by machine, the Schiffli could bind strands to the surface of the net using its shuttle threads, but these were quite easily discernable.

Far more convincing imitations could be made by the Handmachine, first invented in 1829. This worked by passing double-pointed needles in and out of the holes of the net in an ingenious mimicry of human manipulation. Like the Cornely, it was restricted to stouter thread. Its 'darned' motifs were often enlivened by decorative stitches worked by hand. Some Swiss machines with pantographs, accompanied by engineers and designers, were set up in County Londonderry in the north of Ireland in 1913[72], but they were used to embroider fine muslins or lawns, rather than net.

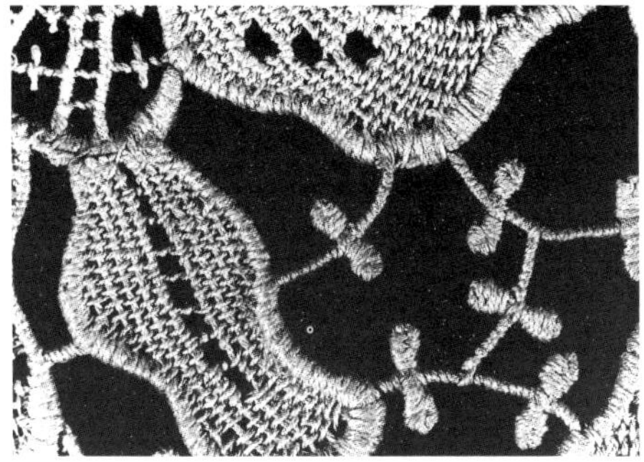

A guipure showing how well the handmachine could darn (Photo: RB)

In 1912, the Vienna-American Export and Import Company of New York posed an aggressive challenge:

> 'As an answer to those complaining that Irish laces must be made in Ireland, since this type of lace made elsewhere is having the greater sale, they should also insist that Vienna Rolls be made in Vienna only, Swiss cheese in Switzerland, and Coney Island frankfurters in Frankfurt . . We will give $10,000 in cash to anyone finding in our line one piece of any type of lace other than Real Irish Lace made in Austria, by Austrian girls'[73].

It is unclear whether this refers to Limerick laces or to Irish crochet, then being made in Hungary[74], and the rage of Paris - its speed of manufacture, relative cheapness and bombastic effect giving it dramatic and unmistakable social impact.

From the late 1890s, heavy and light laces tended to be worn together in a somewhat dizzying alliance. Thus the very popular method of decorating machine nets - with jet, glass beads, sequins (paillets), bugles or tubes, shiny metal thread and multi-coloured 'wood silk' (rayon) - were no obstacle to Limerick's success. Nor were the widely promoted enticements to middle class ladies 'of refined taste and deft fingers' to make for themselves run, tambour and cut cambric work[75].

The year 1913 produced no evidence of a decline in the demand for Limerick and Carrickmacross laces, even for high occasions, yet the fraudulent imitations, frequently passed off as Irish even in Ireland itself, foreshadowed doom, and the introduction of a national trademark was discussed, but not pursued[76].

In design, birds and animals were no longer popular for personal accessories, butterflies alone being tolerated.

In a report from the Department of Agriculture the author, Miss Anderson, considers the training of teachers and the organisation of classes; she rejoices that children under 12 years of age are no longer employed; and reflects on the advantages of the amateur over the professional lacemaker: 'Of course amateurs can follow their own fancy untrammelled by any consideration for the market, but industrial workers cannot neglect any prejudice which affects the commercial side of the industry'[77]. With regard to design, she says, 'Foreign copyists and machinery have combined to hackney and cheapen many of the Irish trade patterns, and it is by the development, variation and improvement, constantly carried out, of patterns that outside competition can be met and successfully checked'.

While the fate of Limerick lace might, in 1913, be regarded as balanced on a knife edge against its accumulated competitors, the outbreak of World War I in the following year placed both sides in jeopardy. During the social revolutions which followed, women who wanted or needed to work preferred the better wages and companionship of factory life to the isolation and poor monetary reward of embroidering nets. Some domestic production, for profit, did however continue until the 1930s, and special commissions for ceremonial occasions were accepted up to the 1960s, such as a gown trimmed with Limerick lace, costing over £200, for the last Court reception in 1958. At the same time, women's magazines were still instructing brides how to make their own veils from silk net embroidered with run stitches in filoselle silk[78].

The making of Limerick lace was revived in the 1980s at the Convent of the Good Shepherd and also at the Tait Business Centre, Dominic Street, Limerick, under the guidance of Father Joe Young. This resurgence was partly for pleasure, partly as a tourist attraction, though present-day living wages are far too high for any substantial pieces to be financially profitable, and the problems of finding materials consistent with those of Limerick in its prime are likely to remain insoluble.

To understand what Limerick laces were it will be helpful to look back at the craft magazines, and the actual laces, of the turn of the century to see what was done, or instructed to be done, at that time.

A circular design of fuchsias, from Kinsale, dated March 1906 (Courtesy: Nat. Mus. Ireland)

The Traditional Materials of Limerick Laces

NETS

In 1802 there were over 1,000 Stocking Frames, some in Spitalfields and some in the Nottingham and Derby area, making silk point or patent nets (see p.6), but though this manufacture continued in England at least until 1819 and far longer in continental Europe, and although black and white knitted nets embroidered with run, tambour or appliqué work survive in many museums, such as Copenhagen, Antwerp, Nottingham, St Gallen, Litchfield and the Daughters of the American Revolution Museum in Washington DC, there is no evidence that Walker used them as a basis for his Limerick laces.

The silk or cotton warp-knitted nets, were quite easily adapted to patterning during manufacture, and by the late 1820s the Jacquard apparatus had been applied to them. Bearing in mind that Walker's venture into Ireland was largely motivated by economic considerations, he would almost certainly have used the relatively inexpensive cotton bobbinet of the 'round mesh' type, its evenly hexagonal holes having every side made of two threads, crossed at the top and bottom and twisted around each other on the four remaining slopes. Costing 5 guineas (£5.25) a yard in 1810, a similar quantity cost only 18d by 1851[79]. Round net, also known as plain net, or two-twist bobbinet, was a perfect machine copy of the Bucks point ground which forms the openwork in bobbin-made Lille and East Midlands laces.

In the early 1830s, shortly after Walker's establishment of the Limerick industry, the Ring Spinner was invented, enabling finer cotton threads to be spun. A lighter more elegant mesh followed, its sloping sides elongated by additional twists into a diamond shape, and its cross-sides so shortened that the holes appeared to have only four sides instead of six. This new net was referred to as Brussels, since much of it was produced by Belgian machines, and it was commonly used for Brussels bobbin and needle lace appliqués. It was also called diamond or square from the shape of the holes. It existed in two forms, a 3-twist untraversed and an extra-twist [4-twist] traversed.

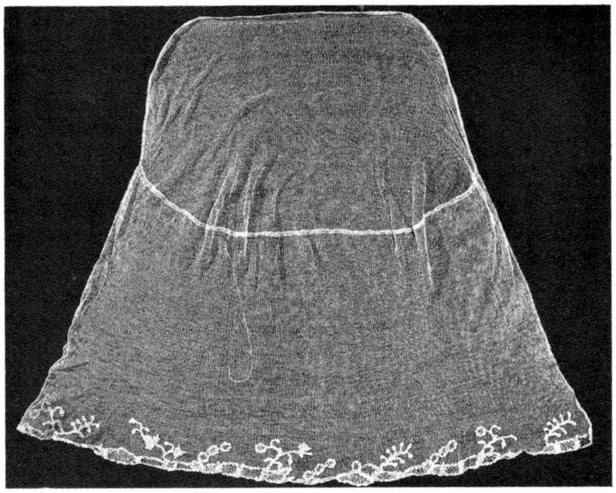

This cap of cotton warp-knit net, with needle lace appliqué, has a draw-cord and a short veil which falls forward over the face, c1809 (Photo: RB)

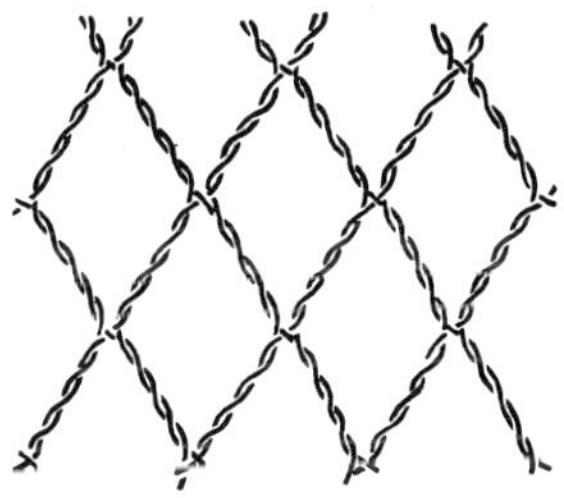

Four-twist bobbinet

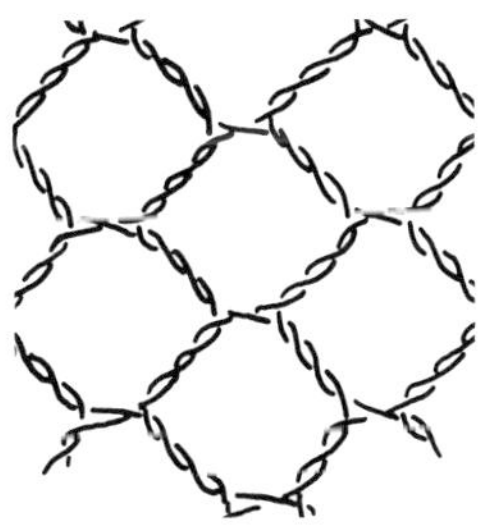

Three-twist net

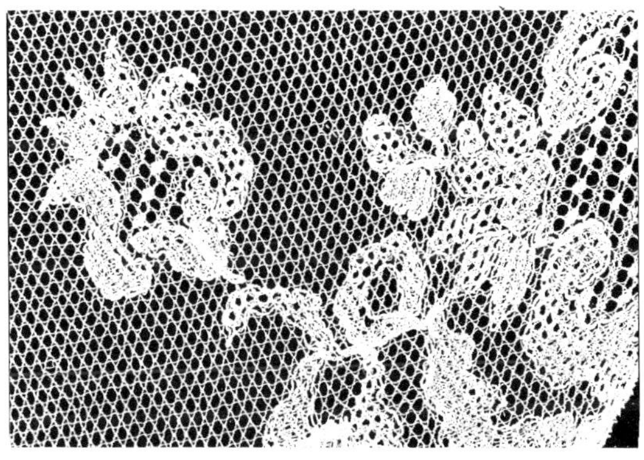

Machine six-point star net with run embroidery

From the 1830s to the 1960s both round and square nets were available to Limerick workers. The square net had a more elaborate method of construction and used an exceptionally fine quality cotton; it was slower to make, and more expensive, but it had a shimmering breath-catching translucency that made it achingly covetable by all who could afford it, or whose social pretensions rendered it needful. It had disadvantages however : these nets 'never retain their width and length, so if a shawl is made square, it will be found after it has been worn a few times, 4 inches too long on one side and as much narrower on the other'[80].

Silk bobbinets developed by Heathcoat at his factory in Paris in 1824 only gradually invaded the English market, but by 1851 he had 300 silk-net machines at Tiverton[81]. In the early 20th century, six-point star or illusion nets in cotton or silk were sent from Nottingham or Tiverton to Ireland for embroidery in floss silks for bridal veils and flounces[82].

Production of Brussels net ceased in the 1960s[83], and it is no longer possible to reproduce faithfully the delirious euphoria of Limerick's ingenious stitch effects upon this incomparable fabric. The same thread-movements, faithfully copied hole by hole on a two-twist net, often lose their harmonious balance and appear distorted.

A choice of cotton nets in 1914: 2. Brussels; 3. Extra-twist Brussels; 4. Bretonne; 8. Tulle; 9. Chantilly

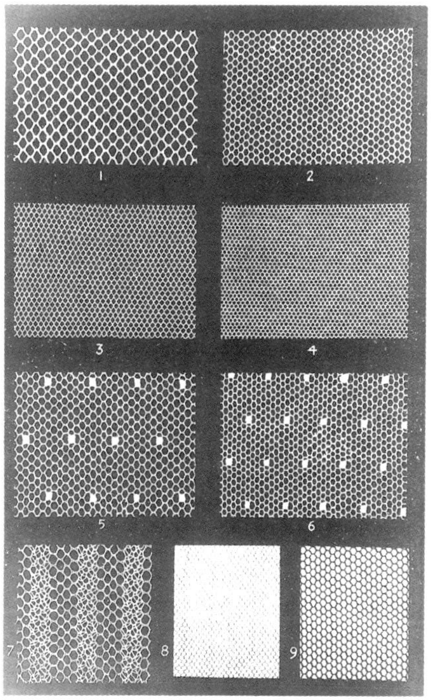

The choice of nets today is restricted. Small-, medium- and large-holed round nets are currently available. The decision about which to use must depend on the eye-efficiency and the needle-manipulation skill of the worker. The smaller the holes, the denser and more misty the net will appear since there is more thread per unit area. The result of this can be seen here in the photographs where small-hole nets allow less light to shine on to the contrasting background, and the general effect is grey.

Synthetic-thread bobbinets also exist, but most synthetic nets are made not by twisting but by warp-knitting on a Raschel machine. They were developed during the 1950s and 60s when Limerick laces had long withdrawn from competitive commercial production. They tend to be stiffer and sharper, are often coloured, and may deteriorate with use.

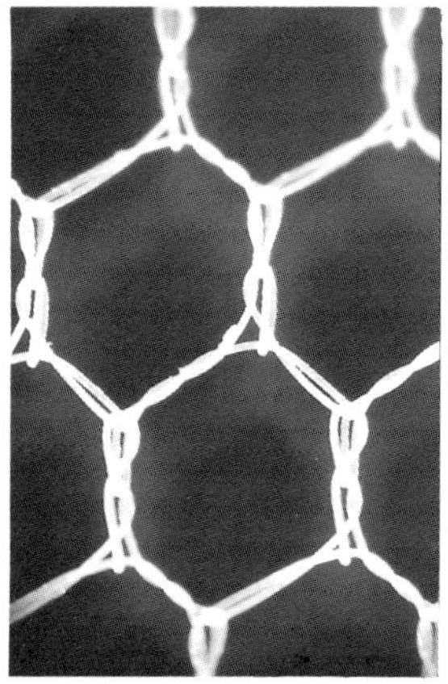

Raschel hex net in synthetic thread

In the late 1960s, stark white terylene nets were sufficiently novel to replace for a while the soft creamy bobbinets, especially for bridal veils, while for Carrickmacross appliqué a nylon organza took the place of the scarce and expensive cotton batiste[84].

Other nets, such as the thin wire-like Mechlin, or tulle uni, (with only one twist in its sides) appear in foreign embroideries of the late 19th century, but not in Limerick.

THREADS

The traditional embroidery threads, like the traditional nets, are no longer available. Instructions, published during the first quarter of the 20th century, for making Limerick run and tambour as a domestic hobby, suggest a 120-count cotton for the outline and a 250-count for the run fillings[85]; and Pethebridge in 1947 suggests 'Carrickmacross lace thread, 150 or 200 for whipping in cambric appliqué'[86].

But there is no need to despair. The old professional Limerick workers were not hidebound by unbreakable rules. With a freedom of attitude restricted only by the need to avoid shocking an expectant market, they might use the sewing thread doubled (see p.42) to increase its density or, more rarely, use thread finer than the mesh sides so that the emphasis was focussed on the holes rather than the embroidery. Their constant aim was not conformity to some rigid pathway but the achievement of a pleasing and beautiful effect.

Dillmont (referring to French embroidered nets, rather than to those of Ireland) advises: 'Coarse net often employed for the modern kinds of work [soft furnishings] requires a coarse material such as DMC coton perlé no.5, DMC floss flax nos.8 and 16, or DMC rayon no.30'[87]. At the beginning of the 20th century floss silk, or flax, or thick loosely-spun cotton was often recommended for tambour work.

Synthetic threads should be used for embroidery on synthetic nets.

CAMBRIC

To what extent linen (flax) threads may have been used for the embroidery is debatable. Miss Penderel Moody, proprietor of the Revival Pillow Lace School of Lower Sloane Street, London, listed in her early 20th-century brochure: 'designs for Limerick and Carrickmacross; tambour needles, 4d a packet; and Limerick thread, écru only, in nos.120 and 250, at 5d (about 2p) per 1000 yards'. She also had linen threads available up to no.300 (equivalent to 100 count cotton); while Harris pure flax thread, 'now up to the 360s', was advertised in *Needlecraft* in 1906.

The cambric used for Irish appliqué and guipure would initially have been a fine linen. Some intriguing figures regarding its manufacture were recorded by Martin, in 1847: 'two acres of flax produce 550 lb of dressed flax . . . 158 females (being paid 7d per day) take twelve months to spin this into 16,500 hanks, each hank holding 12 leas . . . 30 hanks of these yarns are needed to make up one pound weight . . . thus the count of the thread [defined as the number of leas per pound weight] is 360 [equivalent to a 120 cotton count]. Eighteen weavers paid 9 to 10 shillings [equivalent to 40 or 50 pence per week] then take 12 months to convert all the yarn into 210 webs of cambric, each web having enough material for 5 dozen pocket handkerchiefs'[88].

Later in the century, when fine linen cloth was more difficult to come by, cotton fabrics were used, for example batiste, a highly mercerised soft and lightweight fabric, or organdie, mercerised and starched to a stiff transparent finish. A nylon-cotton mix is now sometimes preferred. It is crease-resistant and shrink-proof, thus eliminating the two major hazards of the traditional appliqués and guipures: the difficulty of restoring their smooth surface after washing and the very real risk of the material pulling irreversibly away from its binding outline. On the other hand it may be so sheer that it scarcely conceals the net beneath, fudging the impact of the design.

Materials should always be matched to the effect you wish to achieve, and it is esential to spend some time and care working small samples to check how the ideas in your head will appear in reality.

HOOKS, NEEDLES, SUPPORT

These differ today in no essential way from traditional forms, but sizes must of course be matched with the net, the thread, and any other materials used. Further details will be given in the practical sections.

A tambour lace printing block carved with a design of vine leaves and grapes (Courtesy: Hamilton District Museum)

PATTERNS

Prior to the invention of lithographic printing, in 1837, costly engraved wood blocks were sometimes used to mark the pattern on the net, and printing blocks of this type survive at Nottingham, Limerick and Hamilton in Scotland.

A cheaper method was to take rubbings of laces already made. A sheet of paper was placed over the lace and rubbed lightly with a soft pencil so that the minutely raised outlines and stitch areas showed up as ghostly white shapes.

Not until the late 19th century were designs drawn so artistically and so professionally that they were works of art in their own right. However the working outline had still to be extracted from them to construct a pattern in Indian ink on white or yellow glazed calico.

A rubbing of an art nouveau design of poppies (Courtesy: Phyllis Wood)

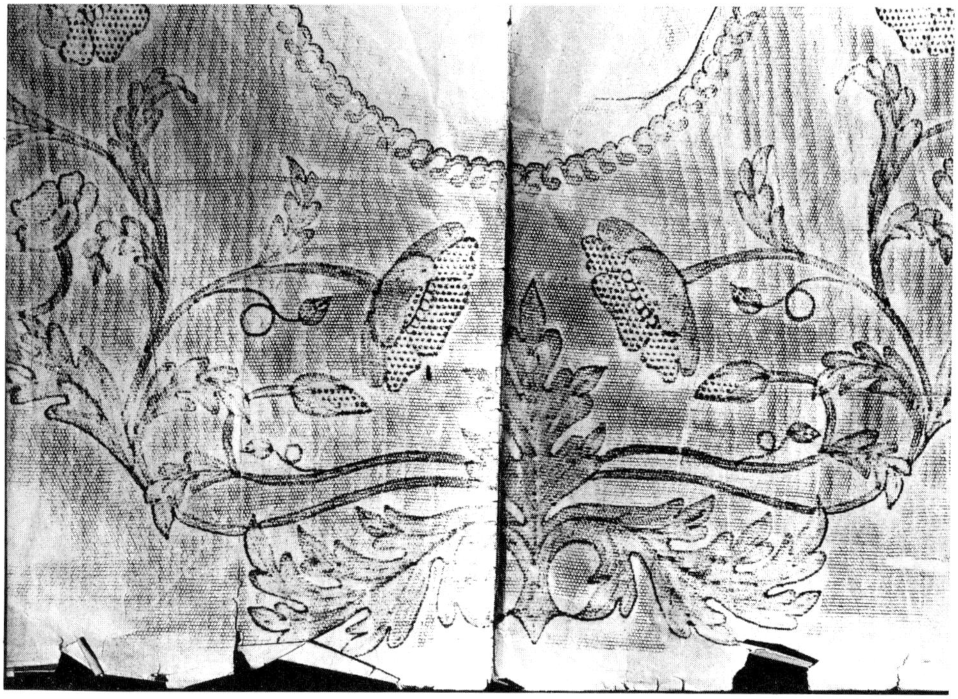

Pouncing was sometimes used in cut cambric work. As described by the *Illustrated Exhibitor* of 1852, Vol I (p.47), the design was drawn very accurately on thick letter paper. 'With a coarse needle prick all the outline, making holes about 1/8in apart. Lay the material to be used completely flat on the table. Place the pricked pattern over it. Place weights to keep them firmly down, then rub all the design over with fine pounce so that it penetrates every hole. [The pounce was a powder made of gum arabic and Reckets blue mixed with blanc de Troyes and a little alcohol. It was applied with a small roll of felt or linen. For use on white material a little powdered charcoal was added]. Remove the paper. Link the dots with a solution of colour mixed with varnish and spirits of turpentine, using a fine sable brush'. This method is useful if the motif is to be repeated quite a number of times. It is also helpful where a precise symmetry is required: the paper pattern can be folded once or twice and the pricking taken through every layer at the same time. The two or four parts will then be identical.

Yet another possibility, particularly useful for tambour work, was to place the inked pattern under the fabric, if it was sufficiently translucent, and 'trace the design with a quill pen or a fine brush dipped in a solution of stone blue in thin gum water'. This could be washed out when the lace was complete. An alternative 'ink' was a mixture of sepia and charcoal.

From 1874, transfer patterns could be bought and hot-ironed on to tracing paper.

Today, photocopying patterns or designs, with relatively easy enlargement or reduction, saves a great deal of labour.

Limerick lace with design left, pattern right

Tidying the cut edge of the net

At the end of the work, remove the pattern if this has not already been done. The net must now be cut to the required shape. Its raw edges will certainly fray with use and so, before cutting begins, every edge must be secured. This can be done in several ways:
(a) visible buttonhole-stitching;
(b) invisible buttonhole-stitching using the finest possible cotton thread;
(c) a double line of close running stitches;
(d) overcasting;
(e) attaching a machine picot border, also known as pearl edging, invented c1825[89] to the net on the face side, using small inconspicuous stitches.
(f) attaching a bobbin-made picot border. This was still available in the early 20th-century, but increasingly hard to find.
(g) attaching a narrow beaded tape which will pass tidily around curves and angles;
(h) for straight borders only, running a double line of stitches along the edge and then, instead of cutting the net back to this point, turning it under to form a narrow hem, and tambouring or running along it again to hold it flat.

If at this point the lace is no longer quite clean, it can be washed in its frame so that it remains taut. Discretion must be used here - stain from wood or plastic may mark it.

Cut around the lace, leaving an over-wide margin, until you have it out of the frame. Then with extreme care cut the net back to the border. In (e) and (f) cut behind the picots so that they are clearly visible. In (g), cut to the edge of the tape.

(a)

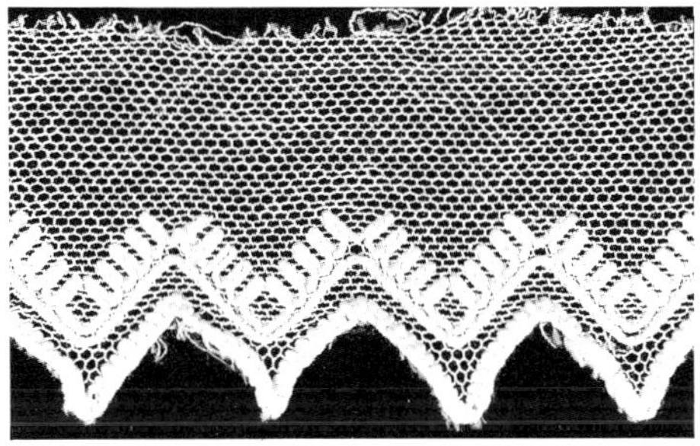

(d)

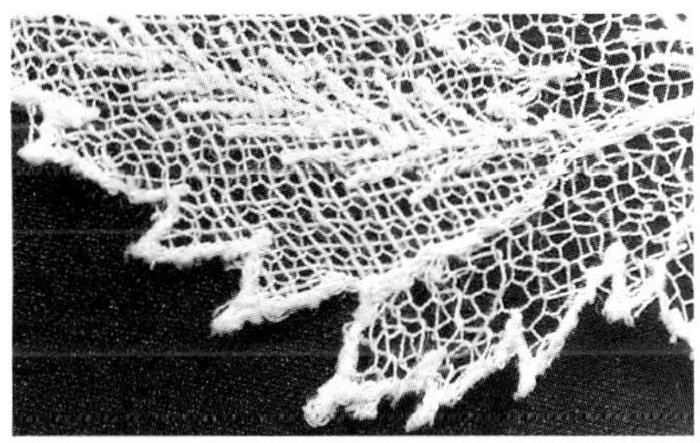

(f)

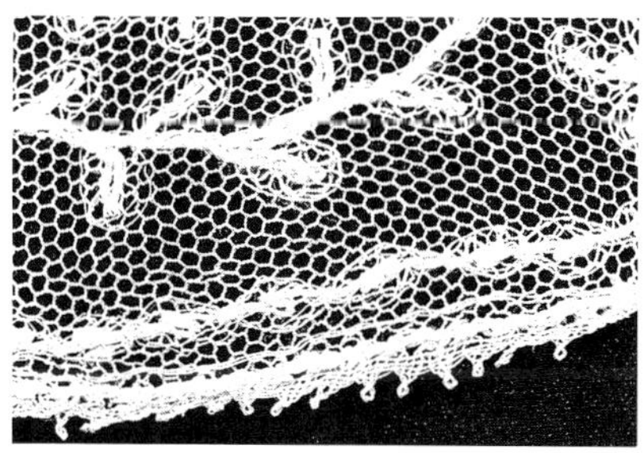

In the next section, practical instructions for making Limerick run are given. The designs and stitches are taken from actual laces of around 100 years ago.
NOTE: designs shown for one Limerick technique can often be adapted with little or no modification for the others.

Above: A pattern for working. Half a collar, with rose, shamrock and thistle (Courtesy: Phyllis Wood)

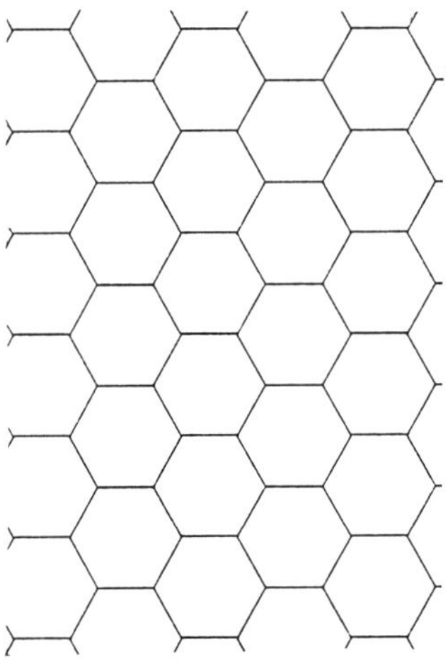

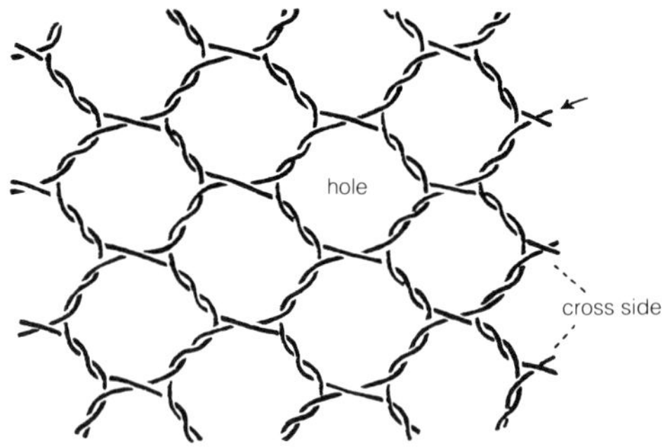

Above: The vertical alignment of the cross-sides
Top right: A two-twist bobbinet
Right: A card set up with two ivy leaves on coloured paper, the stitch numbers marked, and the three layers tacked together

2. INSTRUCTIONS FOR MAKING LIMERICK RUN

Run embroidery is worked by plunging a blunt-tipped sewing needle in and out of the meshes of the net. If only a sharp-pointed needle is available, use it eye-first.

Materials

NET Two-twist bobbinets in several hole sizes are available.

PATTERNS AND SUPPORTS The twisted sides of the meshes slant diagonally across the net, giving an elasticity which needs to be controlled if the work is to keep its shape. For small pieces, the net can be tacked to a card and held in the hand. For larger pieces, it must be stretched within a rigid frame, aligning the cross sides parallel to the vertical axis, since the warp threads run in this direction and are more resistant to stretching.

The pattern can be traced, photocopied or rubbed from the original design. A drawing in black waterproof ink on tinted paper or stiffened cloth is ideal, since it can be re-used several times over. Pencil drawings should be sprayed with a fixative such as Perfix, to avoid blurring the shape or discolouring the net. The older technique of painting in Chinese white on coloured paper can also be used.

The pattern outline expresses the shape of the design elements which must be worked on the net. Any faults in its construction will automatically carry over to the lace, and so must be corrected before the embroidery begins.

For convenience of handling, a border of about 2 inches is left around the outer limit of the pattern. Net, pattern and support are arranged one above the other. If no frame is used, stick or tack the pattern to the card. Cut the net slightly larger than the size required to allow any roughened edges to be trimmed back later. Lay the net over the pattern, carefully aligning the cross-sides. Using a sharp needle and coloured thread, tack the three layers together, taking care to avoid the design areas where run stitches will later be worked.

When using a frame, first align the net as described above. The pattern paper can be strengthened by tacking it to brown paper cut to the same size. Position the pattern beneath the net, pin it in place and tack net and pattern together. Remove the pins. Small frames can, like the card, be hand-held and turned in different directions, to facilitate the working of more complex stitches. Alternatively they can be fixed to a stand which can be adjusted for height and angle, leaving both hands free.

Larger frames are usually square or rectangular. They have an inner margin of stout canvas to which the net, correctly lined up, should be firmly fixed by backstitching. These frames have tensioning cords and ratchets so that the net can be tightened in either direction.

If an extensive piece such as a veil or flounce is attempted, it is helpful to have a frame with a roller. The net actually being embroidered can then be stretched tight, while that already worked, and that waiting to be, is kept smooth and uncrumpled.

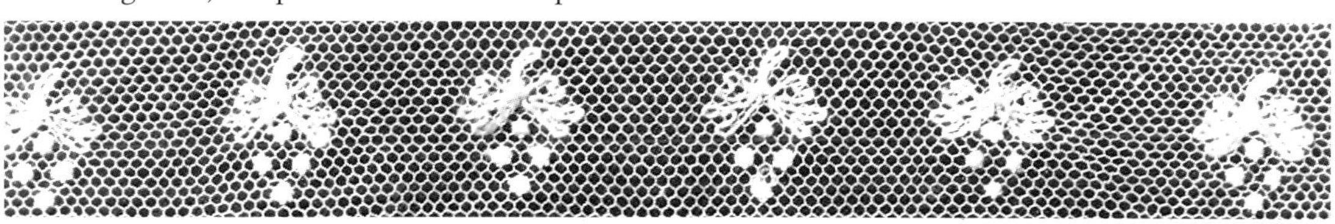

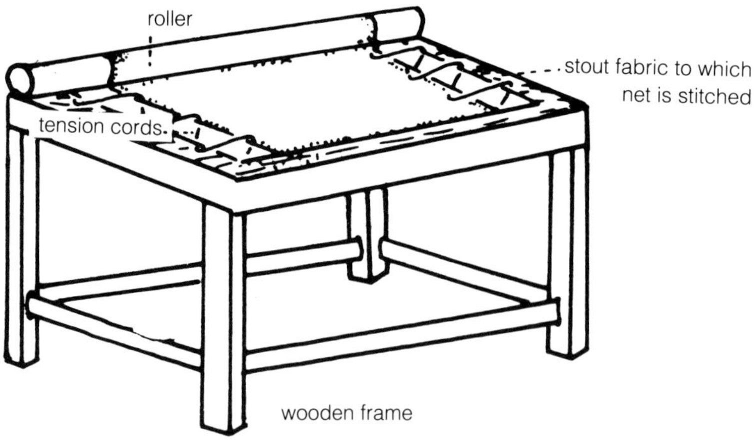

Large embroidery frame set up for run work

NEEDLES AND THREADS

A sharp sewing needle is needed for joining the pattern to the net;
long and short blunt-tipped needles for the embroidery;
thickish thread such as cordonnet special 30 for the outline; thinner thread such as 60s or finer for the infillings.

These counts are for cotton thread. Equivalent linen thread counts are approximately x3.

In spite of the report of the judges of the International Exhibition of 1851, coloured thread other than black does not seem to have been used in traditional Limerick laces. However since the traditional nets and threads are in any case no longer available, there is no reason why experiments in this direction should not be made.

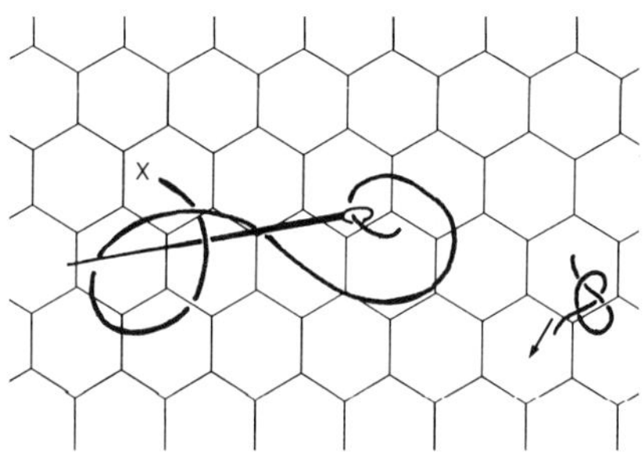

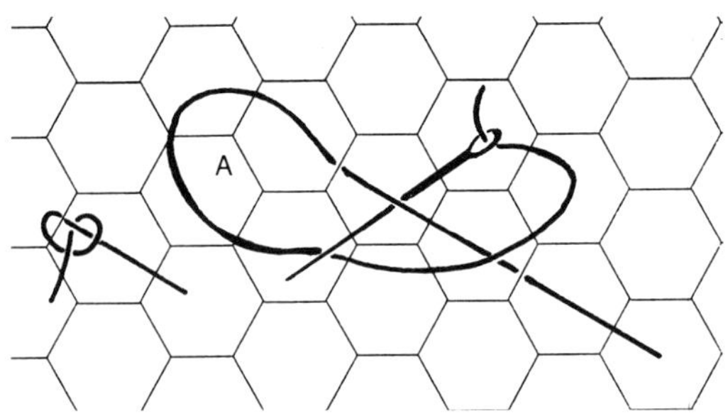

Starting off with a single thread
Starting off with a double thread

Above: Ending
Below: The pattern is moved and retacked beneath the net for every repeat

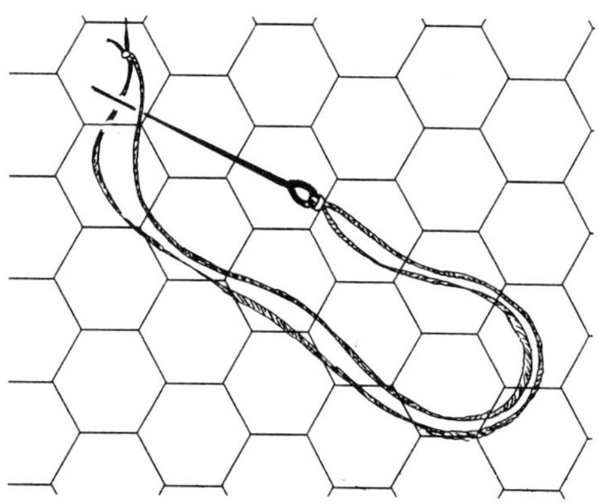

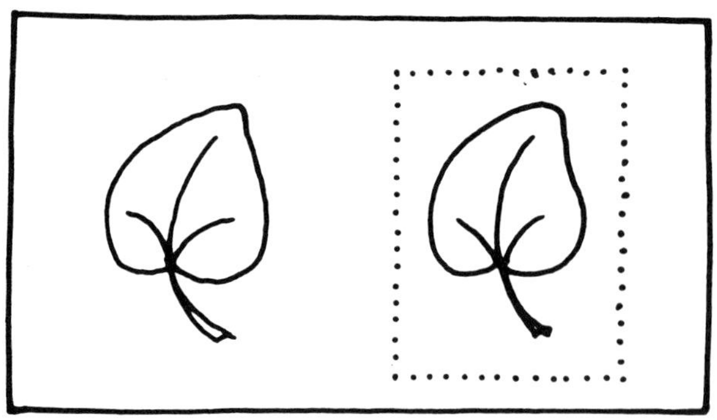

OUTLINING

The thread outline follows the pattern outline marked on the drawing. The thread used to mark it in must be thick enough to make the shapes of the motifs quite clear. Either a single coarse thread, or thinner threads doubled or trebled on themselves, can be used. In Limerick work the outline was commonly double. A silky floss flax was occasionally used.

All the threads are cut into relatively short lengths to avoid either overtwisting or unwinding of the plyed strands. Cut these lengths in advance so that fresh strands are ready to be pulled out as required.

To begin, choose a suitable point along the outline cord and tie the thread to the net with an inconspicuous knot.

If a double thread is used, thread the needle then tie the two lengths together just behind the eye to keep them even, and make a second knot at the far end to hold the extremities together. Pass the needle under a mesh side, then through the loop of the two threads, and tighten.

Another method is to fold the thread in two, pass the free ends through the eye and tie them immediately behind it. Pass the needle under a mesh side, through the terminal loop, and tighten.

Choose a needle size which fits the thread comfortably. For a design with a straightish outline a long needle speeds up the work, for one that twists and turns a small needle is easier to handle.

Follow the pattern outline very carefully, taking the needle and thread in and out of the meshes in a regular manner.

When the first length of thread is finished, secure it by passing the needle under a bar and back through the loop. Tighten.

Alternatively, make two very small buttonhole stitches, then carry the thread through to the back, and leave it. Attach a new length over the point where the first thread ended, to strengthen the join.

Continue in this way until the first design unit is completely sketched out in thread. If the next motif is a repeat of the first, untack the pattern, move it to the required position, re-attach it to the net, and work as before.

Slight overtwisting or unwinding of the thread can be corrected by letting the needle hang and twirl until the balanced state is restored.

Once the outlines of all parts of the design are complete, the drawn pattern is no longer needed. It can be removed by cutting through the tacking threads at the back, using blunt-tipped scissors, but it can if preferred be left in position, especially with small designs where a card has been used for support.

Running in the outline; This thread could just as well be double or even treble

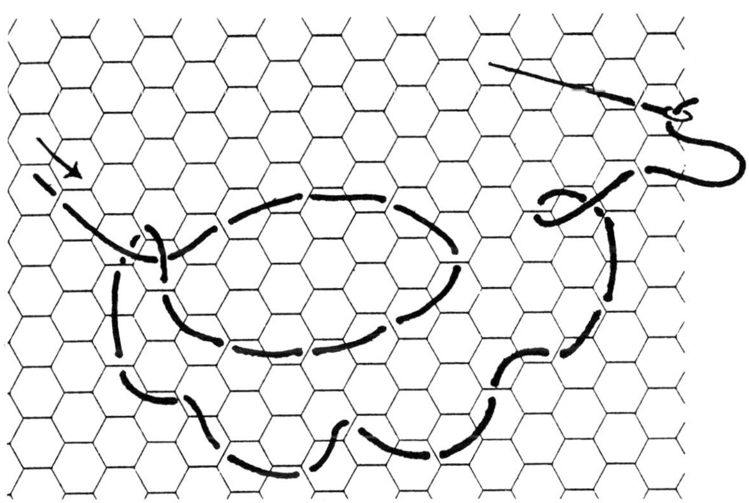

INFILLING

Infilling consists of filling in with embroidery the outline shapes just completed.

The dominant parts of the design, referred to as 'solid areas', have a small range of stitches, while the more decorative areas or 'caskets', are characterized by an almost limitless range.

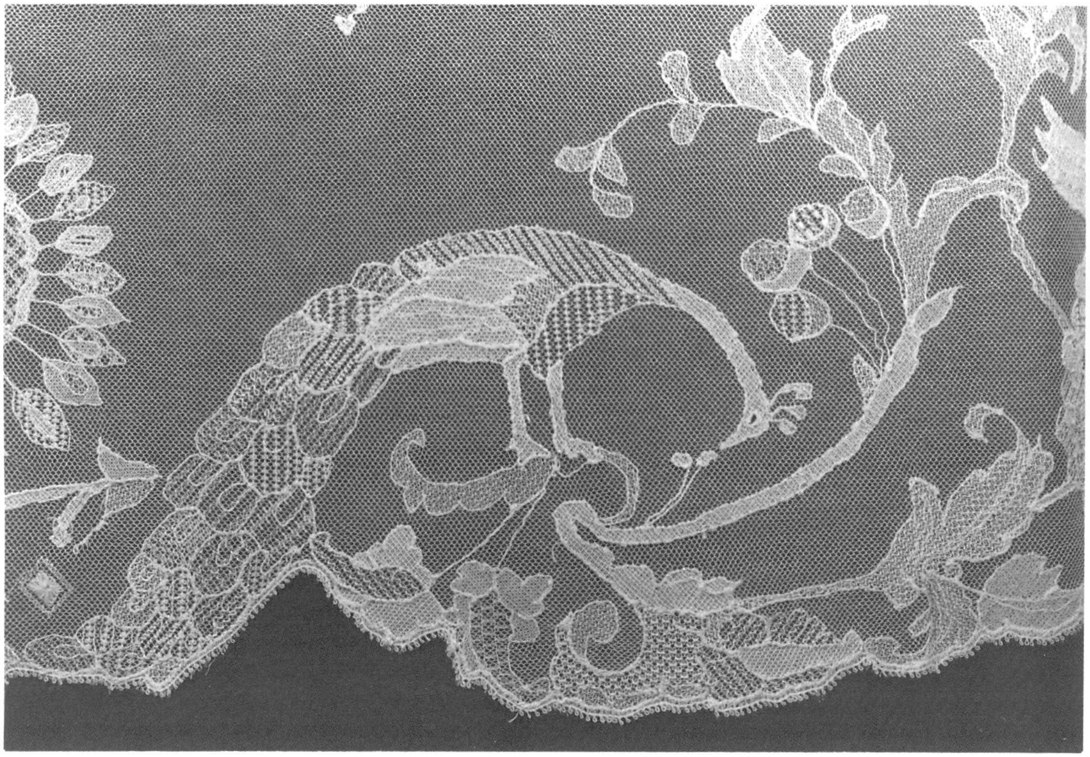

The peacock's trailing tail and the encircling foliage enclose a richly varied sample of Limerick run stitches, without in the least detracting from the delicately composed design (Courtesy: Wolfgang Haüsle)

Single threads are normally used for both areas, though their thickness may vary, and double threads do occasionally occur. Tie a knot either around the outline itself or around a mesh side immediately next to it. Leave a loose end which can be neatly sewn in, or cut back later.

At the end of each row take the embroidery thread around the outline cord, then along it to the position for the following row. This whipping will be almost invisible in the finished work.

The use of short lengths of thread reduces the fluffing out of the fibres caused by repeated friction against the sides of the meshes.

DIRECTION OF WORKING The embroidery can be worked in whatever direction you prefer, and cards or smaller frames can be constantly turned. However, the relation of the embroidery thread to the cross sides of the meshes, shown in the line drawings, must be carefully adhered to. The drawings are intended to illustrate the thread movements. The final appearance of the casket stitches is shown in the accompanying photographs.

STITCH NAMES

In The *Needlecraft* and *Weldons* magazines of c1900, the stitches were named after the comparable forms used in surface embroidery - darning, tent, cross, and so on. As far as possible, we have done the same. In Ireland, and other countries, the stitches are often given colloquial names.

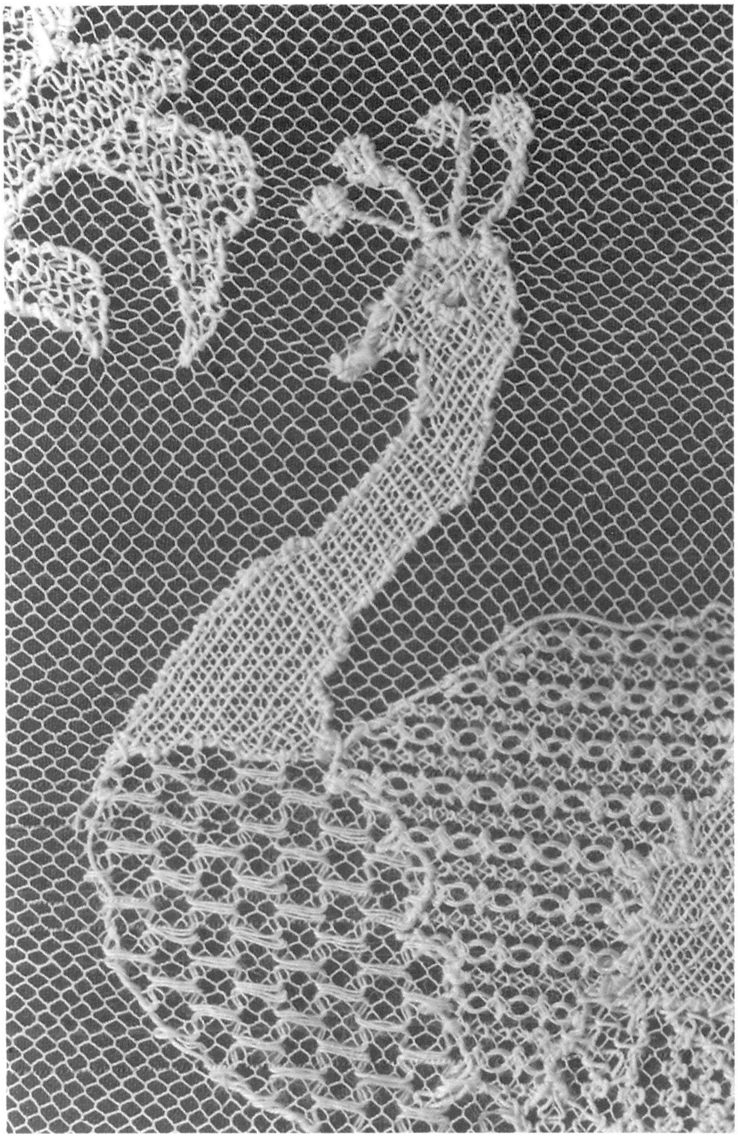

The foliage, top left, is filled with tent-stitch; the head and neck with darning-stitch; the breast and lower body with three decorative casket stitches. The photograph is taken from the reverse side to show the whipping of the embroidery thread around the outline more clearly (Courtesy: Wolfgang Haüsle)

THE SOLID DESIGN AREAS

Thread lengths are begun and ended as described for the outline cord (p.42) and for infilling.
Four main stitches are used:
Run, Darning, Tent and Cross. Each shows minor variations.

Running stitch

Row 1. Work diagonally, taking the needle over and under alternate twist sides.
Row 2. Work back in the same way, along the next line of holes.
Run-stitch may be worked vertically, horizontally or diagonally to the cross-sides, depending on the effect required. The length of the stitches can also be varied, for example take the thread over 2 mesh sides, then under 1.
On p.32, thick running stitches give heaviness to the flower base and leaves. For an even more solid effect, pass the thread of both LR and RL rows through the same holes.

Darning stitch

The term 'darning stitch' is here used in the sense of Mary Thomas's 'surface darning stitch'.
Set 1. Secure the embroidery thread. Stretch it diagonally over a line of holes to the far side of the area, around the outline and back over the next line of holes. Repeat until the space is filled.
Set 2. Working diagonally in the opposing direction, take the needle over each first-set thread and under each mesh side. The two sets of threads cross obliquely in the centre of each hole.
In the old Brussels nets, sets 1 and 2 crossed almost at right angles.

Tent-stitch

Four variations occur quite commonly:

A. Work tent stitch horizontally, as shown, reversing the over-unders in alternate rows.

B. Work as A, but keep the over-unders constant in every row.

C. Row 1. Take the needle horizontally over two twist-sides, then vertically downwards under two cross-sides. Repeat to the end of the row.
Row 2. Return through the next row of holes, keeping the same over-under sequence, and the same direction of slope.

D. Work diagonally as in C, but in alternate rows take the needle over 2 cross-sides and under 2 twist-sides. Repeat these rows until the space is filled.
Note: Tent-stitch may also be used as a casket filling. Its effect can be lightened by leaving lines of holes between the rows of stitches, or two lines of holes between each two rows of stitches.

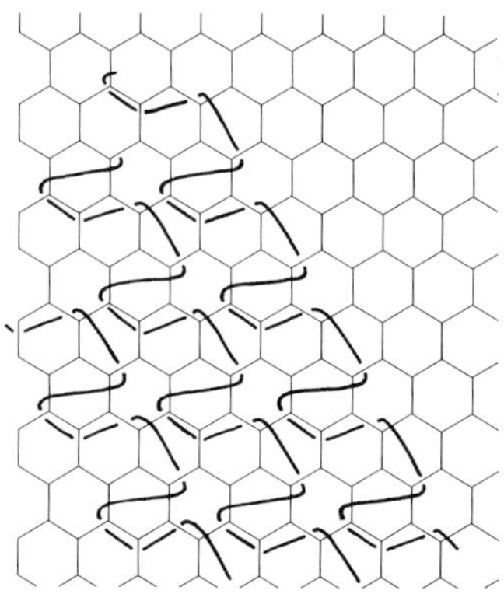

A hybrid of tent- and cobweb-stitch

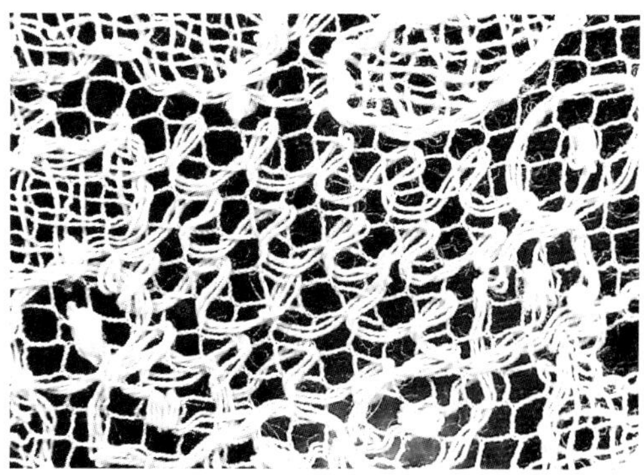

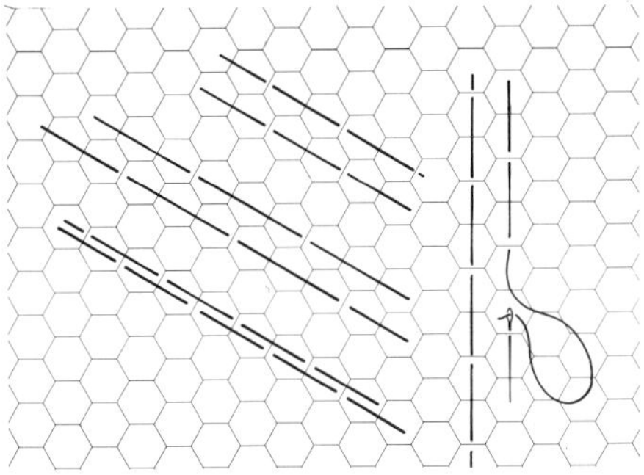

Run-stitch

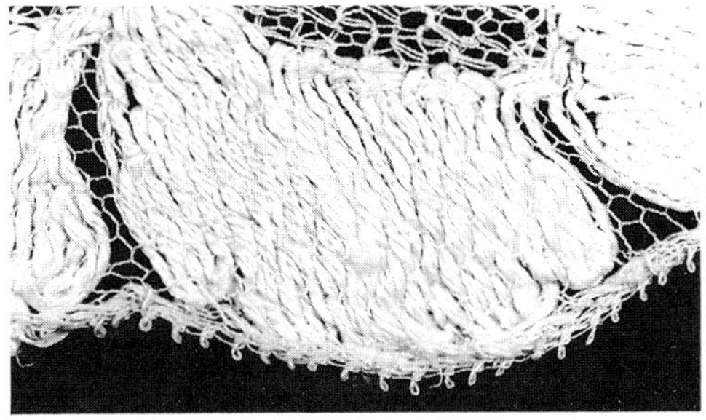

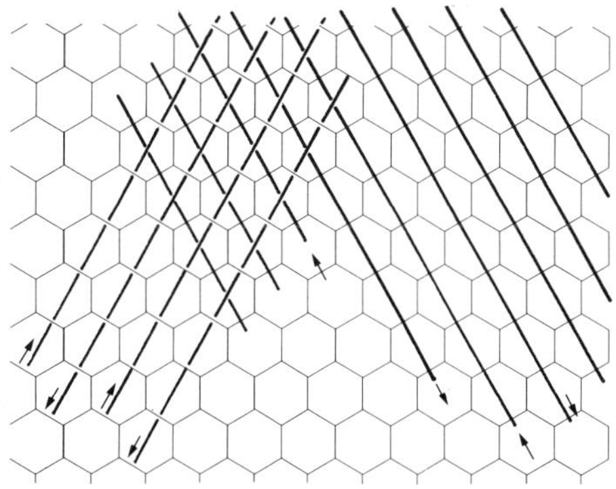

Darning-stitch ->

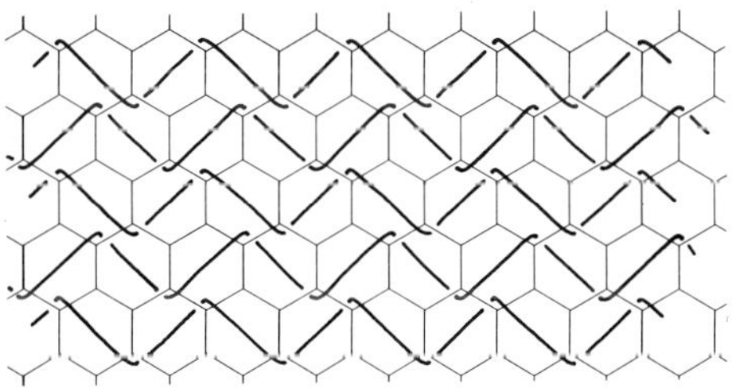

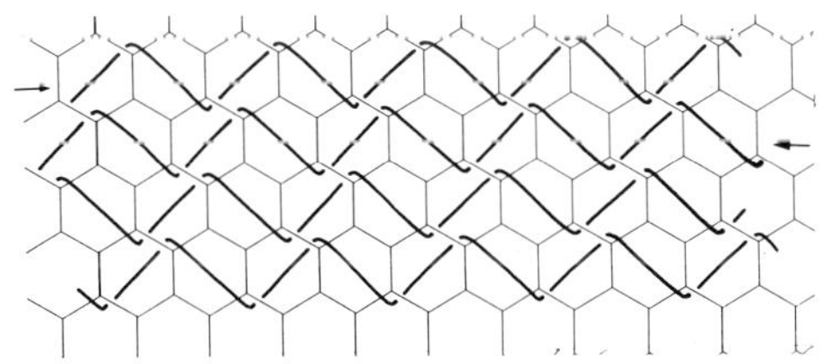

Four tent-stitch variations

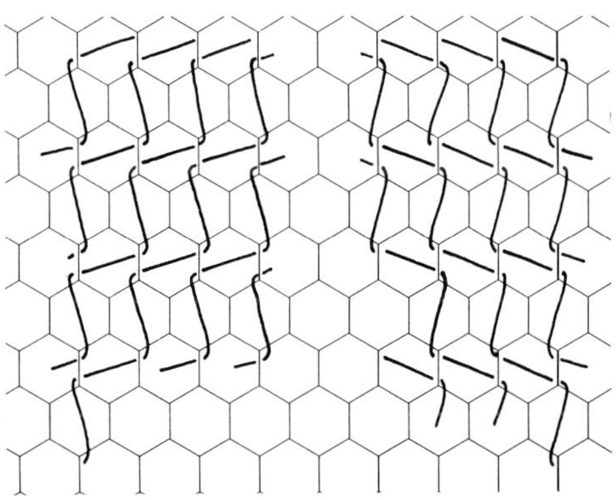

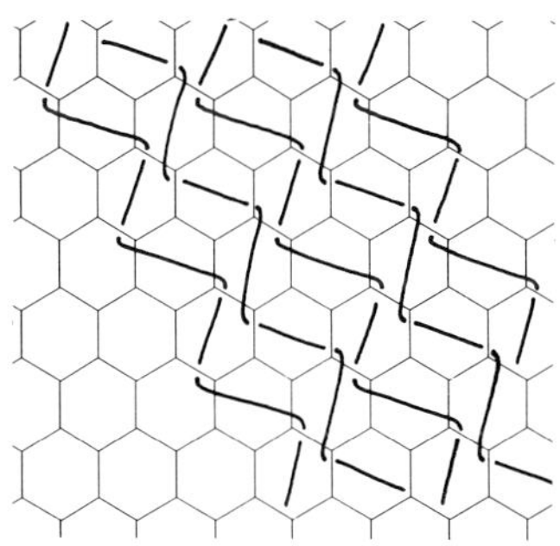

Cross-stitch

This is most easily worked in LR RL rows, with the cross-sides aligned horizontally. The direction of the upper stitch of the cross is usually from bottom left to top right, as in embroidery.
There are two forms:

A. Row 1: LR. Work a row of tent-stitch with the top-left/bottom-right slant passing over the holes, and the connecting slope beneath them.
Row 2 (RL). Work back over the same meshes, taking the needle over the face-side (upper) strokes and under the reverse side (under) strokes. All the uppermost strokes will lie in the same direction.

B. As A, but pack the tent-stitches closely together so that while the face-side threads pass over a hole, the reverse-side threads are hidden beneath the cross sides. An overcrowded appearance can be avoided by leaving a line of holes between the stitch rows.

Cross-stitch can also be worked diagonally.
It often appears as a casket filling, in conjunction with other stitches (see stitch 3, p.33).

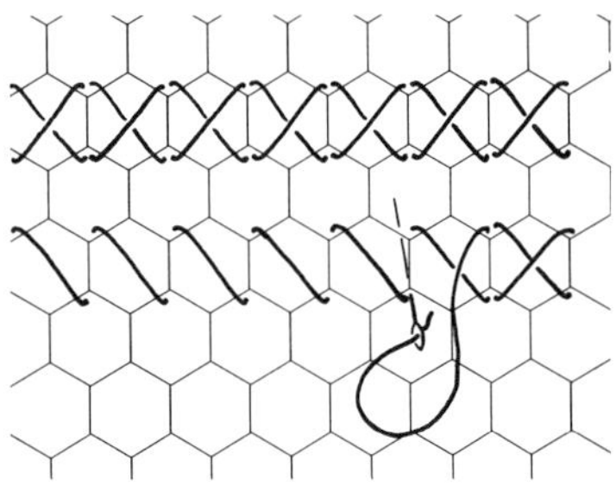
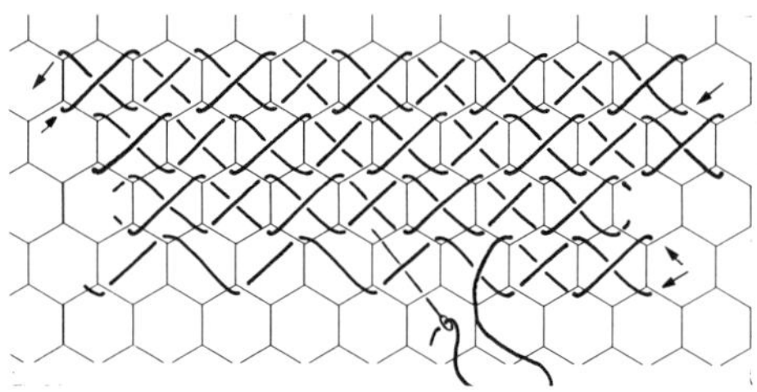
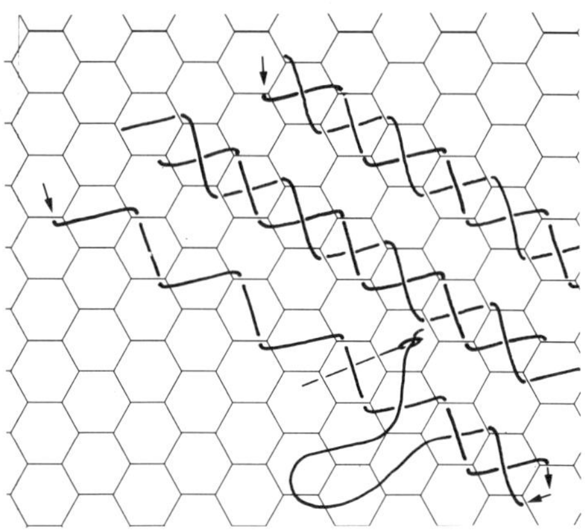
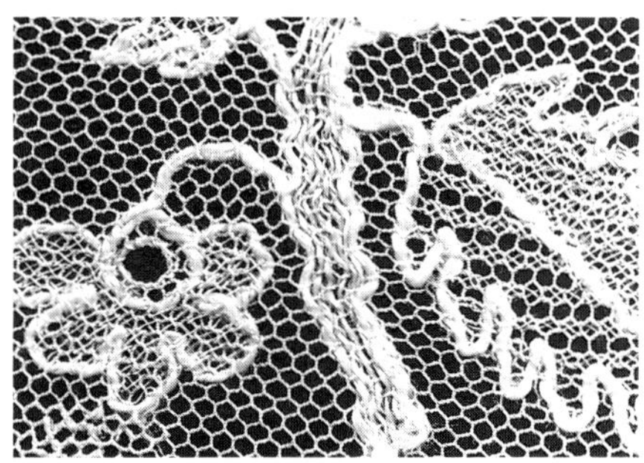
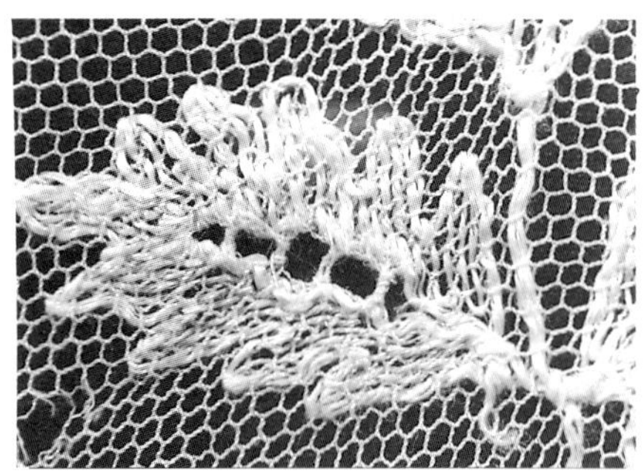

Above: Three cross-stitch variations

Right: A stem thickened with close running-stitches, and a vein widened by holes

LIGHT AND SHADE Diversity can be introduced into the solid areas using stitches of different density, for example cross-stitch in every row on the darker side, and cross-stitch in alternate rows on the lighter side of a leaf; or combinations of cross-stitch (heavy) with tent-stitch (light); or, when a leaf or petal is curled over, darning-stitch for the top and tent-stitch for the exposed underside. Such ploys convey an attractively realistic effect of sunlight and perspective, which can be intensified by using thicker or thinner threads.

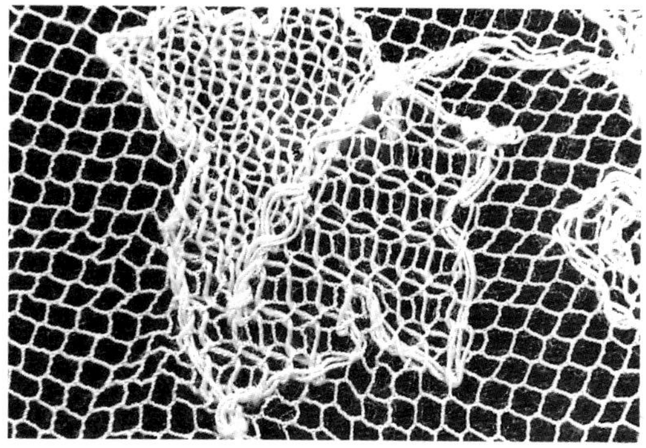

Two light and shade effects

Below right: The method of enlarging a hole

Below left: A series of enlarged holes. Within the casket are smaller holes created by the tension-pull of alternate rows of cobweb- and figure-of-eight stitch

MAKING HOLES Stretching selected mesh holes with a bodkin or similar implement was a common device in 19th- and early 20th-century Limerick laces. Insert the tip into a hole and rotate it gently but firmly easing the sides back to the surrounding meshes. Tightly overcast the edge of these expanded holes with embroidery thread to keep them open.

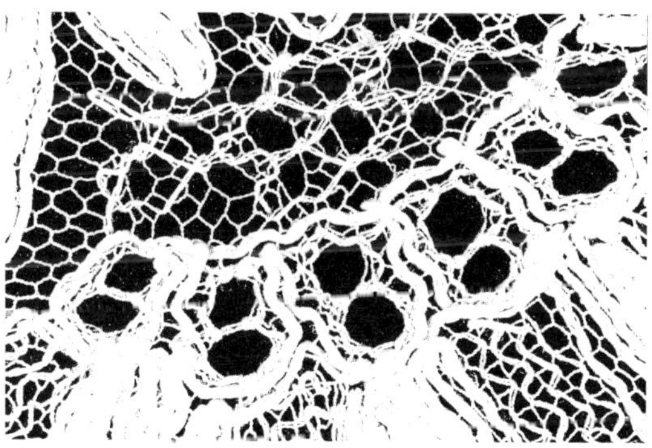

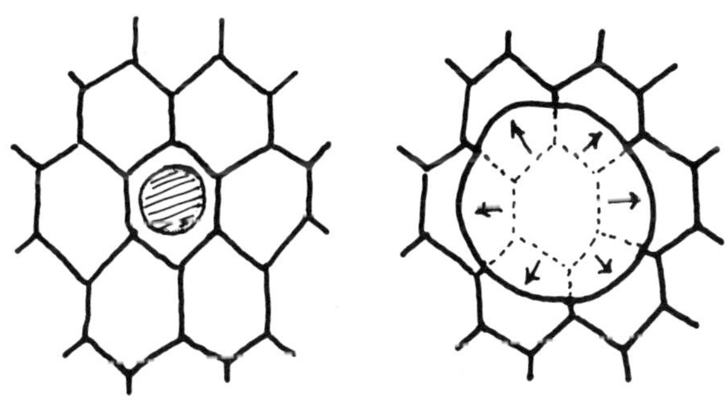

STEMS AND VEINS are usually represented by double strands of outline thread. Occasionally the space inside was sufficiently thick for slender filling stitches to be entered.

Alternative techniques for making stems or veins are:

a. Work run-stitch from base to tip. Return through the same holes, taking the needle in where it previously went out, and out where it previously went in (as in Holbein stitch).

b. In returning from tip to base, take the needle not through the same holes but through an adjacent line.

c. In returning, use overcasting instead of running stitches.

d. Take the thread straight up, catch it around a mesh side, then overcast it on the way back.

e. Veins: select a line of holes along the middle of the leaf. Enlarge alternate holes, making them gradually smaller towards the tip to give a tapering effect. Using a thin embroidery thread, overcast the squashed mesh sides to bind them in place. An outline cord may be run around this type of vein, if desired.

A Schiffli pattern of art deco trees, after a design from Albert Muhler, Atelier Plauen (Courtesy: Dieter Maier)

PRACTISING OUTLINES, AND SOLID DESIGN AREA STITCHES

Study the design shown above. Set up the pattern, net and support as described (p.21). Run in the outline (p.23).

Fill the internal areas with selected combinations of the four types of stitch just described, and consider the result.

Look for balance, harmony, and the integration of density and texture. Whether you find the result pleasing or displeasing, try to discover why.

Take another copy of the pattern and try a different stitch combination. In this way you will learn not only about shapes and how best to emphasize them, but how to divide up the interior spaces to give variation without disintegration.

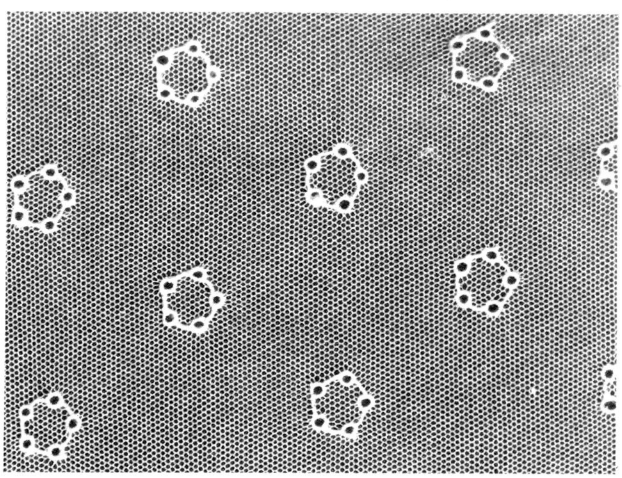

Use this simple all-over pattern to practise making enlarged holes. Overcast each with thickish embroidery thread and continue from one hole to the next with similar stitches. Finish each circle independently

THE CASKETS (WELDON'S CASQUETS)

In the following pages, a selection of 19th and early 20th-century Limerick laces are considered. Numbered maps indicate the position of their decorative stitches. Each stitch is illustrated by enlarged photographs, explanatory line drawings, and written instructions. In every case the cross-sides may be aligned either vertically or horizontally for convenience of working, but make sure that your stitch movements bear the same relation to the net as that shown in the drawings.

SQUARE VEIL WITH FLOWERS

In this small veil the design is outlined with and internally divided by thick loosely-twisted silky strands. The same strands are used to emphasize and highlight selected stems and leaves.

All the infillings are worked in high-count matt cotton no thicker that the net threads. The meshes are small and round, the solid design areas worked in cross-stitch (see p.28), and there are many different casket stitches.

The lighter shading of the leaves is achieved by alternating the cross-stitch rows with lines of empty holes.

The cut border of the net is scalloped, and finished with run strands of outline thread closely overcast. A narrow machine picot is added.

The photograph shows the main areas of the design. The caskets from Areas (1) and (2) are described below.

The veil is 1 metre wide, and slightly more than 1 metre long. The lower panel consists of:

(1) a line of pendant blossoms

(2) a row of leaves and sprays

The panel is enclosed by a border of flower sprigs, 5.8cm deep, and this continues around the entire veil

The upper part of the veil is sprinkled with tiny rosette-like flower heads each of the six petals containing an enlarged hole

Area 1. Pendant blossoms

In each flower the stem and sepal base is filled with close running stitches worked in outline thread.

The central holes of the buds are made as described on p.29, except that their outline surrounds but does not touch them.

In each of the 10 blossoms, the inner caskets of the flower petals are filled with different decorative stitches:

Three of the pendant blossoms. Note the areas where the stitches vary from one flower to the next

Stitch 1. Cobweb-stitch (diagonal)

Since the embroidery thread is finer than the twisted mesh sides, the immediate appearance of the filling is of 6-lobed rosettes clustered around dilated holes.

Slightly enlarge every other hole (o) diagonally across the casket in every second row.

Row 1. Follow these enlargements with cobweb-stitch, arranging the angles between the straight and slanting sides to pass over the intermediate meshes and back under the stretched holes, widening them further.

Row 2. Work back again, in the reverse direction, making a mirror image of row 1.

Repeat these two rows until the casket is filled.

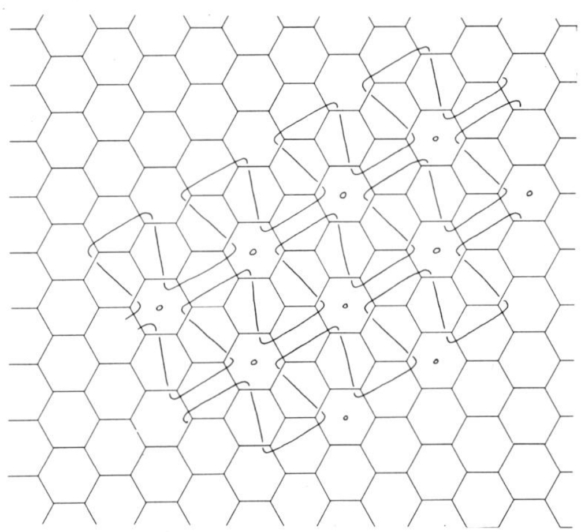

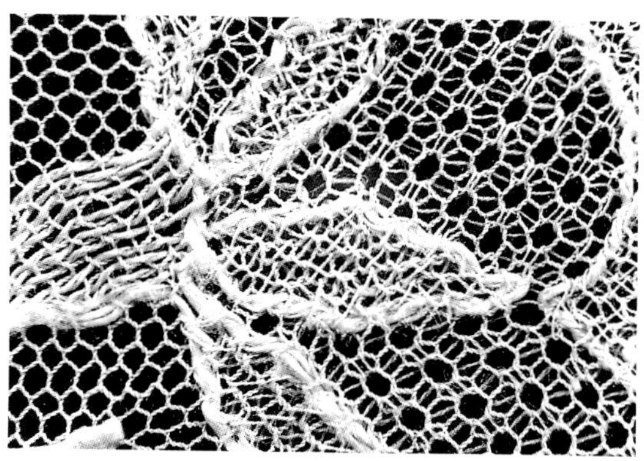

Stitch 2. Cobweb-stitch (vertical)

Elongate alternate holes (o) in a vertical direction. Work the cobweb-stitches in vertical rows, arranging the overs and unders of the thread as before.

The final appearance of this filling is quite different from the diagonal cobweb-stitch.

The photograph shows another example of the same stitch worked on black silk net.

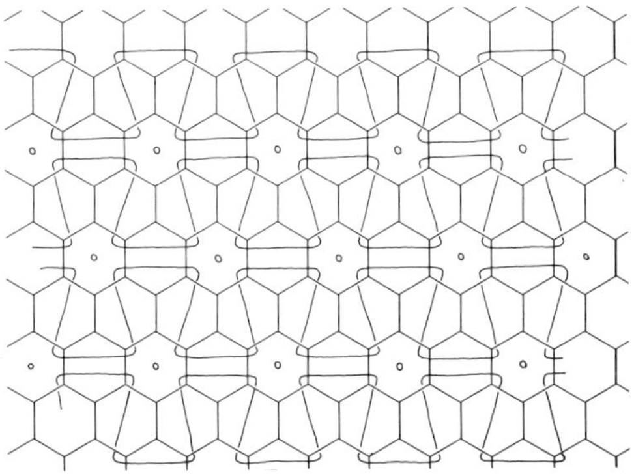

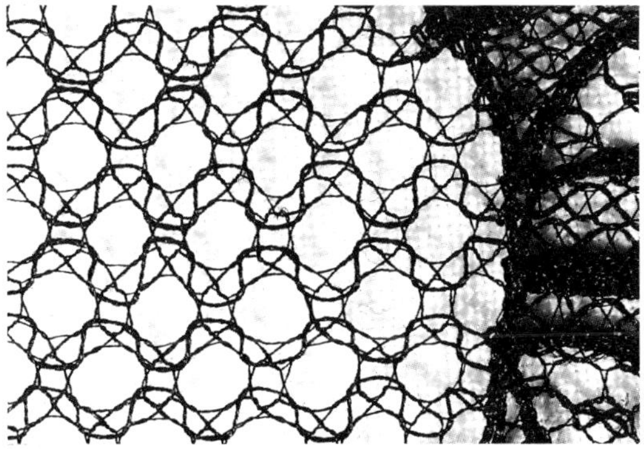

Stitch 3. Cross-stitch and satin-stitch (1)

Diagonal rows of cross-stitch and satin-stitch alternate across the casket, separated from each other by lines of holes.

Rows 1 and 2: Work a row of cross-stitch (p.28).

Miss a line of holes.

Row 3: Wind the thread 3.5 times around alternate holes, carrying it across the back of the intervening mesh.

Miss a line of holes.

Repeat this sequence until the casket is filled.

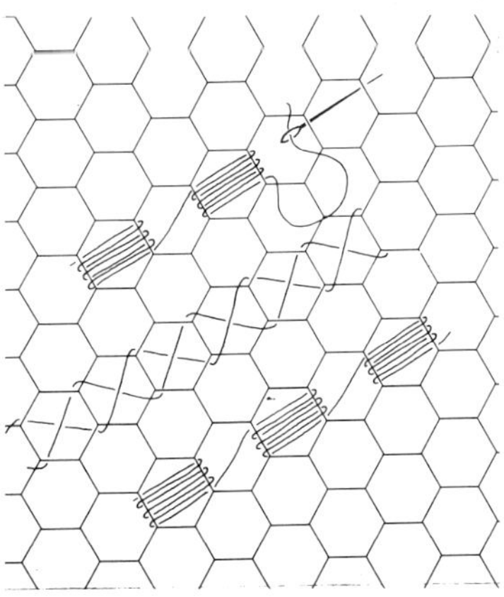

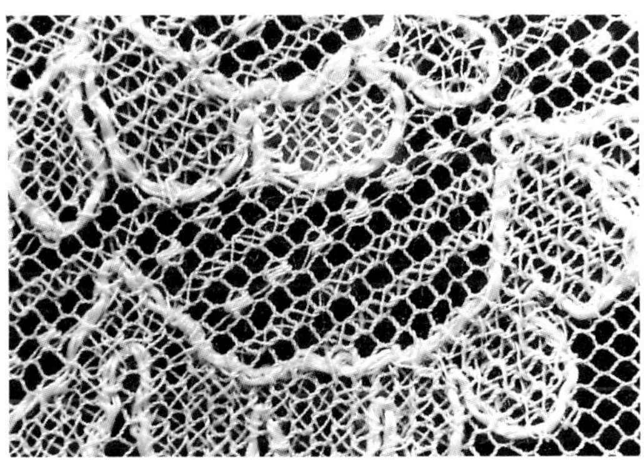

Stitch 4. Cross-stitch and satin-stitch (2)

Work as stitch 3, but in the satin-stitch rows wind the embroidery thread only 1.5 times around alternate meshes, pulling the sides together to enlarge the spaces between them.

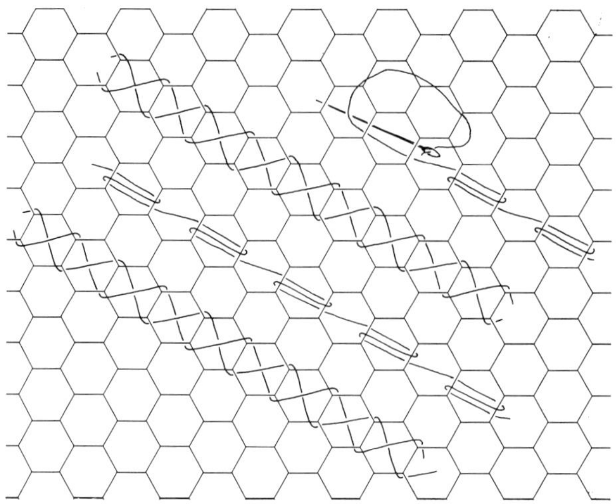

Stitch 5. Zigzag satin-stitch

Row 1: Work a zigzag row of satin stitch, passing the thread 3.5 times around each pair of meshes.
Row 2: Work back again in the opposite direction, to form a mirror-image of row 1.
Repeat these two rows until the casket is filled.

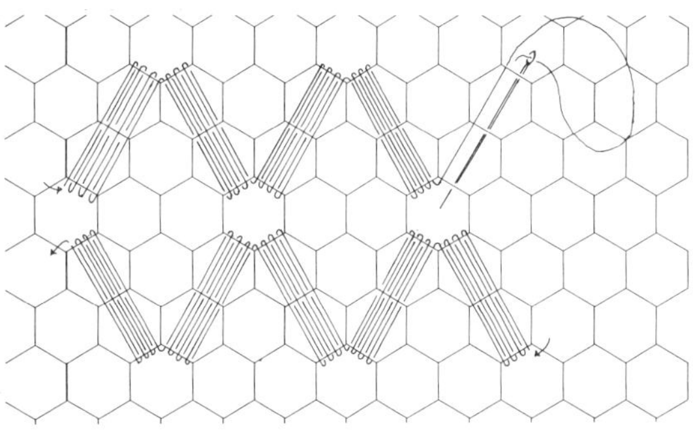

Stitch 5

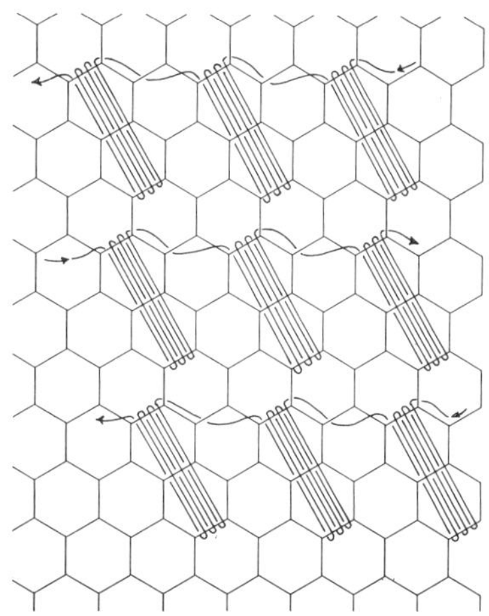

Stitch 6

Stitch 6. Satin-stitch spots (1)

Rows 1 and 2 run up and down between the top and bottom of the casket.
Row 1: Work 3 slanting satin-stitches over pairs of meshes. Leave the next two meshes empty, whipping the thread over and under the sides as you pass from one spot to the next. Repeat to the end of the row.
Row 2: Work back in the reverse direction so that the spots alternate with those of row 1.
Continue until the casket is filled.

Stitch 7. Satin-stitch spots (2)

Row 1: Work diagonal rows of satin-stitch spots, taking the embroidery thread 2.5 times around alternate meshes, carrying the thread from one spot to the next across the reverse side of the missed holes.
Row 2: Work back in the opposite direction to form a mirror-image of row 1.
Repeat these two rows until the casket is filled.

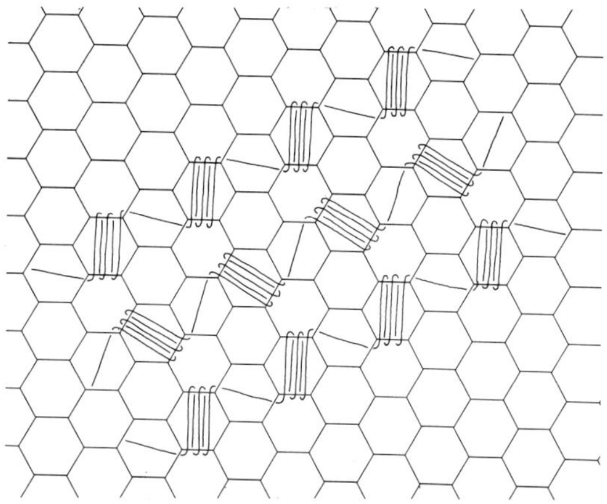

Stitch 8. Cross-stitch lattice

Set 1. Fill the casket with diagonal rows of cross-stitch, spacing the rows 2 holes apart.
Set 2. Work similarly spaced rows along opposing diagonals so that they intersect with set 1 to form a trellis, each unit enclosing 4 holes. At the points of intersection, carry the threads of set 2 across the back of set 1.

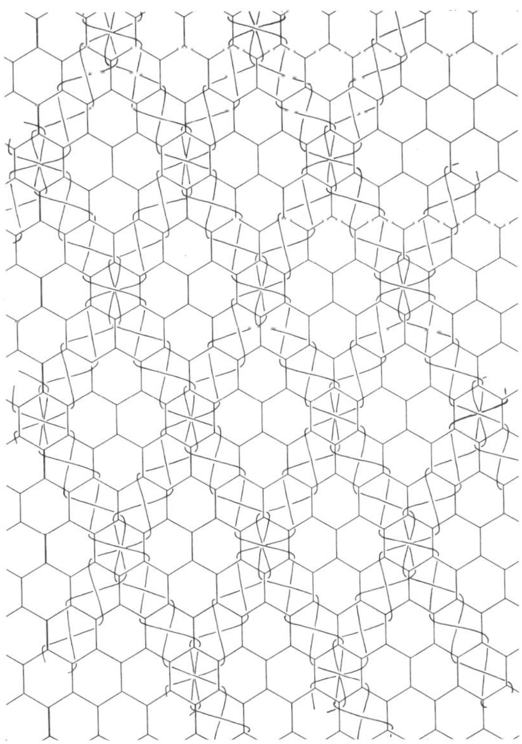

Stitch 9. Tent-stitch with satin-stitch ties

This is most easily worked with the cross-sides aligned horizontally.
Set 1. Work tent-stitch along every fourth row so that 3 rows of empty holes lie in between. Reverse the direction of slant in LR and RL rows to form mirror images.
Continue until the casket is filled.
Set 2. In the mid-row of holes, between the lines of tent-stitch, use 1.5 satin-stitch turns to bind the cross-sides together in groups of 3. The desired effect is shown in the photograph. Assist this movement by enlarging the intermediate holes (o) with a stiletto.
Arrange the ties to alternate in successive rows.

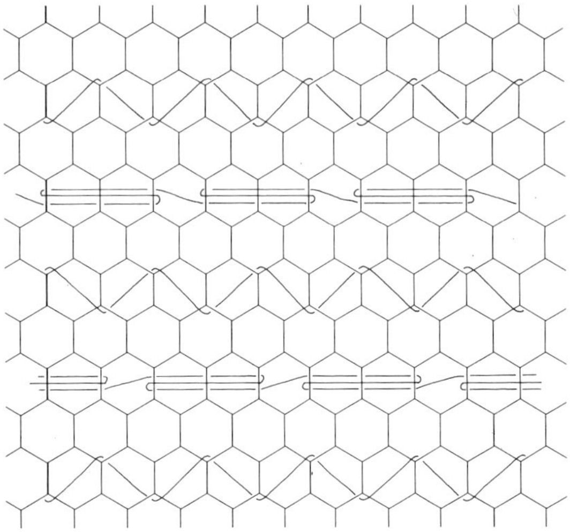

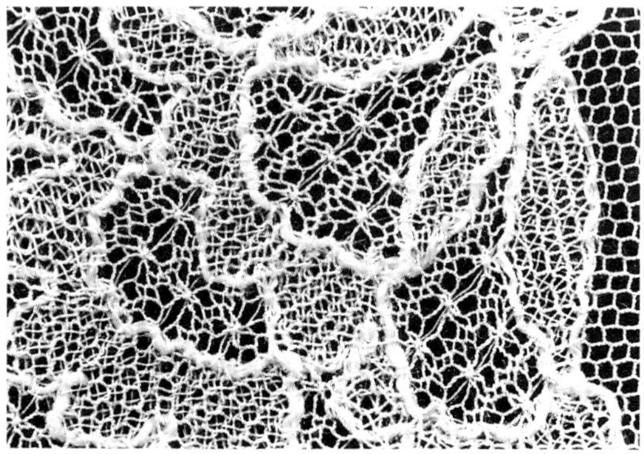

Stitch 10. Crossed figure-of-eight stitch

With the cross-sides aligned vertically, work in LR RL rows, following the over-unders indicated in the line drawings.
Pull on the stitches as you make them so that the mesh sides between them are lengthened.

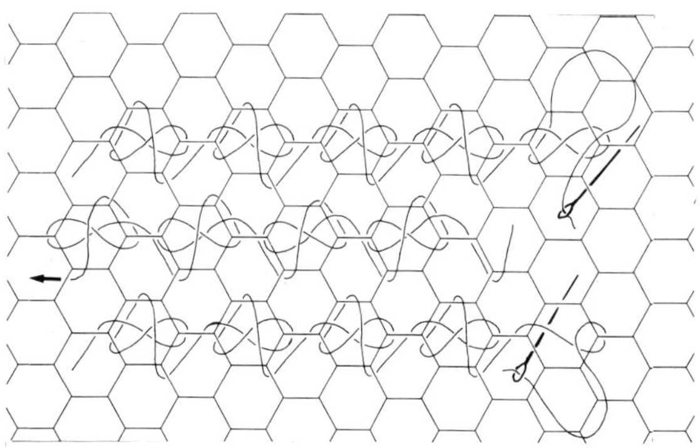

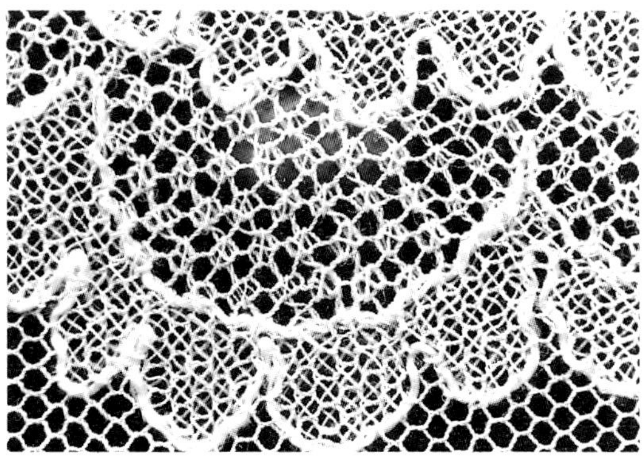

Area 2. Leaves and sprays.

Each spray consists of a pair of large daisies sharing their elegantly curved stems with clusters of alien flowers. Arching over them is a flurry of thick shiny leaves, run in with outline thread. All the stitches are constant from one spray to the next apart from the centres of the daisy heads, ringed by a circlet of holes.

A and B are the areas which change their stitches from one flower to the next. C = circlet of holes. M = midrib

CIRCLET OF HOLES Run in the outline cord in a sinuous manner, going twice around the holes, and passing over and under the mesh sides as indicated. At the end of the first round, reverse direction and return, passing over the cord already laid, at each crossing.
Gently enlarge the encircled holes.

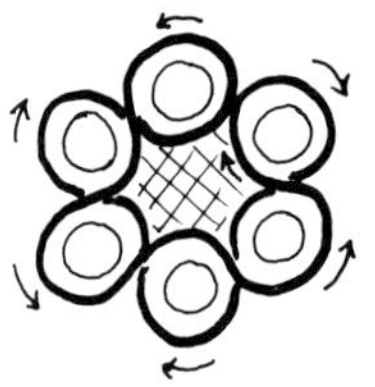

MIDRIBS The holes tapering to a point along the middle of each leaf are worked in the usual way, but only a single outline encloses the entire length.

Stitch 1. Cross-stitch lattice with satin-stitch spots

Stage 1. Construct a lattice as described for stitch 8 (p.35), but spacing the stitch rows 3 holes apart so that each unit encloses 9 meshes.
Stage 2. In the centre hole of each unit, take the thread 3 times diagonally around the twist sides. Carry the thread horizontally (LR) across the back of the work to the centre hole of the next diamond, until you reach the end of the row.
Return RL across the next row of central holes, sloping the satin-stitch spots in the reverse direction.

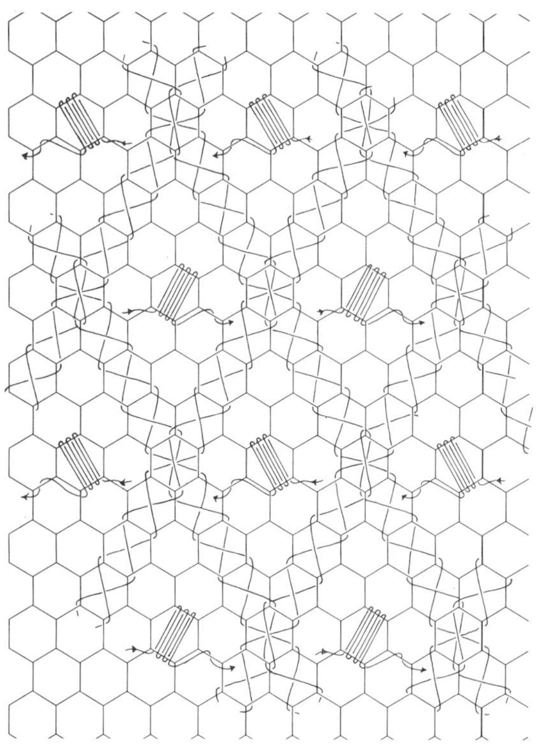

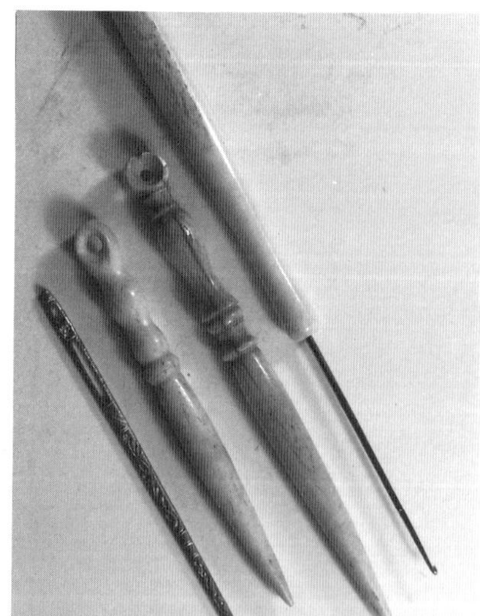

Embroidering implements: a bodkin, two bone stilettos and a hook fixed in an ivory handle that could be used for tambour work

Stitch 2. Cross-stitch chevrons

Row 1: Work tent-stitch (p.26) over a sequence of 5 holes, moving diagonally to the right.
Take the thread downwards under the next hole so that you can reverse its direction. Move diagonally to the left, working tent stitch over the next 5 holes.
Take the thread under the next hole, and reverse the direction again.
Repeat these zigzag movements until you reach the far side of the casket.
Row 2: Work back again in the reverse direction, converting the tent-stitch into cross-stitch. At the turning points of the chevrons, cross the threads in the final hole, as shown, to make a sharp angle.
Repeat these two rows until the casket is filled. Leave a line of holes between the cross-stitch rows.

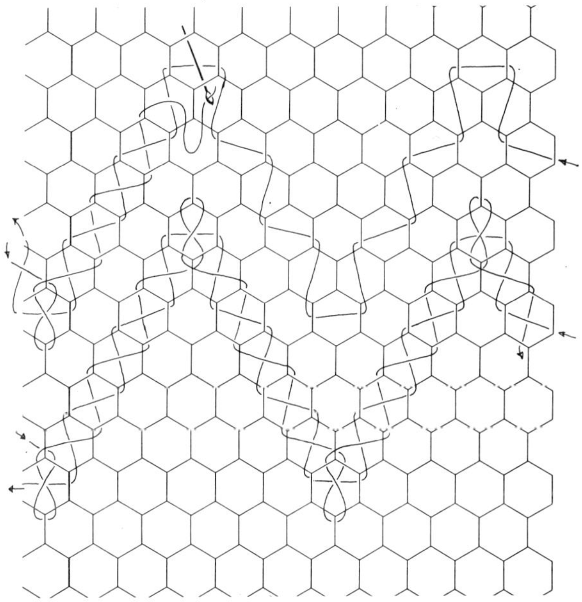

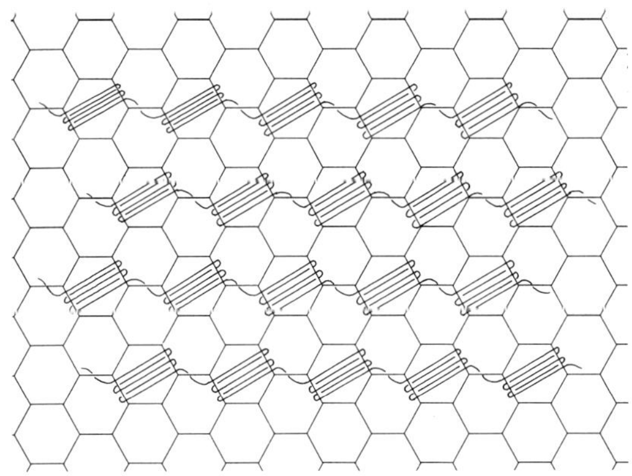

Stitch 3. Satin-stitch spots (compare stitch 7, p.35)

Row 1: LR. Take the thread slantingly 2.5 times around the twist sides of every mesh, in the manner of satin-stitch. Carry the thread under the intervening cross-sides.
Miss a line of holes.
Row 2: RL. Work in a similar manner, slanting the stitches in the same direction. The spots should alternate with those of row 1.
Repeat rows 1 and 2 until the casket is filled.

Stitch 4. Cross-stitch stripes and satin-stitch rosettes

Fill the casket with diagonal rows of cross-stitch spaced 5 holes apart.
In the middle line of empty holes, work a series of satin-stitch rosettes.
*Proceeding LR, take the thread across hole (1) to the centre, wind it twice around 2 twist-sides then, working clockwise, twice around each mesh until you reach (6). At the last stitch, carry the thread across the reverse side to the exit point.
Whip along 2 meshes, and repeat from *.
Rosettes can be worked equally well in a RL direction.
If the thread runs out, take it through to the back of the net to be sewn in later, and join on a new length.

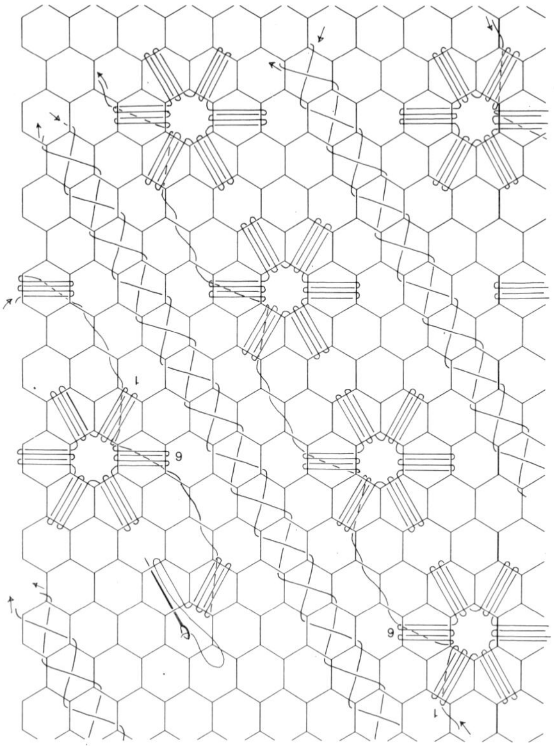

Stitch 4

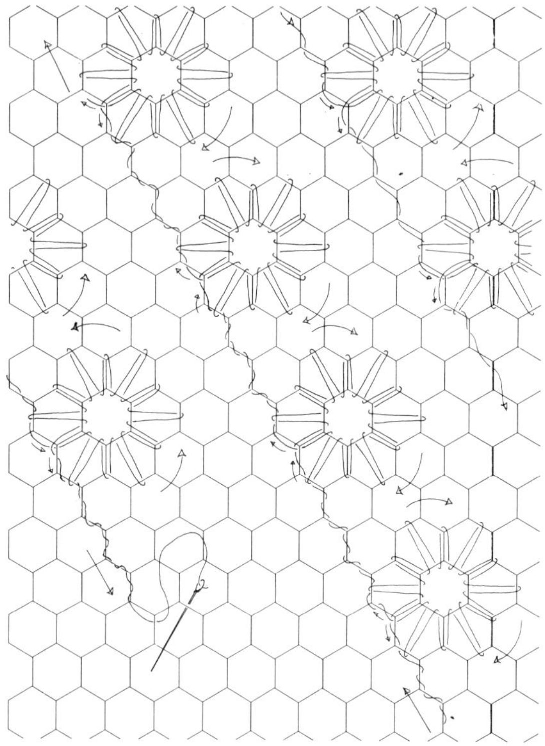

Stitch 5

Stitch 5. Satin-stitch starfish

Work 12 single satin-stitches across the 6 meshes surrounding a central hole. Repeat diagonally across the casket, spacing the starfishes 1 hole apart, and the rows 2 holes apart.
The thread is passed from one starfish to the next by whipping it over the intervening mesh sides.
In the LR rows, the starfish are worked anti-clockwise, in the RL rows clockwise (see arrows).
The working of this stitch in the antique veil is somewhat irregular.

Stitch 6. Cross-stitch stripes with multiple ties

Work 3 rows of cross-stitch horizontally. Miss a line of holes.
Work 5 rows of tight binding stitches similar to those of stitch 9 (p.36), but encircle only 1 mesh at a time instead of 2. Miss a line of holes.
Repeat this sequence until the casket is filled.

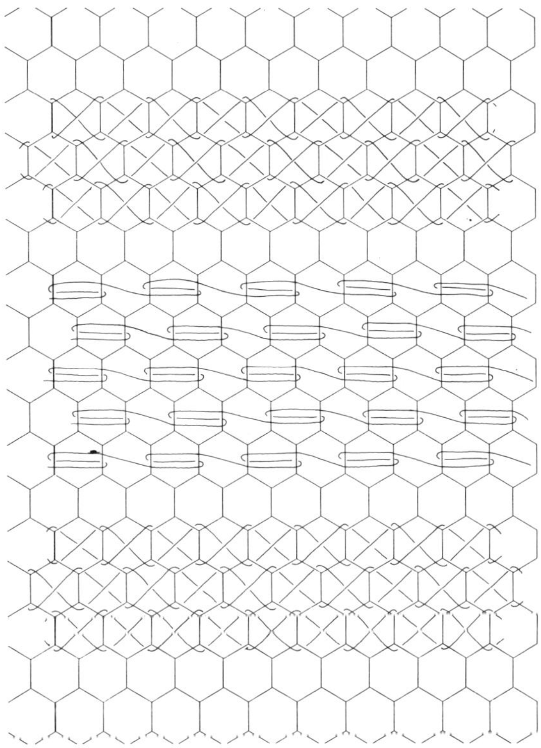

Left: stitch 6 inside its flower centre. Note the cross-stitch fillings of the solid areas, and the outline cord rimming the circlet of enlarged holes

Right: detail, from the reverse side

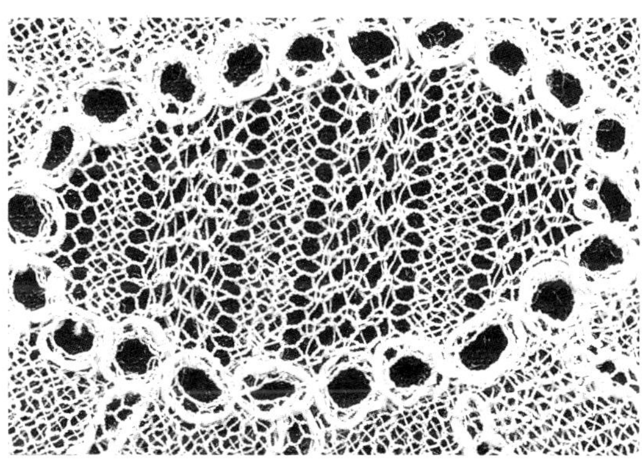

PLASTRON WITH CHOKER COLLAR

This plastron or over-bodice presents an interesting contrast to the Square Veil with Flowers (pp.31-41). The net is a 4-twist diamond mesh. The outline cord is double and quite thick. The solid design areas consist entirely of darning stitch worked in a single cotton thread barely thicker than the twisted mesh sides. The casket stitches dominate the design. Worked with double strands of sewing thread, their dramatic patterns stand out boldly against the pale light setting of flowers and leaves.

Note: In the line drawings the double threads are shown as single. The plastron has been distinguished into four areas to show the distribution of the decorative stitches more clearly: Front, Border, Yoke, Collar.

Collar (choker)

Yoke (across shoulder)

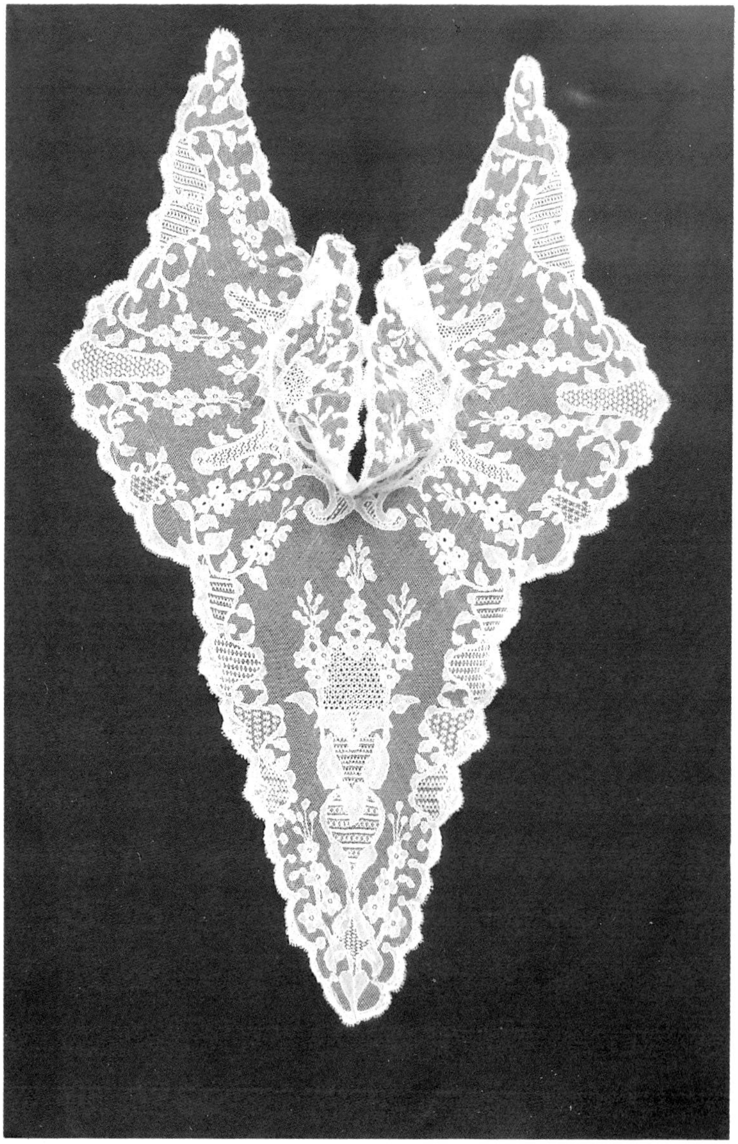

Front (centre)

Border (along the edge)

Stitch map of the plastron front

Stitch map of the plastron border

Area 1. Plastron front

Stitch 1. Interlocking tent-stitch

Throughout the casket, 3 sets of tent-stitch pass horizontally and diagonally, intersecting to give the appearance of stout overlapping rosettes centred upon expanded holes.

Set 1. Row 1: Work tent-stitch along a horizontal row.
Leave a line of holes.
Row 2: Work tent-stitch back again, in mirror image form.
Leave a line of holes.
Repeat this sequence until the casket is filled.

Set 2. Work diagonal rows of tent-stitch from top L to bottom R, each mirror-imaging the row before. Leave a line of empty holes between each row.

Set 3. Work similar diagonal rows of tent-stitch from top R to bottom L.

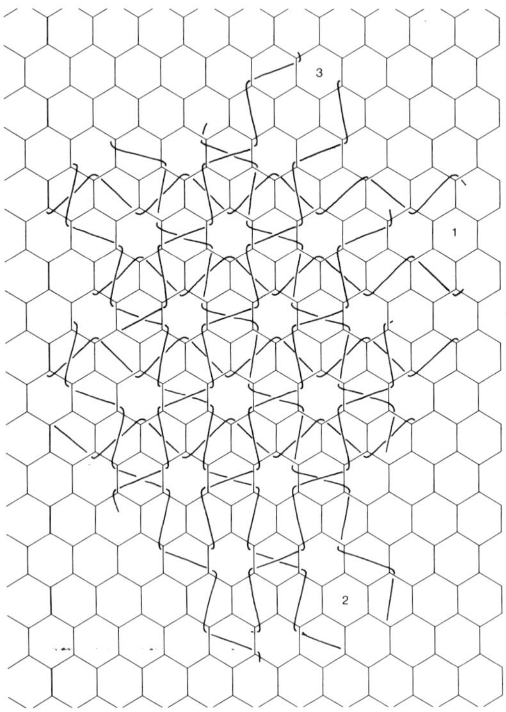

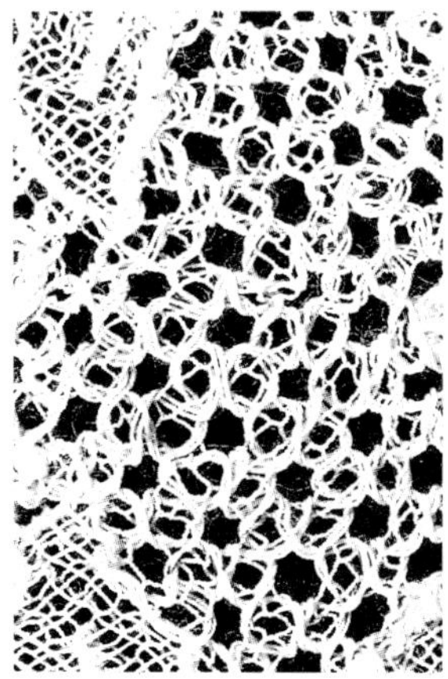

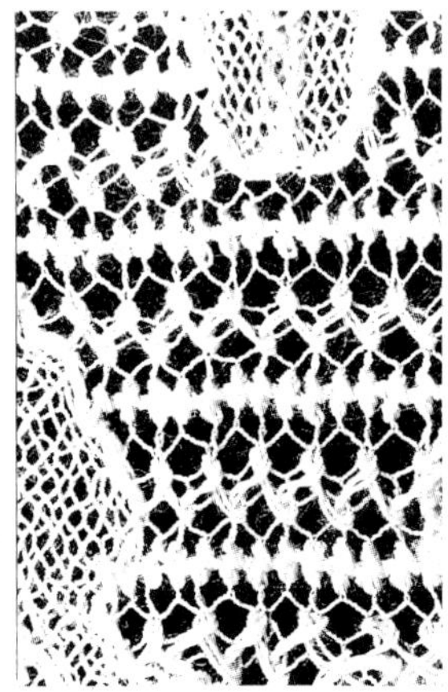

Top left and middle, stitch 1

Top right and below, stitch 2

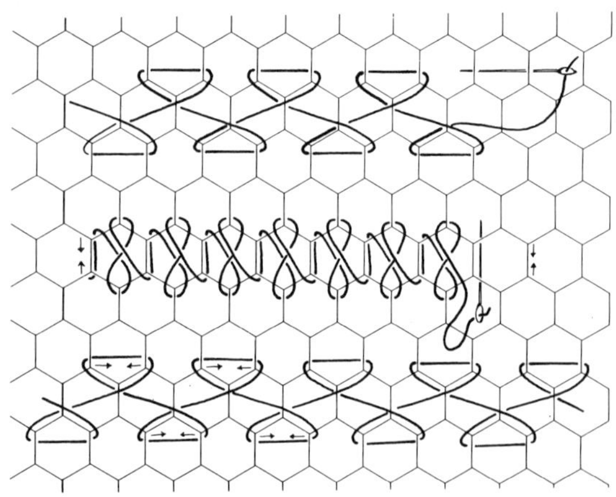

Stitch 2. Herringbone and figure-of-eight stitch

Both stitches pull the mesh walls which they encircle inwards, until they touch.

In herringbone, the compression is from side to side. In figure-of-eight, the compression is vertical (see arrows).

Stitch 3. Large cross-stitch with figure-of-eight ovals

Rows of over-size cross-stitch alternate with rows of large ovals, each row separated from the next by a line of holes.

Set 1. Cross-stitch rows: Work each tent-stitch unit over 2 rows in height and across 3 in width. Three movements over the same meshes - LR, RL, LR - are needed for a full cross-stitch row to be completed. The top L to bottom R threads lie on the surface throughout.

It may be helpful to turn the work round between rows so that the stitch movements can be repeated in the reverse direction, and in effect you are working continually from L to R.

Set 2. Ovals: each big oval encloses 4 holes.

Row 1: LR. Work a series of loops somewhat similar to extra-large detached buttonhole stitches. Catch down the top of each oval (A) and the bottom of each cross-over portion (B) by taking the thread beneath a cross-side.

Row 2: RL. Return with the thread movements reversed so that the stitches overlie those of row 1, completing the appearance of ovals and figures-of-eight.

Stitch 3

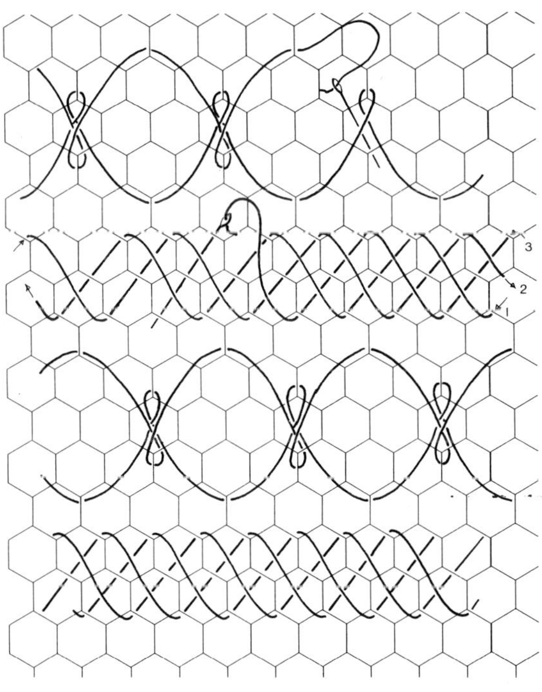

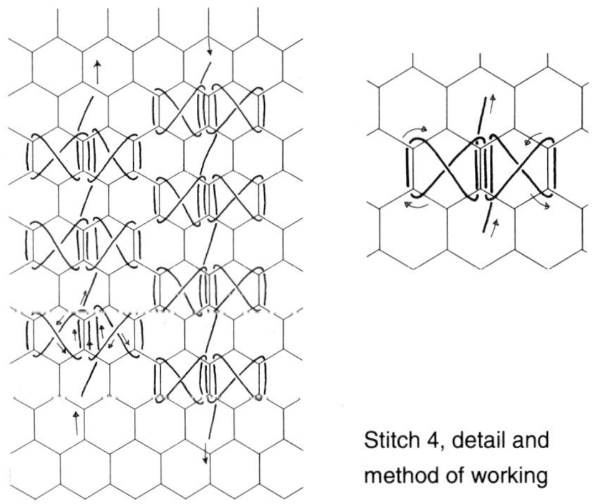

Stitch 4, detail and method of working

Stitch 4. Double figure-of-eight

In each row, the encircled mesh sides are squashed together until the hole within them vanishes. At the same time the cross-sides, between the stitches, are elongated.

Row 1: LR. The course of the embroidery thread is indicated by arrows. The effect of the horizontal compression is shown in the photograph.

Row 2: RL. This is most easily worked by turning the lace around so that you again proceed from L to R, in effect repeating the previous rows. The direction of the crossing threads will then appear reversed.

*Area 2. **Plastron border.***

Four different casket stitches are found along the plastron edge. Enclosure (5) contains tent-stitch.

Stitch 1. Cobweb- and fern-stitch

Row 1: Using the double embroidery thread, work a single row of cobweb-stitches (see st.2, p.33) across the width of the casket.
At the end of the row, whip the thread twice around the outline, and past 3 lines of empty holes to the position for row 2.
Row 2: In alternate holes work 3 radiating satin-stitches, arranging the tip of the tallest to lie just below the indent of a row 1 cobweb-stitch.
Row 3: Work a row of cobweb-stitch to mirror-image row 1.
Row 4: Work another row of fern-stitch, making the triple fronds alternate with those of row 2.
Repeat rows 1 to 4 until the casket is filled.

Stitch 2. Double figure-of-eight stitch

This is an enlarged and modified version of stitch 4 (p.45).
Row 1: Work LR, following the sequence of thread movements shown in the drawing.
Row 2: Turn the net round so that the original R side is now at the L. Work back in the opposite direction, repeating row 1. All the over and under thread-crossings will appear reversed.
Repeat these 2 rows until the casket is filled.
The holes (x) are slightly enlarged by the pull of the stitches.

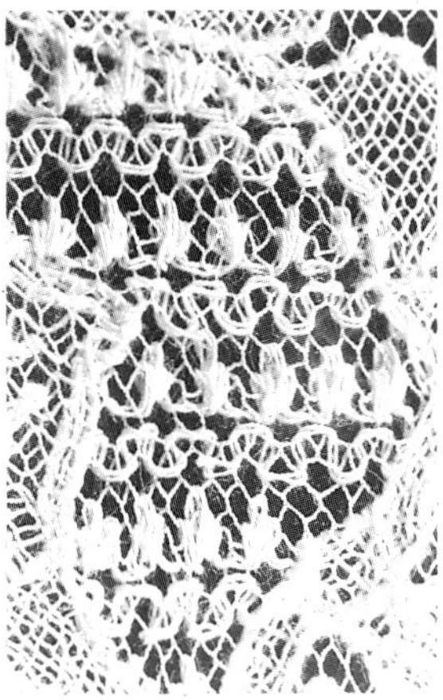

Stitch 1

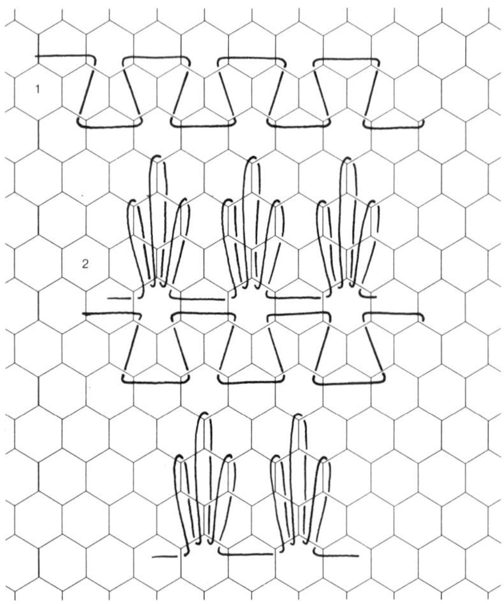

Stitch 2

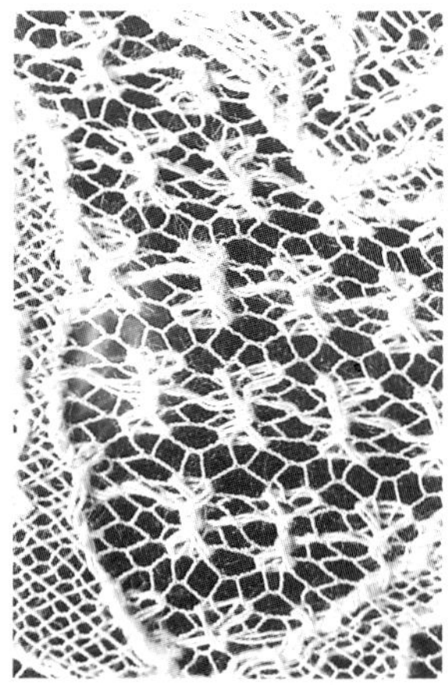

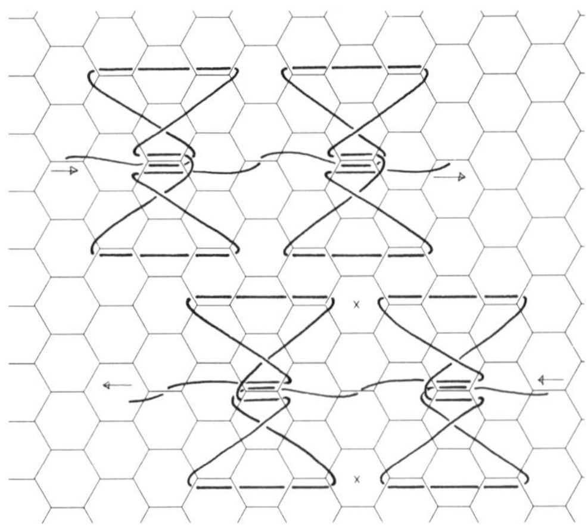

Stitch 3. Figure-of-eight and cobweb-stitch

Row 1: LR. Work the figures-of-eight as shown, passing over the twist-sides and under the cross-sides, following the direction indicated by the arrows.

Row 2: RL. Work a row of cobweb-stitches, passing the horizontal sides over the central crossing of each figure-of-eight, and the slanting sides under the meshes.

Repeat these two rows until the space is filled.

The figures-of-eight of successive rows are vertically in line with each other.

Note: In the line drawing, rows 1 and 2 are shown separately at the L and R sides, before being superimposed.

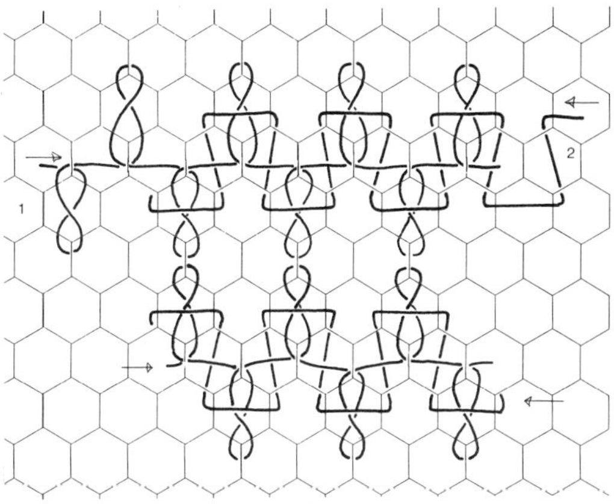

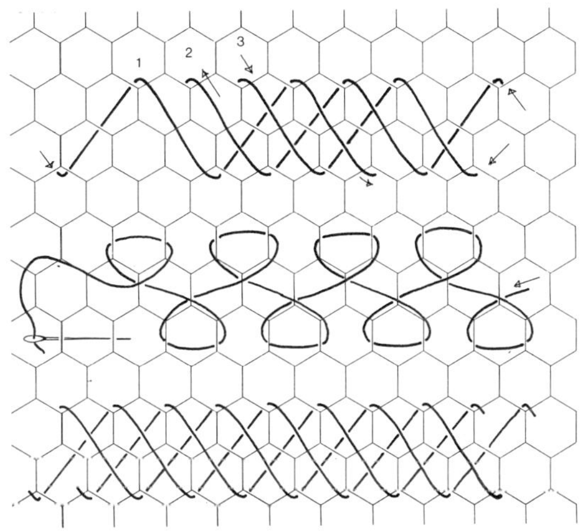

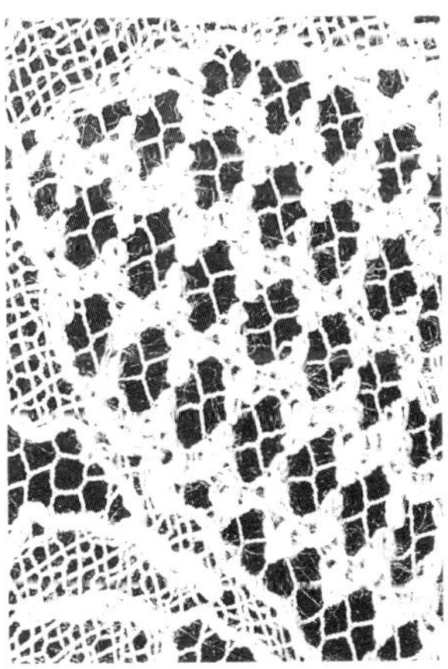

Left: Stitch 3

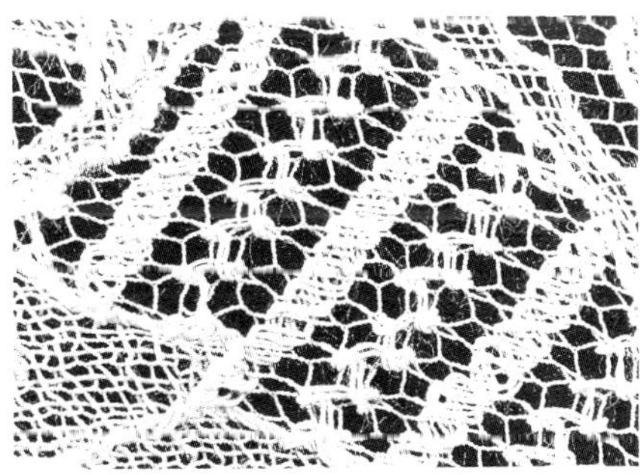

Right: Stitch 4

Stitch 4. Cross-stitch and herringbone-stitch

*Rows 1, 2 and 3: LR, RL, LR. Work large cross-stitches - over 3 mesh sides and under 3 mesh sides as in st.3, p.45. The tightening of the threads causes the meshes which they enclose to be pulled vertically together.

Miss a line of holes.

Row 4: RL. The herringbone-stitches are worked as shown. With each stitch, tighten the upper and lower loops to obliterate the space within them.

Miss a line of holes.

Repeat from * until the space is filled.

Note: At the top of the line drawing, the 3 rows of tent- stitch are shown separately.

Area 3. Plastron yoke.

Stitch map

Stitch 1. Cross-stitch and detached buttonhole stitch

Row 1: Work a row of loops like very large detached buttonhole stitches. Tighten their closure portions, which encircle 2 cross-sides, to constrict the hole within, at the same time enlarging the hole (x) between them.
Row 2: Turn the work round so that the former R side is now the L. Repeat row 1.
Rows 3, 4, 5: Work outsize cross-stitches as in st.4, p.47.
Repeat this sequence until the casket is filled. In the repetition of rows 1 and 2, the cross-over direction will be reversed.

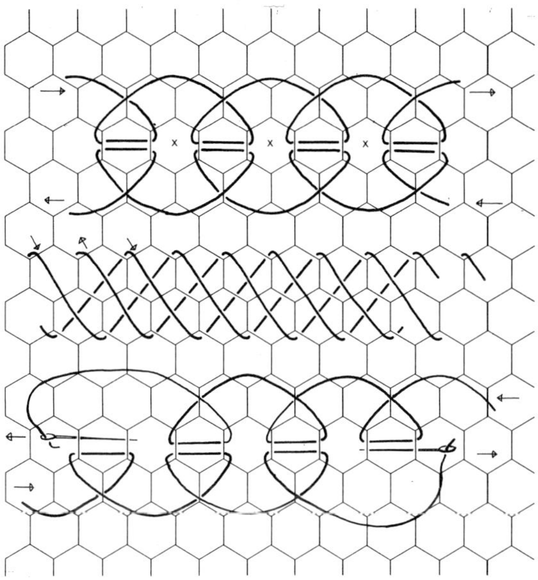

Stitch 2. Figure-of-eight

This large stitch is worked over 3 horizontal and 2 vertical meshes.
Row 1: Follow the direction of the arrows, noting where to pass over and under the mesh sides. The upper and lower loops of the '8' pull the meshes together, slightly enlarging the holes on either side.
Row 2: Turn the work round and repeat row 1, making the figures-of-eight alternate with those of row 1. The direction of the thread-crossings will appear reversed.
Repeat rows 1 and 2 until the casket is filled. Turn the work round between every row so that the direction of the embroidery is constantly from L to R.

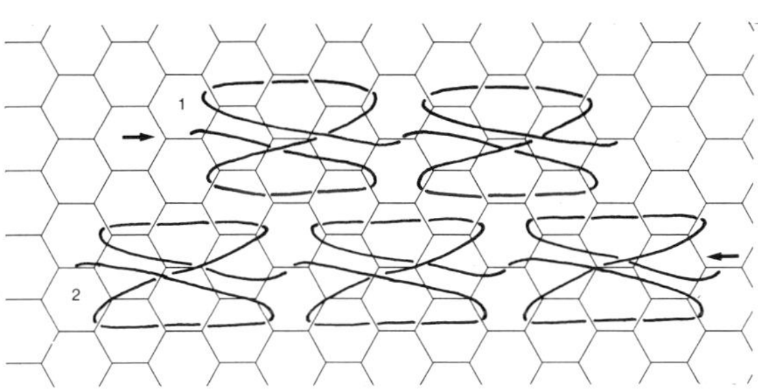

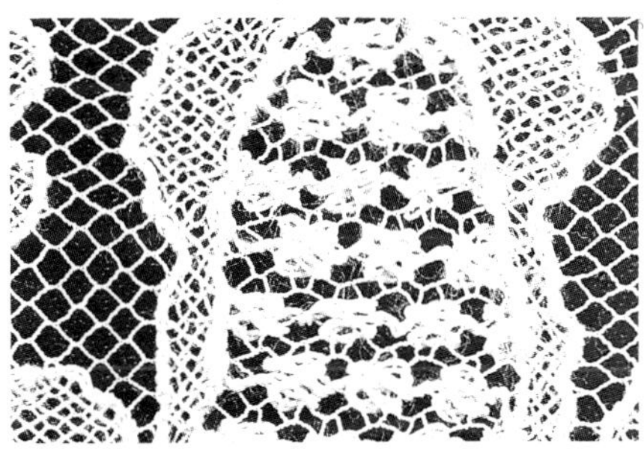

Stitch 3. Figure-of-eight trelliswork

The stitches are aligned so that the enclosures lie diagonally to the casket. In the working, however, the cross-sides are aligned horizontally and, in relation to this, one set of figure-of-eight rows runs vertically and the other set diagonally.

Set 1. Row 1: Work a vertical row of figure-of-eight stitches.

Leave two lines of holes.

Row 2: Turn the piece round and follow the arrows, working parallel to row 1. The thread crossings will appear reversed.

Fill the casket with Set 1.

Set 2. Repeat the thread movements of set 1 in a diagonal direction. The rows will now intersect to form a trelliswork of equal-sized diamonds, each enclosing 7 holes. The central hole (o) is slightly enlarged in a vertical direction (see photograph). The four corners of the diamonds are kept neat by taking the thread of set 2 beneath the stitch already worked there. The shape of the set 2 stitches is slightly different from those of set 1 because of their different relation to the cross-sides.

Repeat the rows of set 2 until the trellis fills the casket.

Stitch 3

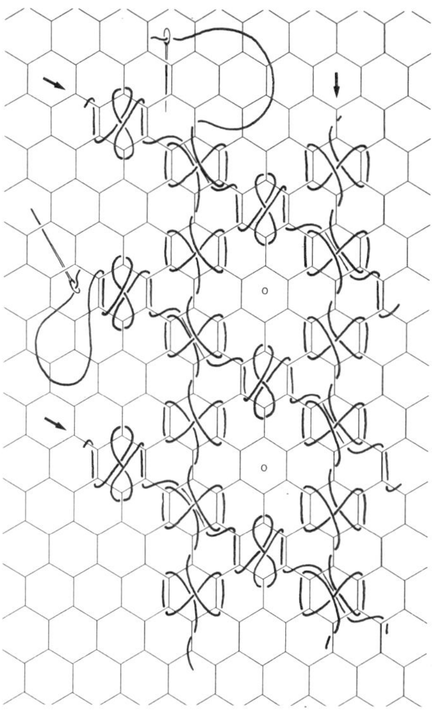

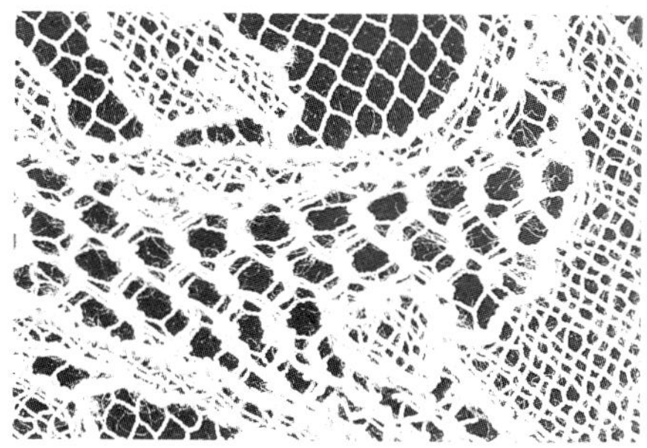

Stitch 4

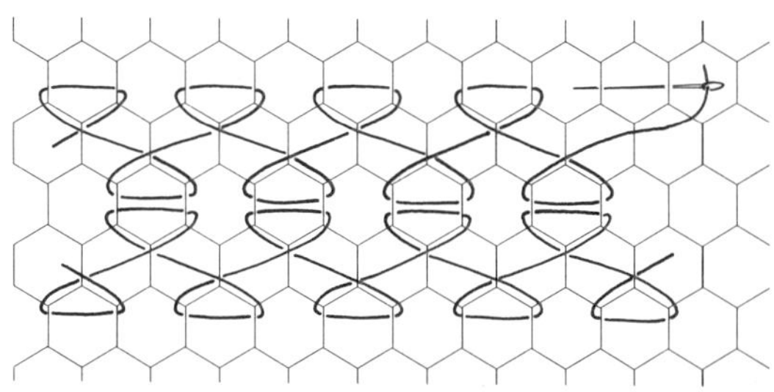

Stitch 4. Herringbone-stitch

The working is similar to st.4, p.47, but here herringbone fills the whole casket.

Turn the work round between successive rows, to preserve the LR direction of the embroidery. Again, constriction by both the upper and lower loops gives a compact appearance to the stitches, and at the same time increases the openness of the intervening spaces.

Stitch 5. Six-point stars

The stars are not worked as complete units, but in double rows, LR and RL, each row completing half the star.

Row 1: Work a single slanted satin-stitch over each of 3 adjoining meshes so that they radiate from a common centre. Pass over the next hole, taking the thread under one twist-side for stability. Work another half-star of 3 stitches, and continue in this way until you reach the end of the row.

Row 2: Turn the work round and return, completing the stars by working a mirror-image of row 1.

Repeat rows 1 and 2 until the casket is filled, spacing the lines of stars close together, and arranging them to alternate with each other.

Left: Stitch 5

Below: Stitch 6

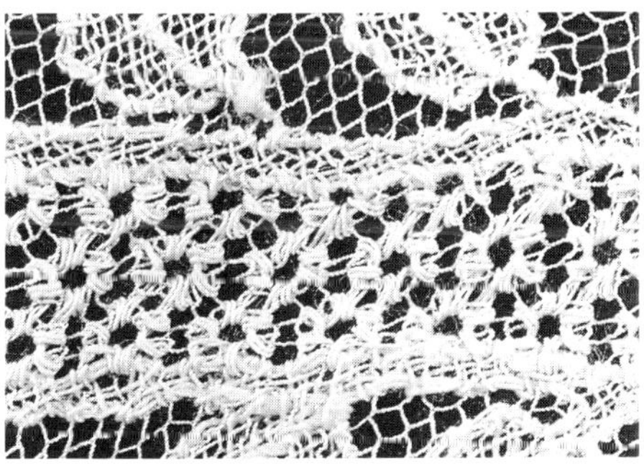

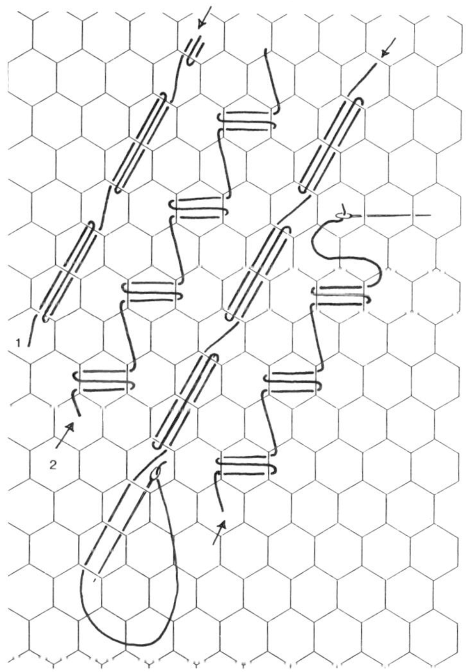

Stitch 6. Satin-stitch ties

*Row 1: Working diagonally, encircle each pair of meshes, passing the thread around the twist-sides 1.5 times. Tighten until the holes on either side of the stitch appear stretched (see stitch 9, p.36 and stitch 6, p.41).

Leave a line of holes.

Row 2: Working diagonally, in parallel to row 1, encircle the cross-sides of alternate meshes, tightening the thread until the central hole is almost obliterated. Note that the walls are crushed together in a direction at right angles to the compression of row 1.

Leave a line of holes.

Repeat from * until the casket is filled.

All the row 2 stitches are horizontally in line with each other across the casket.

Area 4. Plastron choker.

There is only one decorative stitch. It lies in the tiny caskets marked (1), and is worked in the same way as stitch 1 (p.44).

Actual size: 47 x 240 mm

* * * *

Collar with shamrocks and periwinkles

COLLAR WITH SHAMROCKS AND PERIWINKLES

The shamrock leaves and periwinkle flowers are worked in **darning-stitch** (1), and most of the periwinkle leaves in **tent-stitch** (2), a combination of the two giving an effect of shading.

A fine round-meshed net is used. The embroidery thread is single, no wider than a mesh side. The outline thread is of a thicker cotton used double.

Left: Stitches 1 and 2

Below: Stitch map

Stitch 3. Figure-of-eight

This fills the centres of the periwinkle flowers.

Follow the over-under movements shown in the drawing. Each loop of the '8' pulls on the two cross-sides it encircles, making the surrounding twist-sides stand out like radiating spokes.

At the end of each row turn the work round, so that the embroidery is again in a LR direction. The over-unders of the thread movements will appear reversed.

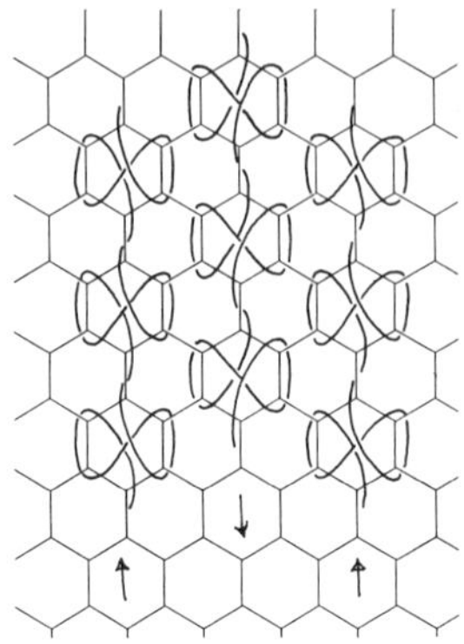

Stitch 3. See photograph p.53

Stitch 4. Interlocking tent-stitch

This is similar to stitch 1 (p.44), but worked on a larger scale.

Study the layout of the stitches shown in the line drawing then, using a stiletto, gently enlarge the holes (o) until they occupy the space of four normal holes.

Work tent-stitch in three different directions - horizontal, diagonal top L/bottom R, and diagonal top R/bottom L - arranging them so that in both diagonals the points of alternate rows face towards each other, i.e. are mirror-imaged.

Each tent-stitch side crosses 3 meshes, and the stitches from the three directions converge at the enlarged holes. The over-under sequence will depend on which set of stitches you work first: the rows worked last will lie on top of any other embroidery threads that they cross.

The fineness of the embroidery thread means that the meshes are more obvious than the stitches themselves. This is less apparent in the line drawing which omits the dilated centre.

Left: the appearance of st.4 in the lace

Right: one set of diagonals

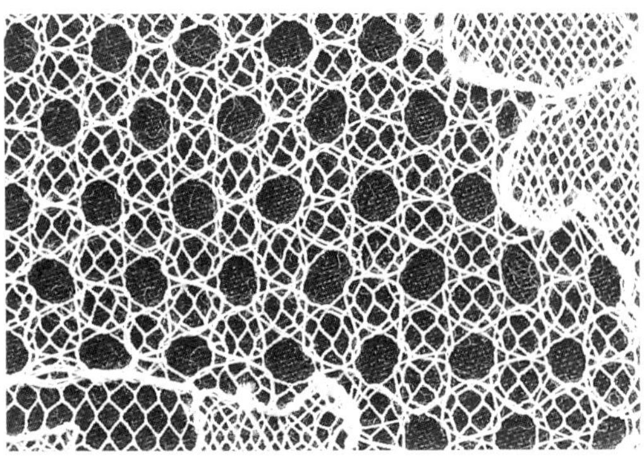

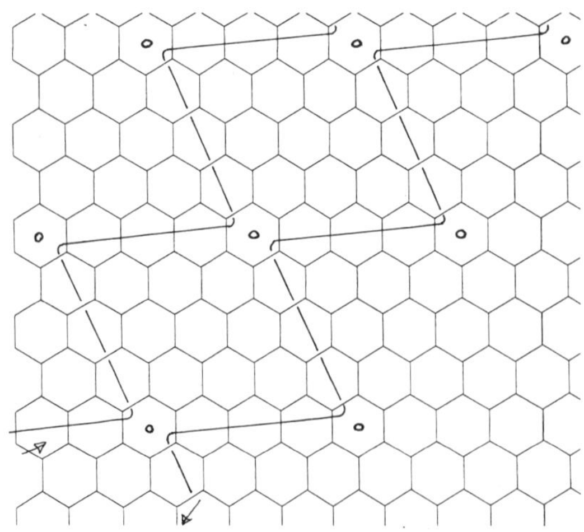

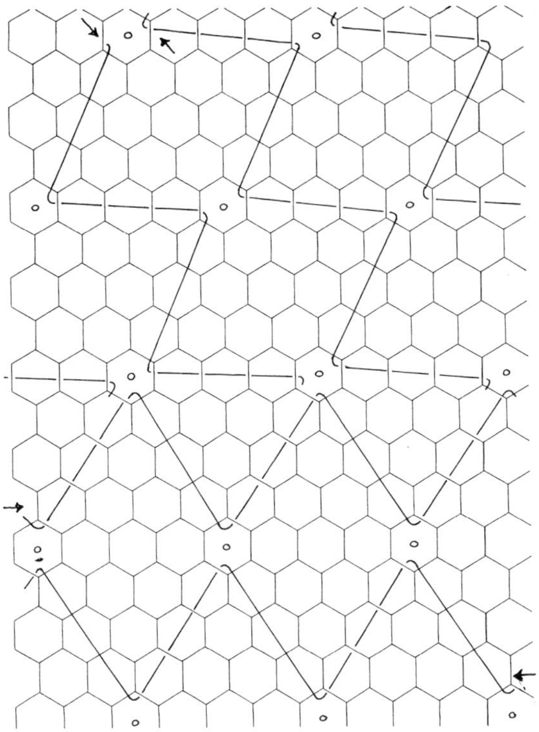

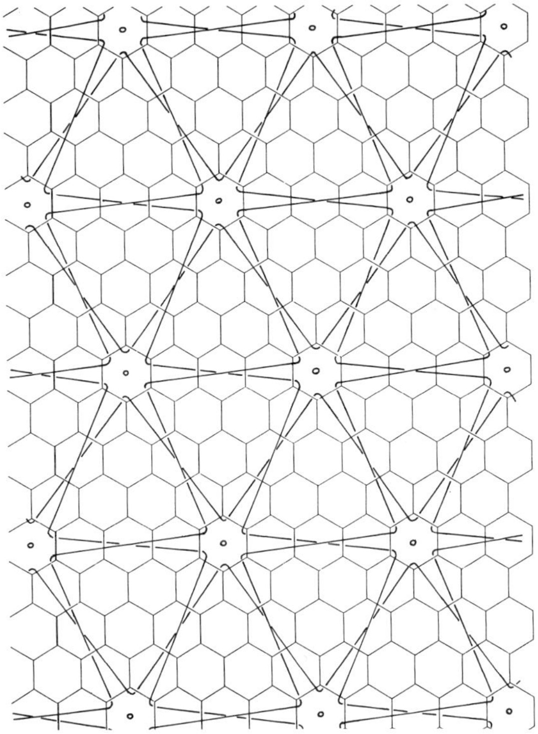

Above: the other set of diagonal stitches

Below: the horizontal set

The completed working

Stitch 5. Diagonal herringbone

Row 1: LR. Work the rows of stitches diagonally across the casket, so that each stitch passes over two meshes.
Leave 2 lines of holes.
Row 2: Turn the net round and work back, again progressing from L to R. The under and over crossings will appear reversed.

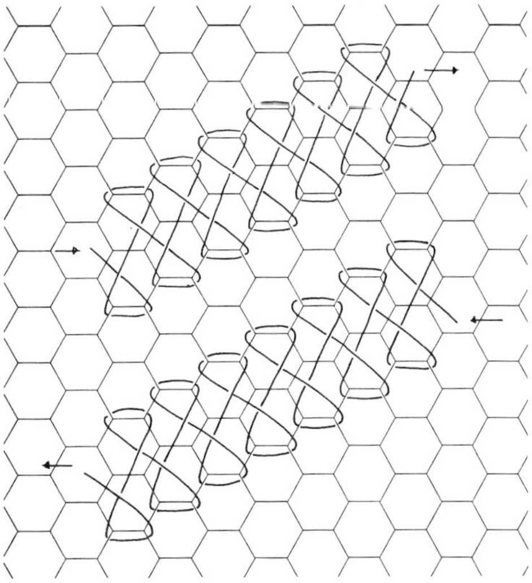

Stitch 6. Double figure-of-eight stitch

The thread movements are the same as for stitch 2 (p.46), but note the striking difference in appearance which results from the use of a much finer thread.

Stitch 6

Stitch 8, detail

Stitch 7. Diagonal tent-stitch

This is used for lighter shading in some petals of the periwinkle flowers. An empty line of holes is left between the stitch rows (see p.26).

Stitch 8. Scattered snow crystals

These are dispersed in a neat and regular manner through the net. Double threads are used to give tactile strength and visual impact.
Follow the route indicated by the arrows, until you are back at the starting point. As you complete each individual loop, carefully tighten it to pull the encircled sides together, until the completed stitch formation takes on the appearance of a hexagonal crystal.
Finish off the beginning and end threads by running them through the back of the spokes.

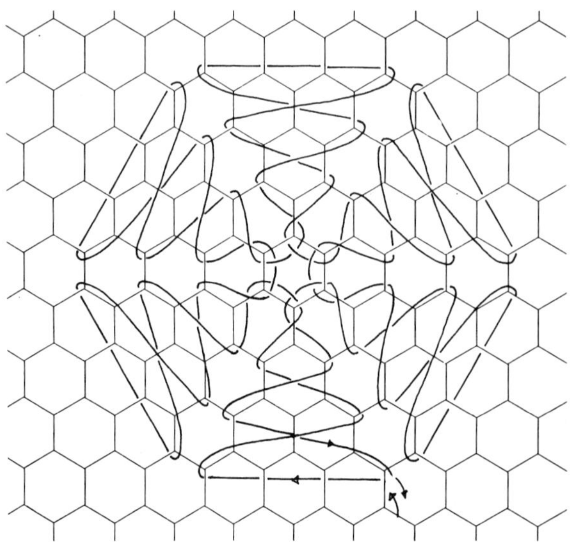

The hole at the tip of the lateral sprigs is worked in the usual manner (p.29).

SHAWL COLLAR WITH LONG TIES

This slightly art nouveau design shows gentian and periwinkle flowers mingled with exotic fruits.

The double-strand outline is sharply clear. The solid design areas contain a light-and-shade mixture of cross-stitch and darning-stitch.

The embroidery thread varies in thickness from the same to twice the width of the mesh sides. It is used single for all the infillings.

An interesting feature is the open loops which lie within the bases of the smaller leaves. They invade like seedless placentas the fruit caskets, and give a scalloped border to the larger petals.

Another unusual feature is the appearance of chain-stitch as a decorative filling (see stitches 1, 2 and 3). If you are not yet familiar with tambouring, these caskets can be worked with a sewing needle.

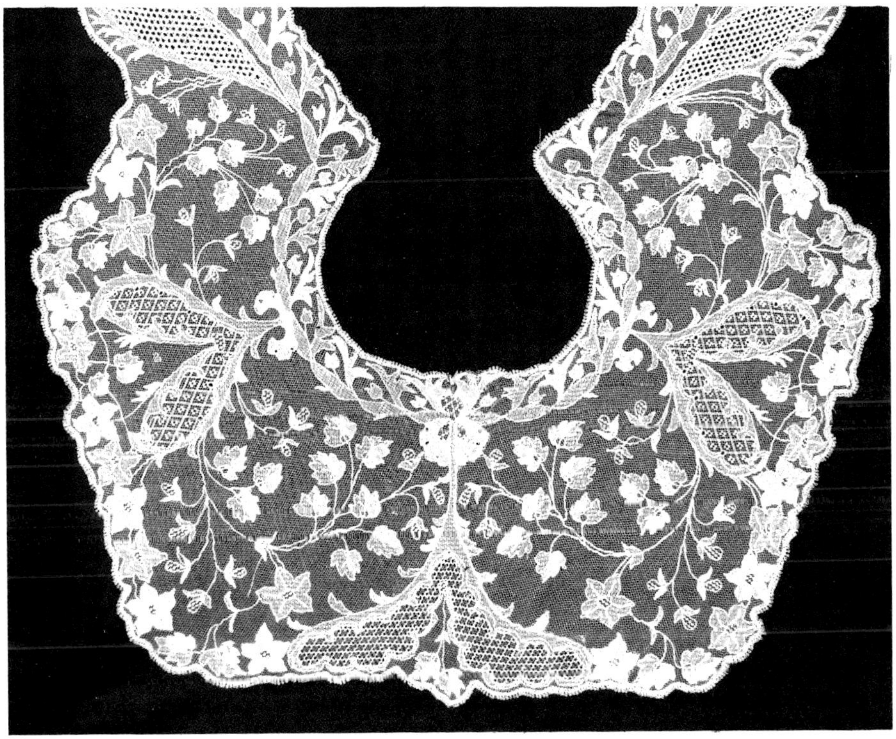

Above: the collar

Below: part of one of the ties
The decorative fillings vary within the ovate pods

Stitch map of the end of one of the ties

Stitch 1. Chain-stitched strings of beads

*Beginning at the lower side of the casket, work 2 or 3 chain-stitches in an upwards direction.

Work 4 chain-stitches over the next hole, taking them round in a tight circle so that they are almost superimposed on each other.

Work 1 more stitch, to A, then make a circle over the surrounding 12 meshes, until you are back at A.

Repeat from * to the end of the row.

Work all the rows in a similar manner, but alternating the positions of the circles. Rotate the net as described earlier (p.46), so that each row is worked in an upward direction.

Continue in this way until the casket is filled.

Stitch 2. Chain cobweb-stitch with satin-stitch spots

Preliminary. Working horizontally, enlarge every second hole (o) with a stiletto. Leave 3 empty lines, and repeat, so that the dilated spaces alternate with the first set of holes.

Set 1. Satin stitch: passing diagonally, take the thread 3 times vertically around 3 meshes, as shown. Carry the thread beneath the net from the bottom of the lower spot to the top of the one above. Tighten the stitches to reduce the size of the enclosed area.

Repeat this row at the intervals indicated until the casket is filled.

Set 2. Using chain-stitch, work a line of cobweb-stitches horizontally across the net, so that they link the large holes together. Each cobweb-stitch is made up of horizontal and vertical chain-stitches big enough to pass around several meshes. Tighten the thread to create the appearance shown in the photograph.

Repeat this row, arranging the cobwebs so that every angle pulls on the enlarged hole.

Alternate rows are worked in the reverse direction.

Stitch 3. Chain-stitch lattice

Set 1. Work rows of chain-stitch diagonally across the net, following the mesh sides to give a lobed appearance.

Space the rows 3 holes apart.

Set 2. Work a second set of rows along the opposing diagonal, spacing them 3 holes apart, so that each enclosed unit contains 9 holes. At the points of intersection, the set 2 chain-stitch rows will lie immediately on top of the set 1.

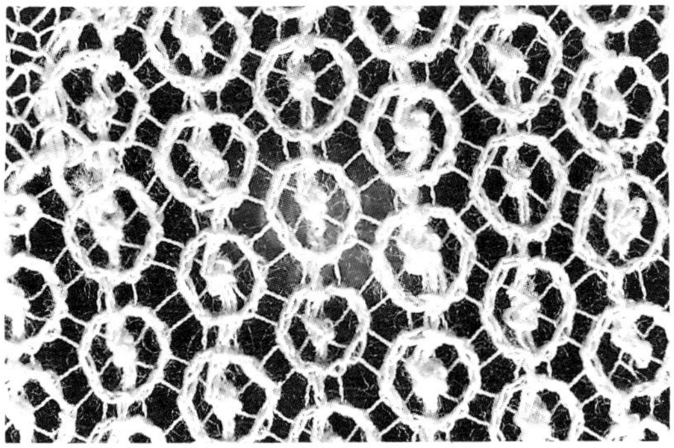

St. 1

St. 1

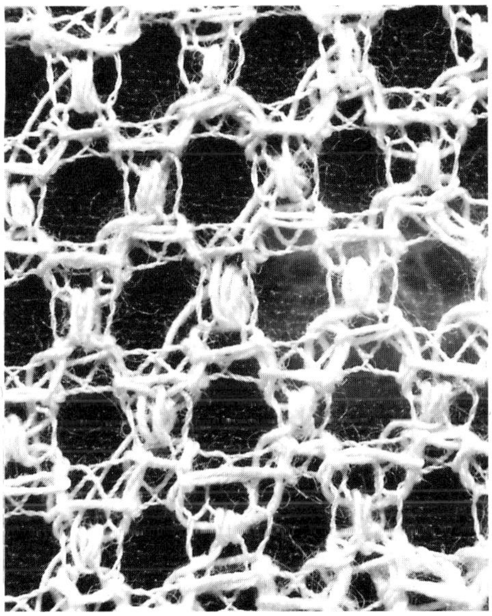

St. 2

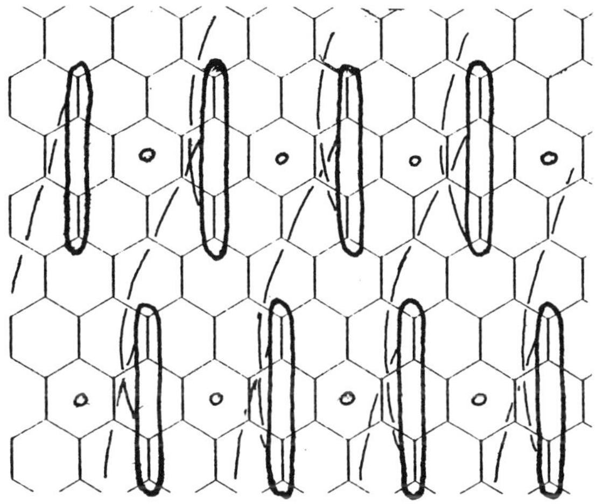

In st.2 the chain-stitched cobweb stitches cannot be fitted properly into a two-twist net, the cross-sides are too long and too inflexible. The sketch (right above) shows the satin-stitch spots and how they are connected

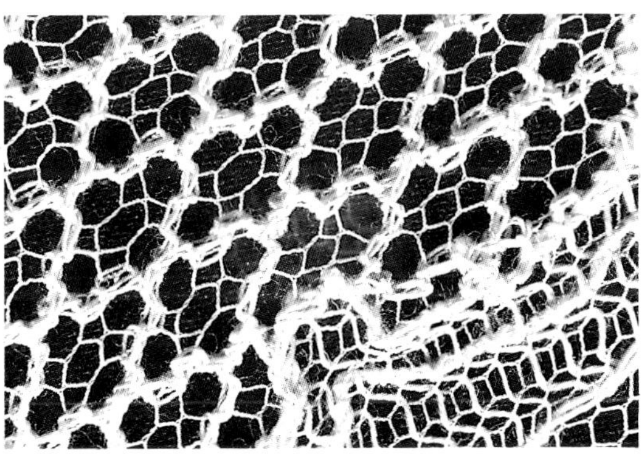

St. 3

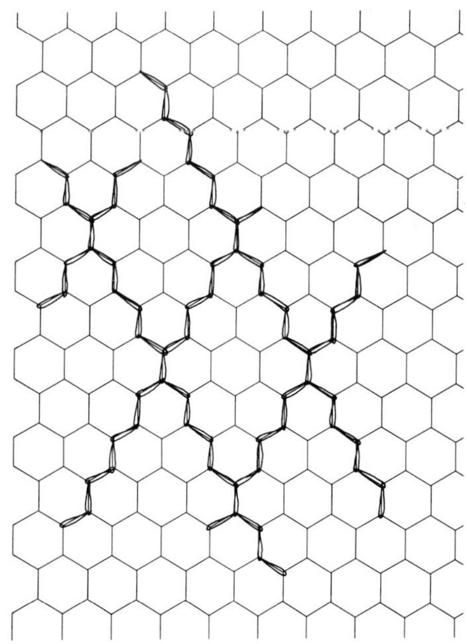

Stitch 4. Chain-stitch rings

Enlarge the holes marked (o) with a stiletto. Overcast the sides of the dilated holes, pulling them back very tightly against the surrounding meshes (see p.29). Work down each row in a snake-like manner, curving from side to side to create alternate right and left halves of the rings.
At the end of the row, turn the lace around so that you are again working in a LR direction, and complete the remaining half of each circle
The rings of successive rows alternate with each other.
When the casket is filled, return to your starting point. Superimpose a ring of chain-stitches on top of the overcasting. Follow the same route as before so that the sinuous lines cross each other as they go from one hole to the next.

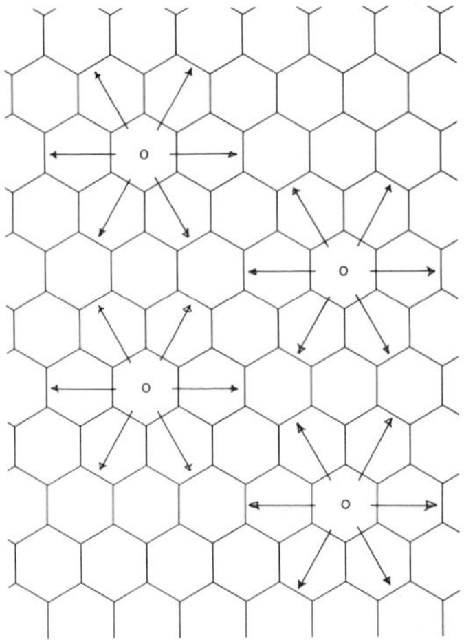

St.4: enlarging the holes

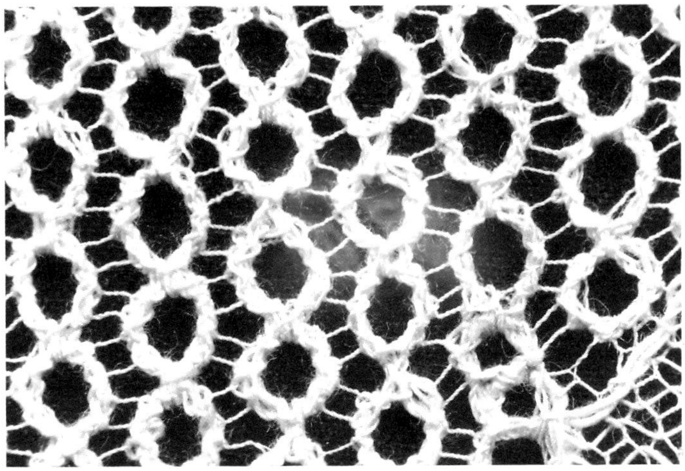

Stitch 5. Cross-stitch lattice with satin-stitch crosses

Work a cross-stitch lattice, placing the parallel rows of each set 5 holes apart, instead of 3 (compare stitch 1, p.38). The crossing of the sets forms the angles of the diamonds.
Each diamond encloses 25 holes. Work a satin-stitch cross around the central hole.
Note: the crosses are worked individually, and the thread is left hanging at the back to be sewn in when the casket is complete, or at the end of the work.

Stitch 6. Ten-point stars

This stitch is similar to stitch 5 (p.51), except that there are 5 rays to each half-unit instead of 3. Pull the stitches fairly tightly to enlarge the central hole of each star, and to emphasize the effect of radiating spokes formed by the stretched twist-sides of the surrounding meshes.
The stars in adjacent rows alternate with each other.

Stitch 7. Chevrons of cobweb and figure-of-eight stitch

The thread movements are exactly the same as in stitch 3 (p.47). The very striking difference in appearance is due firstly to the use of a single instead of a double embroidery thread, and secondly to the arrangement of the rows. Here, the zigzags are parallel, while in the Plastron the points of the Vs face each other.

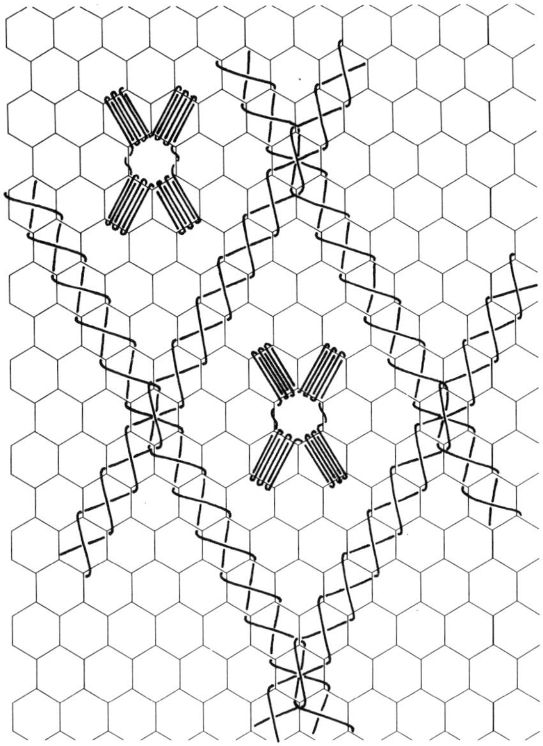

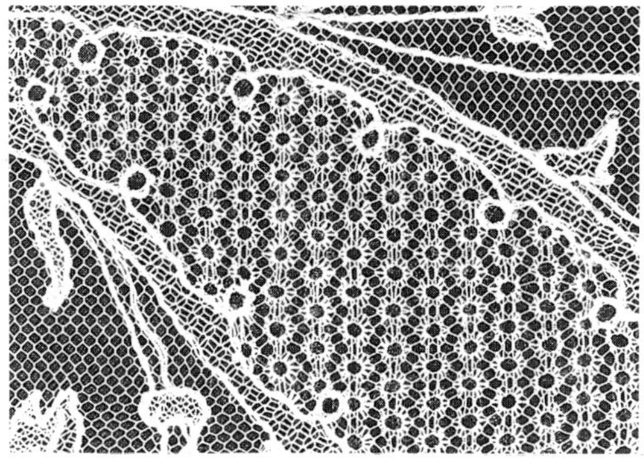

Stitch 5

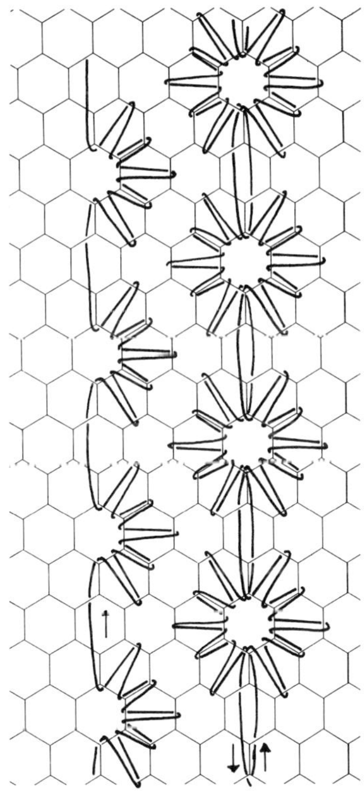

Stitch 6

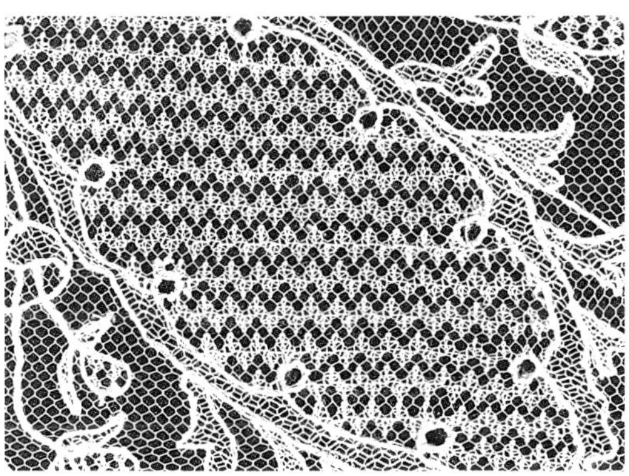

Stitch 7

COLLAR WITH HIGHLIGHTS

In this example, the lightness of the solid and decorative areas, worked in thread only marginally thicker than the mesh sides, might result in the design sinking inconspicuously into the net, but any tendency to blandness is counteracted by the brightness of the thickly two-plyed outline thread which not only delimits the design areas clearly but, packed tightly through the meshes, forms startling highlights in stems, veins and tendrils.

In addition, a thin aura is cast around the motifs by a single strand of embroidery thread which eerily echoes their shapes.

The solid design areas are mainly worked in darning- and tent-stitch, though cross-stitch appears in a few petals.

The figure opposite shows a stitch-map of the collar. Instructions are given only for those stitches which have not been described before.

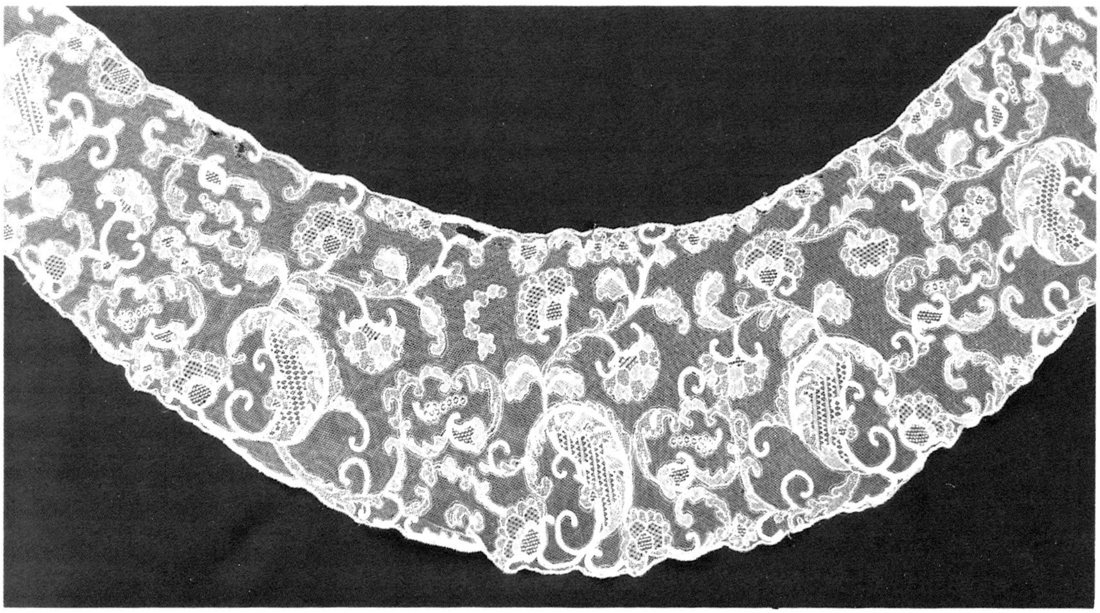

Above: the back of the collar

Left: the rounded end

Above: stitch map of the back

Left: detail to show the highlight effect

Below: a motif isolated

Stitch 1. Figure-of-eight and cobweb (see stitch 3, p.47).

Stitch 2. Crossed figure-of-eight (stitch 10, p.36).

Stitch 3. Interlocking tent-stitch (stitch 1, p.44), **with the central holes enlarged**

Stitch 4. Stripes of cross-stitch and satin-stitch spots

Set 1. Work diagonal rows of satin-stitch spots as shown, slanting the thread 2.5 times around the twist-sides.
The spots alternate in consecutive rows.
Leave 3 lines of holes.
Set 2. Fill each middle line with a row of cross-stitch. Turn the work round between rows so that the cross-over of the threads appears reversed.

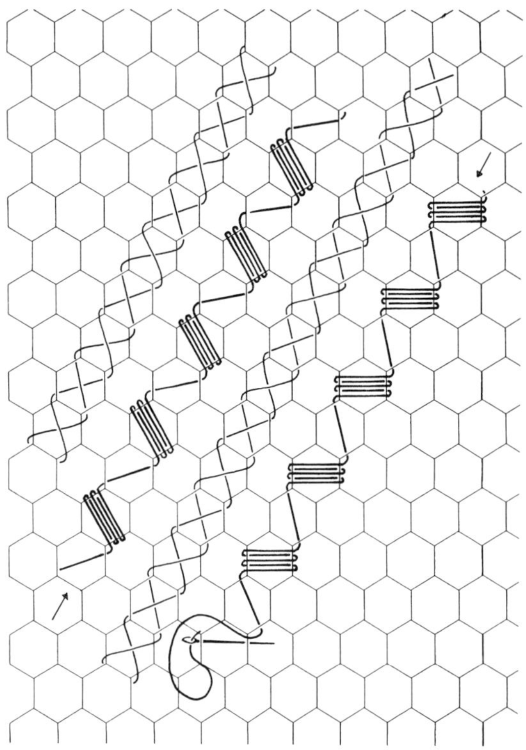

Stitch 4

Stitch 5

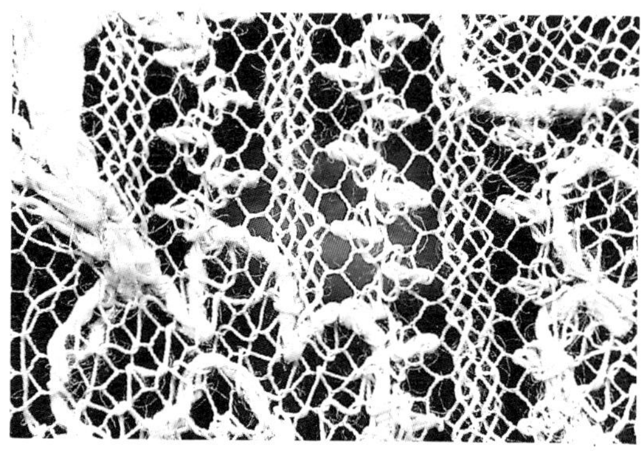

Stitch 5. Superimposed cobweb and figure-of-eight stitch alternating with paired rows of cross-stitch

The points of the figure-of-eight stitches face towards each other (see stitch 3, p.47).

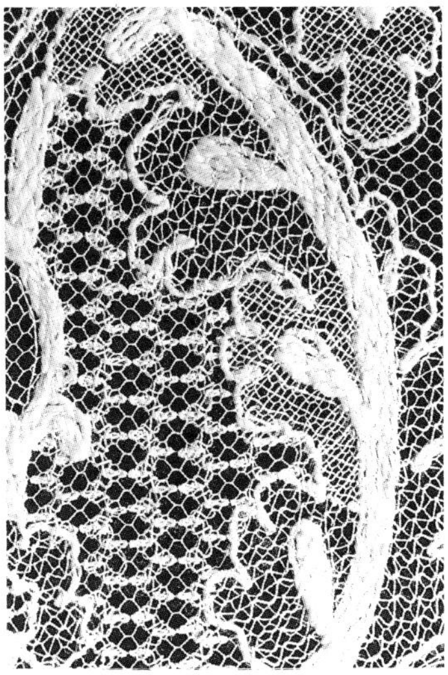

Stitch 1, with darning- and tent-stitch in the solid areas

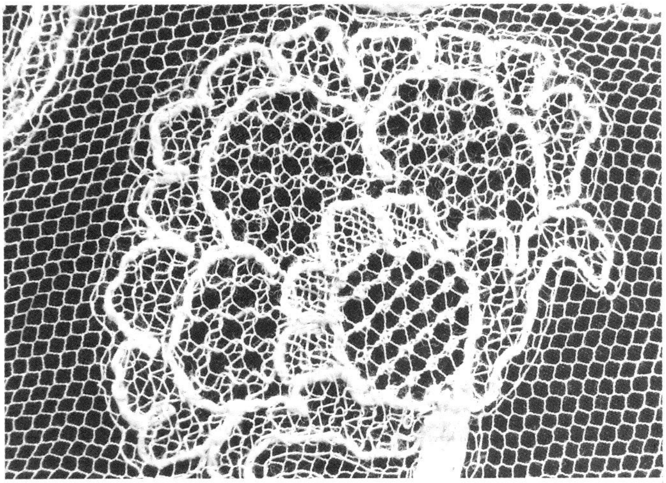

Stitches 2 and 3, with cross- and tent-stitch in the solid areas

Stitch 6. Double-crossed figure-of-eight stitch

Follow the direction of the arrows.

As each figure-of-eight is completed, take the thread a further 1.5 times around its middle, pulling the side of the enclosed holes inwards, and giving the surrounding meshes the appearance of radii.

Stitch 6

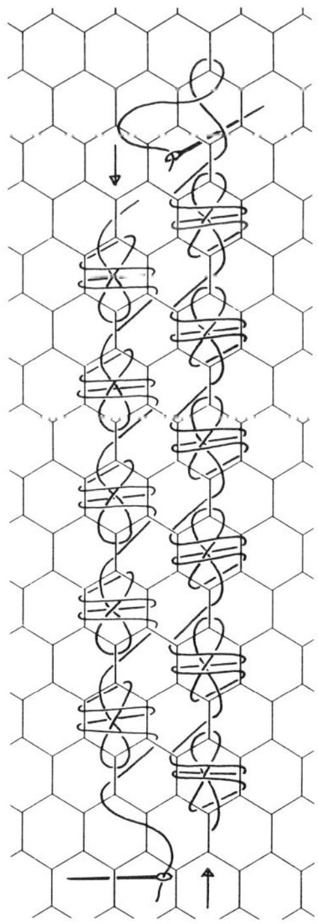

Stitch 7. Grapes with pips

In these bunches of grapes, the shapes are formed by curving the thick outline thread to right and left around them. The return pathway crosses over the first to complete the circles.

The large central pips are worked with a fine embroidery thread. Cross 3 pairs of satin-stitch over each other, and whip the embroidery thread from one pip to the next, as shown.

Run the thin thread around the grapes to form a halo.

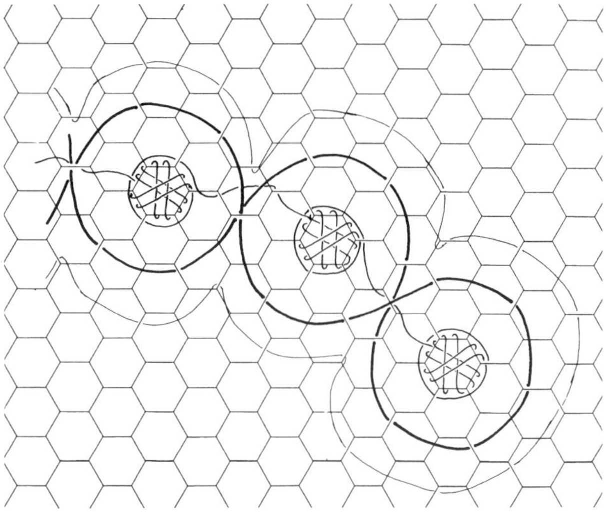

Left: the three terminal grapes. Note the halo

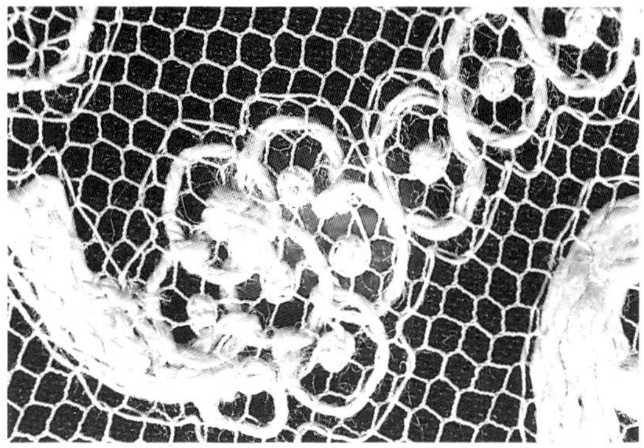

* * * *

Triangular shawl

TRIANGULAR SHAWL

This attractive shawl of unusual design has a strange chinoiserie effect of cloud-like blossoms and sprigs like branching twigs.

The net has small diamond-shaped meshes.

The outline thread is double, and thick relative to the embroidery thread - which is also used double, its individual strands being finer than the mesh sides.

Stitch map

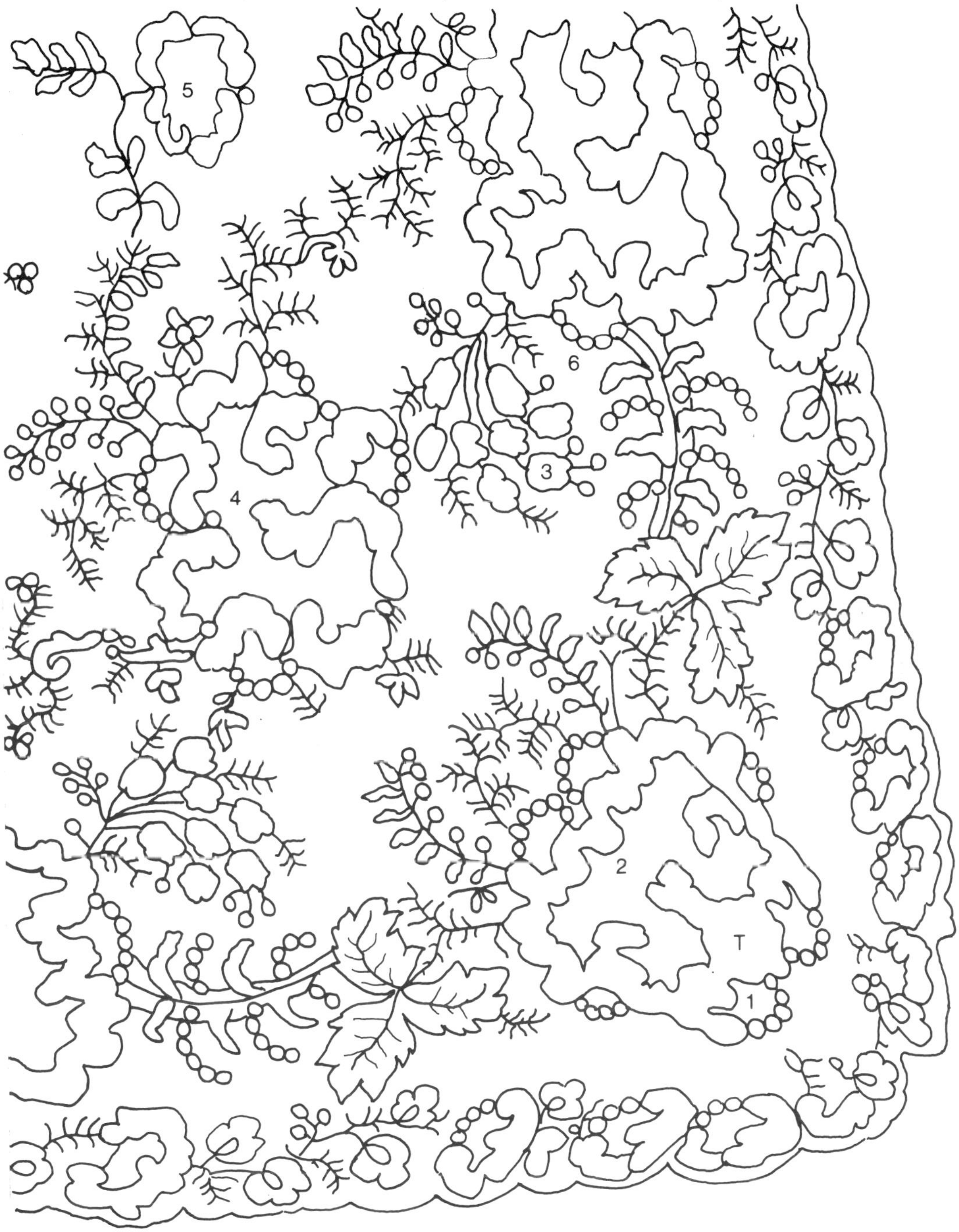

Stitch 1. Crossed figure-of-eight stitch (see stitch 10, p.36).

Stitch 2. Stripes of cross-stitch with satin-stitch spots

Set 1. Work diagonal rows of satin-stitch spots, alternating them over 2 lines of holes, and taking the thread 3 times around each encircled mesh. Leave 3 lines of holes between the spot rows.

Set 2. Along the middle line of empty holes work a row of cross-stitch.

Illustrations on opposite page

Stitch 3. Interlocking tent-stitch (see stitch 1, p.44).

Stitch 4. Stripes of holes and cross-stitch blocks

Set 1. Work a diagonal row of cross-stitch blocks, each covering 4 meshes, and with 3 lines of holes between them. The thread which links the blocks curves over them in the manner of a hood.

Leave a line of empty holes.

Set 2. Enlarge every fourth hole (o) along a diagonal line. Working outwards from these holes, lash the mesh sides against the surrounding net, as indicated by the arrows, to produce a series of circular spaces.

Repeat sets 1 and 2 alternately until the casket is filled. If preferred, complete set 1 before beginning set 2.

Note: In the photograph, the stitch is worked on a diamond- mesh net, and the holes appear oval.

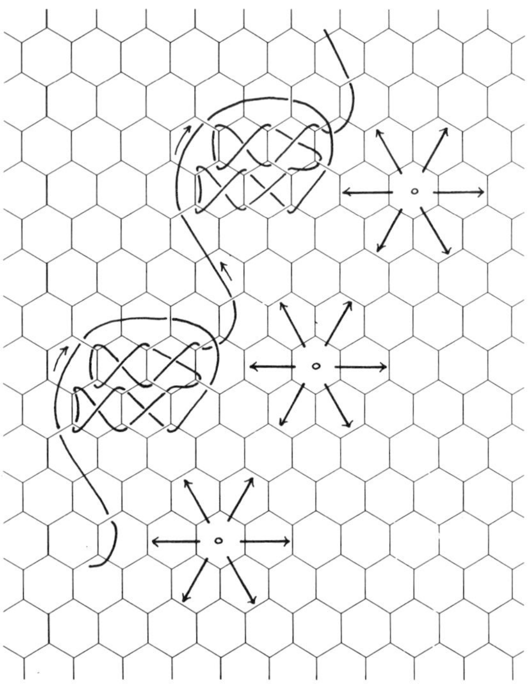

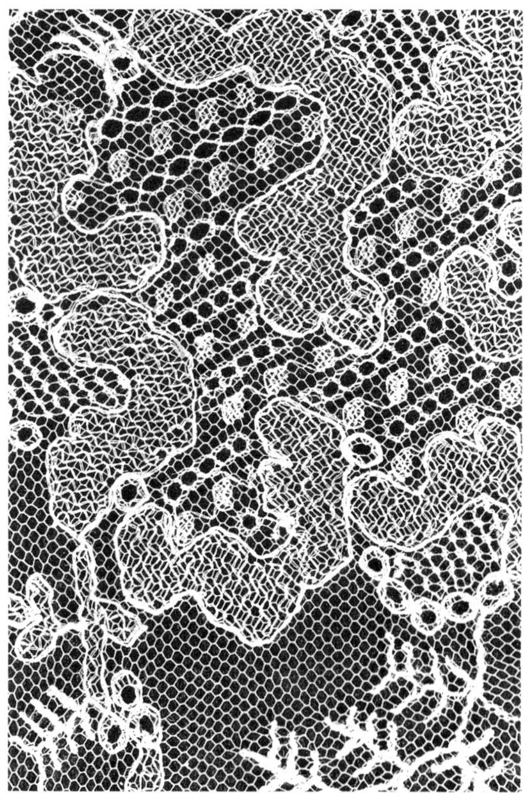

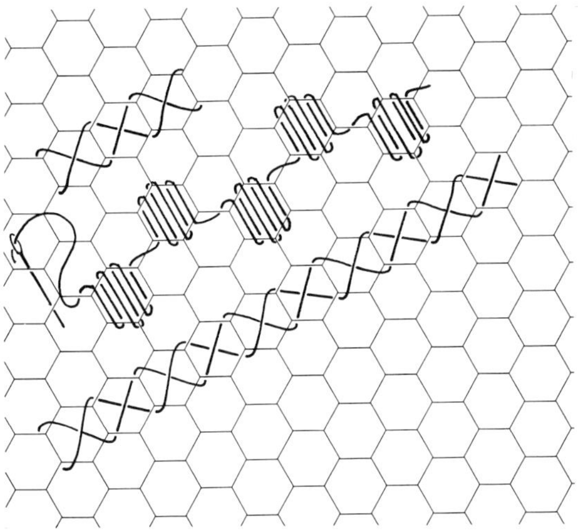

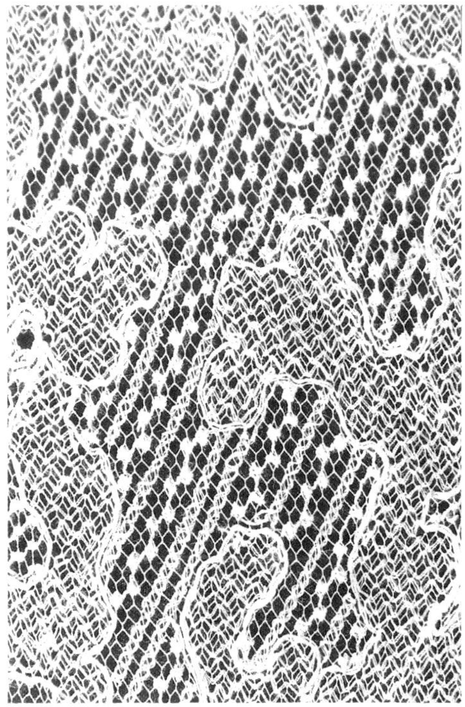

Stitch 2

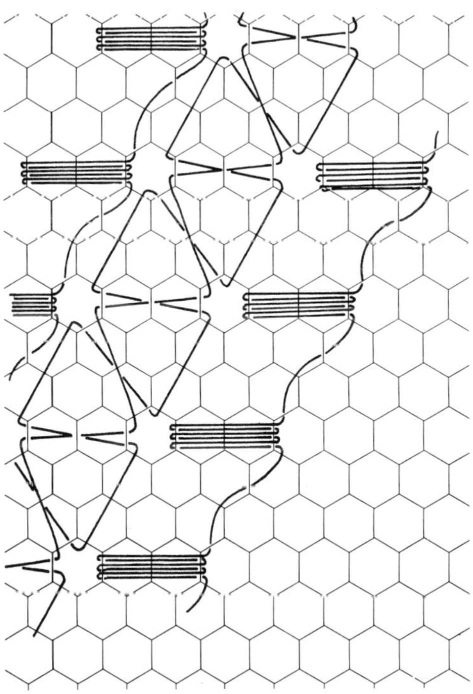

Stitch 5

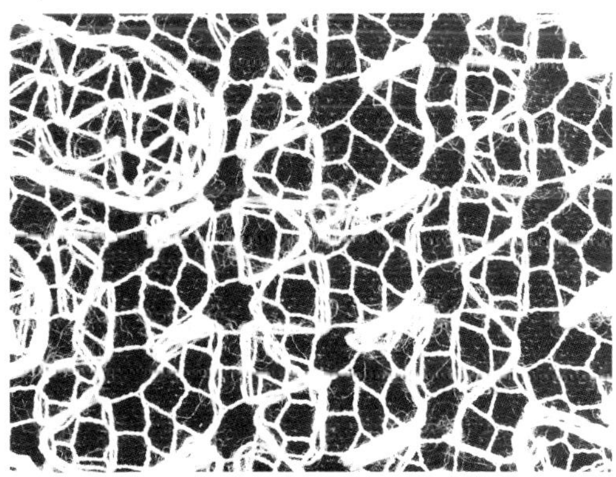

Stitch 5. Triangles of tent- and cobweb-stitch with satin-stitch spots

Set 1. The diagonal rows of triangles are formed by the intersection of large tent-stitches and large cobweb-stitches. Follow the path shown in the line drawing.

Set 2. The diagonal rows between consist of satin-stitch spots with the thread bound 3 times around a pair of meshes. Pull on the satin-stitches so that the holes on either side are enlarged. Space the spots 2 holes apart, and position them opposite the cobweb-stitch indents. Whip the thread over the mesh sides from one spot to the next.

Alternate these two sets until the casket is filled.

Stitch 6. Hollow circles

Enlarge every other hole (o) along a vertical line.

Set 1. Using the double outline cord encircle these holes, snaking around alternate sides to the end, then back again (see stitch 7, p.66).

Set 2. Now using the embroidery thread doubled, work overcasting stitches in a star-like manner from the central hole to just within the outline cord. Draw the mesh sides outwards to meet those of the surrounding holes. Work along one side first, to the end of the row, then turn the work round to complete the overcasting. Cross the embroidery threads at the top and bottom of each circle.

One of the twig-like extensions characteristic of this lace is shown at the end of the line of circles, to indicate the method of working.

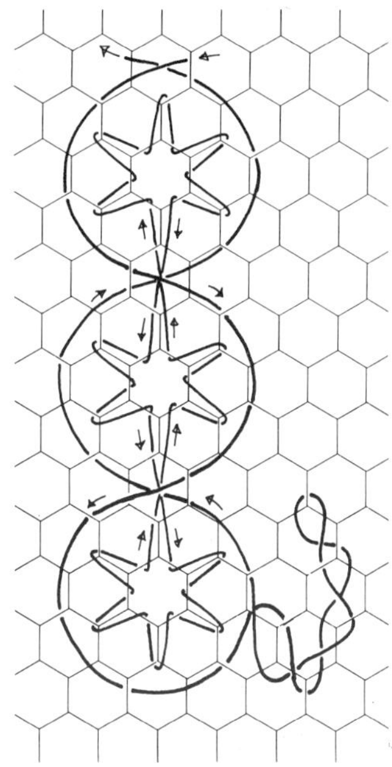

The finished appearance of the hollow circles, and how to work them. Note the tent-stitch on the left of the photograph, and st.5 on the right

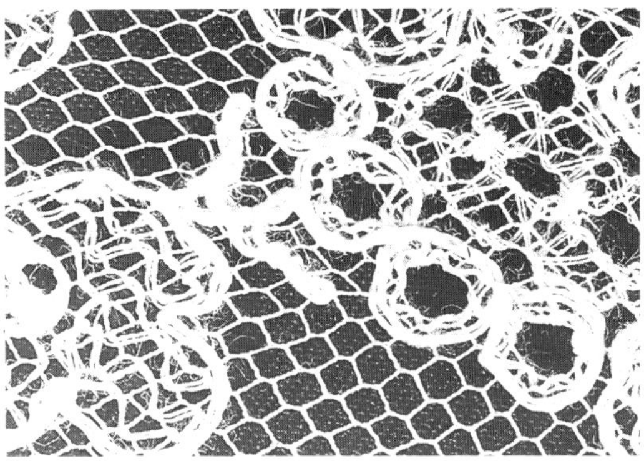

* * * *

This insertion could be filled entirely with the same stitch, or it could be used as a sampler. The outline can be simplified by omitting the loops and curls (based on a Schiffli 'macramé' design, 1913)

DESIGNS TO WORK

First practise the stitches, arranging them perhaps in the form of a sampler. Then try a small design. A simple outline, and just two or three different stitches, is an encouraging way to begin. A number of suggestions, from simple to complex, are illustrated in the following pages.

Based on the Limerick examples already described, select your net, your threads, the type of outline, the solid design area stitches, and the casket stitches you would like to try. Everything you do will be helpful practise.

Run embroideries from other parts of the world are considered in ch. 5.

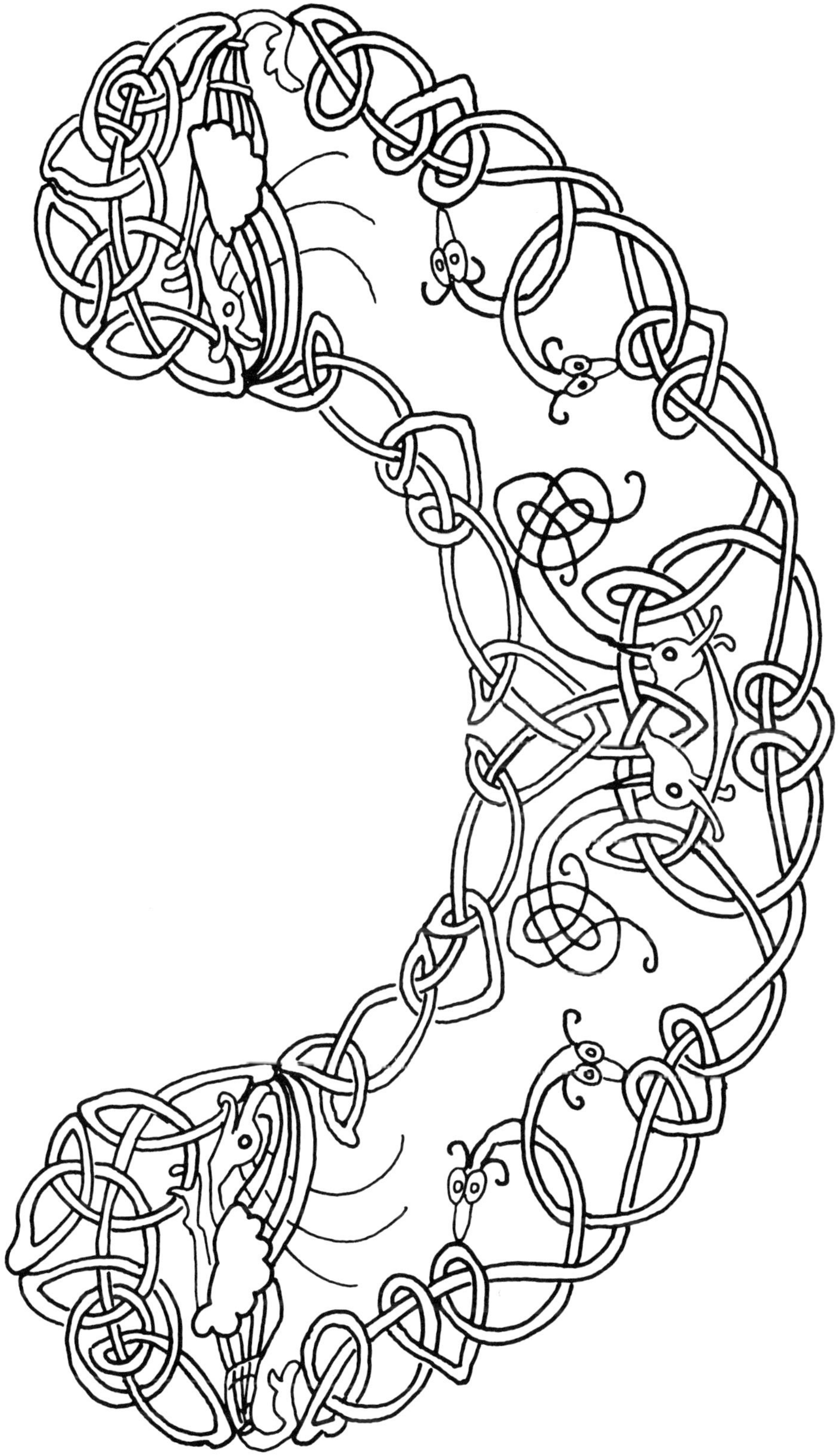

An unusual run embroidery on round bobbinet, simulating the design and general appearence of pulled threadwork. Fragments of net are visible along the lower border. The very fine thread used for the run-stitch of the solid areas, gives an impression of woven muslin. (Courtesy: Dieter Maier).

Design of a fan leaf in 'guipure with insertion of gauze'. (Courtesy: St Clare's Convent, Kenmare).

A Cornely machine. Above is the reel of cord for attachment to the face side of the net. Below, the feed thread passes through a tension spring then up and along the front of the eye-pointed needle (the red line is visible there). The handle, centre, is used to move the net as the embroidery is worked.

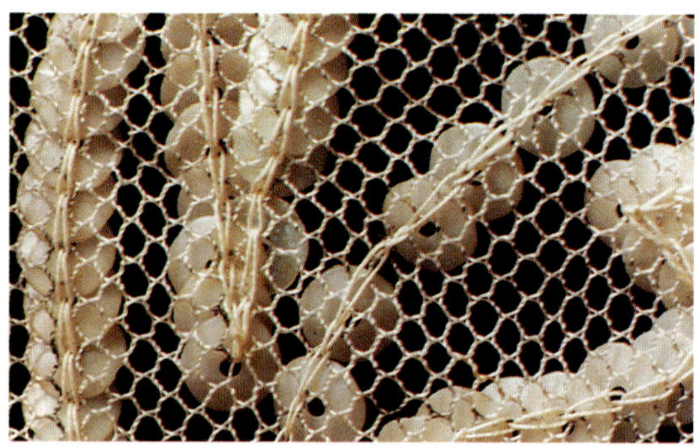

Paillets of mother of pearl attached to the net by chain-stitching (reverse side).

Beaded net, c 1900.

3. INSTRUCTIONS FOR MAKING LIMERICK TAMBOUR

Tambour is chain-stitch embroidery using a hook instead of a needle, and with the thread fed from below instead of poised above the net. The hook can progress far more quickly across the fabric, and since it draws the thread continuously from a ball or reel it eliminates the frequent stopping and starting as lengths of thread need to be renewed. Though the stitches produced by both needle and hook are identical, they are sometimes distinguished as 'chain-stitch' (worked with a needle) and 'tambour stitch' (worked with a hook). For example, in 1882: 'The real tambour stitch is now superseded by chain stitch . . . chain stitch used at present in tambour work is the ordinary chain stitch' [90].

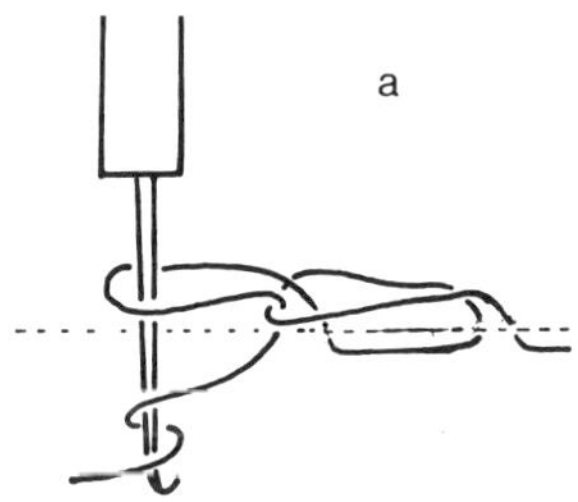

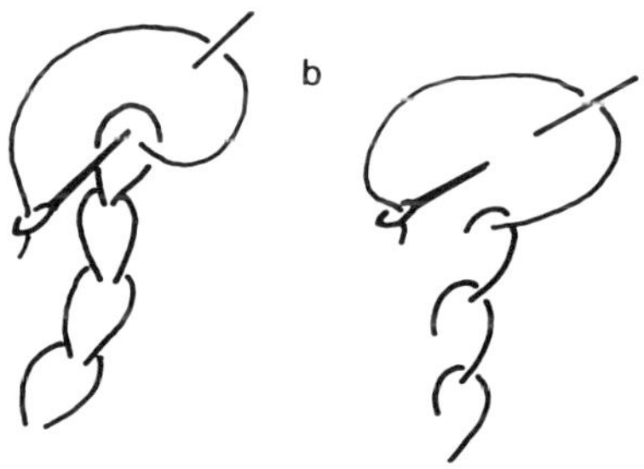

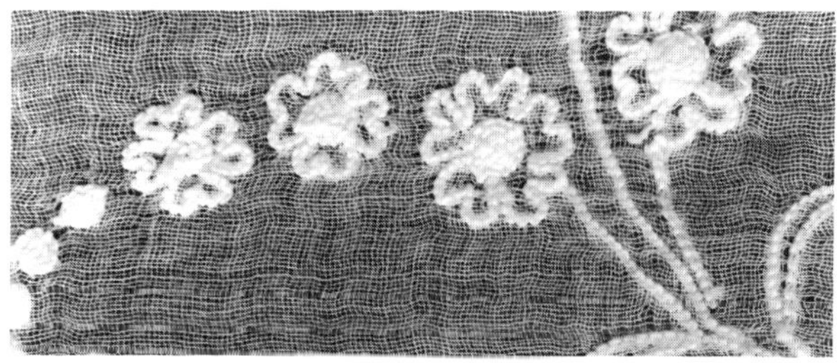

(a) Tambour stitch made with a hook. The two ends of the new loop begin within the previous loop. Note the feed thread below, coiled around the hook, to produce a 'chain-stitch' above the net and a back-stitch below.

(b) Chain-stitch made with a needle. The more flexible positioning of the needle's entry and exit allow the stitch to be either open or closed.

Back-stitch. The reverse side of tambour embroidery, here on muslin.

This regression from hook to needle applied mainly to amateur ladies using thickish thread for soft furnishings, rather than to professionals who had, during the 18th century replaced the needle with the hook to accelerate production, and who continued to use it.

While in run laces corners are easier to negotiate than curvers, designs for tambour work should flow in graceful swirls, and sharp angles which require knots to hold them in place are better avoided.

Materials

NET

Since the 3- or 4-twist 'Brussels' nets are no longer available, the choice lies between the 2-twist bobbinets where all six sides of the meshes are of equal length and the holes are isodiametric or 'round', and some form of Raschel (warp-knitted) nets of which a number of different types exist.

Hole-size and type of net must be chosen according to which best suits your eyesight, your time schedule, and the effect you hope to achieve. Remember that the smaller the holes, the denser the net will appear. The almost invisible fineness of the c1900 nets, made of over 300-count cotton, has long since vanished.

PATTERNS AND SUPPORTS
As in run laces, the net must be firmly supported to prevent its pulling out of shape and also, in tambour, to prevent the stitches being pulled too tightly, and so irremediably puckering the work. In tambour, unlike run embroidery, both hands are actively involved, one above the net, holding the hook, the other below, offering the strand of thread.
This means that a hand-held card cannot be used.

The frame can be similar to that used in run work but it must be self-supported. Its shape can be round, oval or square, and larger forms may have a roller attached.

For domestic tambour work the *Needlecraft Journal*, has a thoughtful suggestion: any reader 'who may not be very robust' should seat herself in 'one of the large armchairs seen in most drawing-rooms. She may rest the frame on the straight arms of the chair and work away to her heart's content'. But there is a word of caution, for if she is too comfortable 'progress will not be so rapid' [91].

It is important that the constant stabbing and withdrawing of the hook does not cause the net to slip. Square and rectangular frames have built-in tensioning devices, and round frames often have screws by which the outer ring can be tightened over the inner. As a precautionary measure, the inner ring of the round frame may be bound with a thin layer of some semi-rough tape or fabric to give a better grip.

Girls at the Industrial School, Golden Bridge, Dublin, learning tambour embroidry on net. Note the reel of thread at the side of the frame, the tensioning pegs and the pattern hung at the front (ref. Cole, 1890).

SETTING UP Stretch the net tight as the skin of a drum over the frame, taking care that the cross sides of the meshes are aligned with the horizontal or vertical axis (see p.20). There do not appear to be any documented records of how precisely the early Limerick tambour workers transferred the pattern to the net.

The fact that the hands are both above and below the net, makes it impossible for the pattern to be fixed underneath as it is for run work. So by some means the pattern must be marked on the net itself. For example:

(a) Place the clearly-drawn pattern beneath the net and draw over it with a hard-lead pencil which will not smudge. When the drawing is complete, take the pattern away. The thin lines on the net will be hidden by the stitches as the lace is worked.

(b) As (a) but instead of pencil use a solution of stone blue and thin gum water [92]. This is the ancient equivalent of a water-soluble marker.

(c) Blocks of sycamore wood carved with the design were sometimes used to print on to the net using pink or blue dyes [93]. The colour washes out during 'finishing'. This method from the past is very expensive unless the blocks can be used many times over. Possibly only very high quality pieces, commissioned for an aristocratic client, would justify their one-off production.
(d) *Weldon's Practical Limerick Lace* [94] suggests the pattern be tacked beneath the net along its upper edge, allowed to hang until needed, then raised for scrutiny, and allowed to fall again. This is possibly Lindsey's suggestion that for experienced tambourers the design was 'drawn on paper and placed before the eye of the worker' [95].

A flow diagram, showing by arrows the pathway which the hook should follow along the maze-like outline, with the minimum of duplication and the maximum of continuity, is not only helpful but necessary for any complex design.

Symmetry, of the two sides of a collar for example, can be achieved in various ways:
(a) Draw the half collar, then photocopy it on to tracing paper and reverse it to make the complete piece.
(b) An older technique is to draw the half-collar, fold the paper along the middle line, prick the outline through to the other side, and join up the holes to make the second half.
(c) In simpler tambour designs where conformity to a fixed pattern is not essential, but balancing the two sides is desirable, the net can be divided up by parallel lines of coloured thread run at regular intervals in a vertical and horizontal direction, and the hook movements balanced on the two sides.
(d) In vermicilli patterns, which were particularly popular around 1912, the design is worked freehand and an artistic eye rather than a pre-drawn outline is required [96].

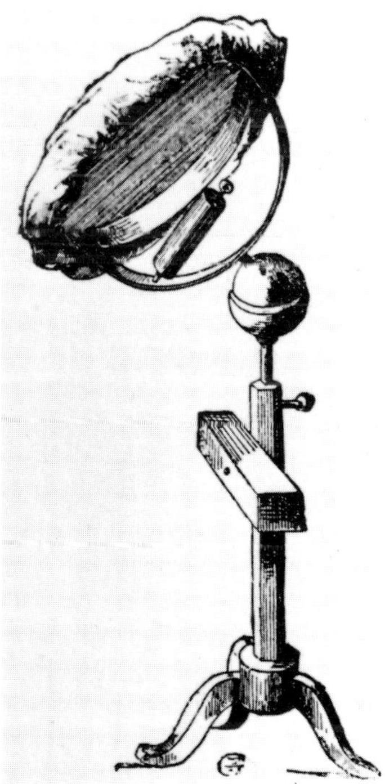

Left: an 18th-century tambour frame (after Diderot, 1763).

Below: flow chart. The arrows indicate the direction of working. The inner circle of the flowers can be made in continuity with the petals (see top right), or separately.

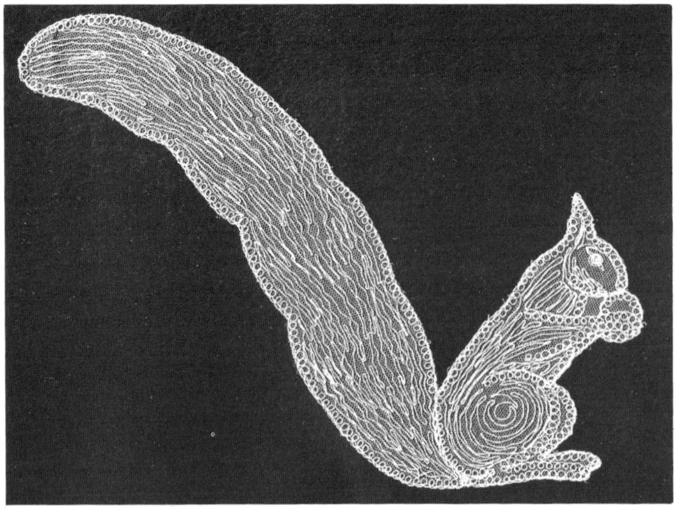

Free-flow tambour embroidery within a looped outline.

THREAD

A thin tightly-twisted cotton, or a firmly twisted silk of similar thickness, works very well. Use them direct from the spool or ball.

It is vitally important, for ease of working and for a harmonious effect, that net, thread and tambour hook are all compatible in size. Spend time experimenting with samples before embarking on any major project.

Suggested thicknesses for use today are:

Outline	Mercer crochet	60	Inside	Cordonnet special	80
	DMC mercerised cotton	80		Retors d'Alsace	50
	Mercer crochet	30		Retors d'Alsace	60
	Mercer crochet	80		Fil a dentelle	70
				Retors d'Alsace	60

Of the traditional threads, 80-count cotton for the outline and 120-count for the infillings are still available. So are the thicker flosses for really coarse nets and big hooks, which were used c1900 for soft furnishings. The infillings can if necessary be lightened by using looser stitches more widely spaced.

The spool of supply thread must be fixed so that it can rotate freely to provide a continuous feed. Place it over a nail fixed to the large table-like frame or, if the frame is smaller, put the spool in a box with a hole in the lid, or in a polythene bag with the opening loosely tied.

Although in theory the thread can be drawn non-stop from start to finish of the piece, in practise the motions of the hook as the stitches are made cause it to overtwist - or untwist, depending on the tightness of the spin. If an even crispness is to be maintained, this distortion must be corrected. If it is not great, allow the reel - with the thread secured to stop it unwinding - to hang and twirl until the balance is restored. Alternatively cut the thread, remove the tortured length, and rejoin. The resultant knots on the reverse side are an indication of hand- as opposed to machine-tambouring.

THE HOOK

The tambour hook, also called a tambour needle or 'hooked needle' [97], is in essence a minute crochet hook made of steel and secured in a handle of ivory, bone, mother-of-pearl or wood by means of a screw.

It scoops up the thread through the holes of the net, and carries it forwards as a loop, creating a single line of chain-stitches over the upper surface. The long tapering cap at the top of the handle can be unscrewed to cover the hook when not in use. In addition, the inside of the handle is hollowed out to make a storage place for spares. In use, the hook is always adjusted to be in line with the screw which secures the shaft to the handle. This enables the worker to keep track of which way the hook is facing (see p.79).

Hooks vary from fine to coarse, to fit with different thicknesses of thread.

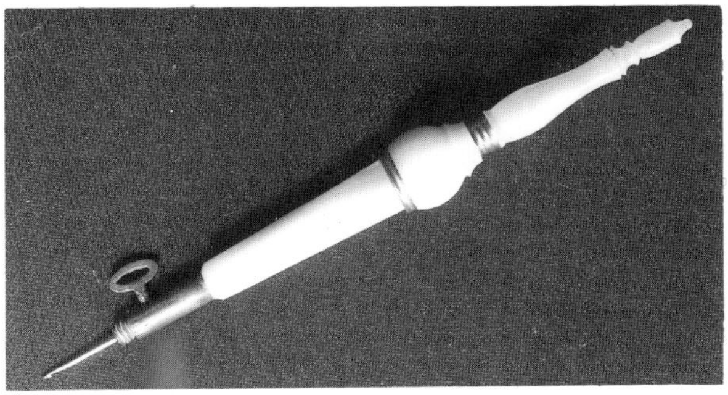

Left: a Victorian tambour hook. Unavailable on modern hooks would be the protective cap (right) and the inner storage area (centre).

Right: a tambour thimble (after Dillmont).

THIMBLE

A special thimble made of thin brass sheeting, and open on one side to give slight elasticity under pressure, may serve to support and guide the hook, and so accelerate the work. The front of the thimble, placed over the finger-nail is notched, and against this notch the shaft slides up and down as the hook is raised and lowered. Although Dillmont, writing in the later 19th century, illustrates the thimble as part of the eqipment, her instructions are aimed at domestic needleworkers. Professional tambourers discarded the thimble as early as 1827: 'In the course of time the thimble was thrown aside, and the needle is now held nearly in the position of a writing pen' [98].

Wielding the hook

Except when beads or sequins are to be added, tambour laces are worked face side up so that the chain stitches will show on the surface.

In the following instructions, reverse right and left if necessary.

The tambour needle is held in the right hand between the thumb and first finger, either straight-up or slanting slightly towards the worker. The middle finger is pressed against the shaft (as if holding a pen), so that the hook is kept steady as it passes in and out of the net.

The feeding thread is held beneath the net, by the thumb and fore finger. Pull it firmly, but without vicious strength, in a forwards direction (away from the worker).

Pass the hook down through the net with the hook facing forwards. Wind the thread 1.5 times anti-clockwise (to the left) around the hook, then pull it slightly forwards to prevent the loops slipping.

Rotate the shaft through 180°, also in an anti-clockwise direction. The hook now faces backwards, towards the worker.

Pull the loop upwards, pressing the shaft against the forward side of the mesh to prevent snagging. If necessary, rest the third and fourth fingers lightly against the net to resist its upwards pull.

When the hook is above the net, rotate the shaft to the right (clockwise) so that it faces forwards again. Draw the captured loop across one or more meshes, working away from yourself. At the same time, briefly release the feed thread to prevent the stitch being over-tightened.

As a beginner, you may need to watch you left hand to see where it is in relation to the thread and the hook, but an automatic rhythm is soon acquired.

There is nothing mandatory about the precise method of holding the shaft, nor about the angle at which the hook is passed up and down through the net. Find your own best position, as long as it results in a smooth regular line of stitches.

Hook movements during stitch-making. The tilting of the hook has been exaggerated.

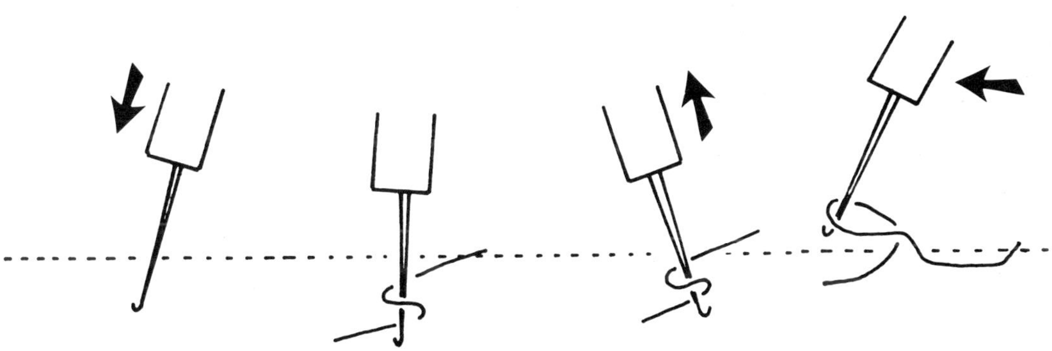

Starting off

Tambour stitches unravel with horrifying facility from the end towards the beginning, and it is essential that first and final stitches be firmly secured.

To begin, poise the needle over the net. Leaving a short free length, twist the feed thread into a loop, pick up the loop on the hook, and pull it through to the upper surface as described above (p.79).

As you make the next stitch, fold the free end (A) forwards alongside the feed thread. Twist both around the hook and pull them upwards through the net. Work another stitch in this way. Now abandon the free end and continue with the feed thread alone.

Rules:

Always make sure the net is tightly tensioned. This gives the hook some resistance to work against.

Don't pull the stitches so tightly that they pucker the net. Distortions are irreversible.

Keep the stitches as near as possible even in size.

Keep the feed thread in the left hand pulled forwards, and adequately tensioned.

The direction of working must be away from you, except in some fancy stitches.

Practise the tambour movements using a large-meshed net. This will develop muscle-memory and the ability to measure distances by eye without conscious calculation.

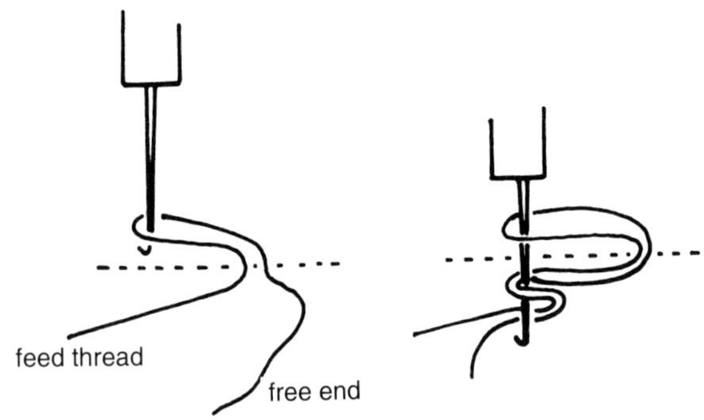

Left: picking up the free end to strengthen the starting point.

Below: two methods of tying a knot to fasten off.

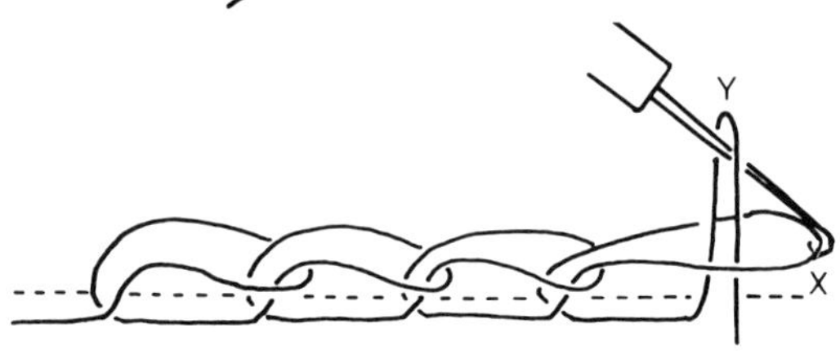

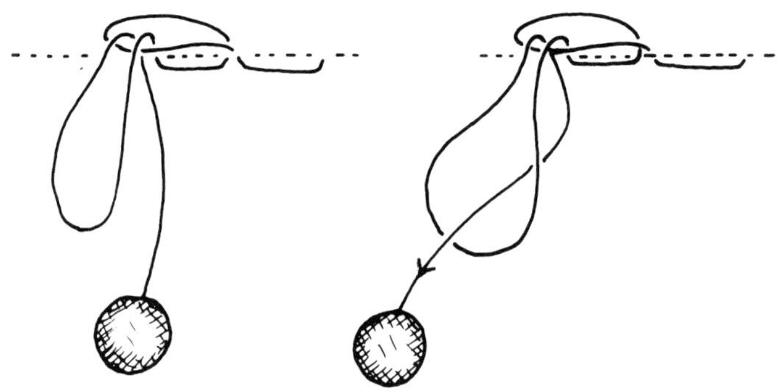

Outlining

The outline thread is always thicker than that of the infillings and is always used single. Start off as described above, then work a line of stitches around the outline until you are back at the beginning. If you want to make a second line, go round again, keeping close to the first.

A knot can be used to fasten off. Enlarge the final stitch, and remove the hook. Take it beneath the net, then up through a hole, catch the loop, and pull it down to the reverse side. Enlarge it further until you can pass the reel through it, then carefully tighten the knot until it is firm.

Alternative techniques are:

(a) Cut the feed thread several inches below the net then, using a crewel or tapestry needle, pull it through to the face side. Pass it through the loop of the last stitch. Take the needle down again and run the thread in and out of the back of the chain-stitches until it is secure.

(b) Enlarge the final stitch (X), as in (a). Pull the next stitch (Y) through it, with the hook pointing forwards. Rotate the shaft so that the hook points backwards (towards the worker). Pick up the front of loop X and pull it through loop Y. Tighten the feed thread to bind X and Y together.

INFILLING

As with Run laces, the working of the design is distinguished into solid areas and decorative caskets.

The solid areas may be heavily emphasized by dense lines of chain stitch arranged in circles, spirals, or long parallel rows. In the last, all the stitches can be made to point in the same direction by cutting the thread at the right-hand side and starting again at the left-hand side of every row.

For more decorative solid areas line the thick closely-worked outlines with a row of loose chain stitches in thinner thread so that they hug the inner border like a shadow.

Highlight effects can be created in a similar manner. Grapes and wheatears were perennial pattern favourites in Catholic countries, but in 1903, according to *McCall's Magazine*, grapes were 'the most fashionable of the season's offerings' [99].

The basic stitch of tambour laces is always the same, and their decorative infillings depend not on a wide range of variations but simply on changes of stitch size, arrangement, direction of working and thread diameter. They can be manipulated into spots, zigzags, square-toothed battlements, solid diamond shapes with stitches radiating from a common centre, simple petal forms, or half-fans.

Whenever the line of tambouring has to turn sharply back on itself, knot the loop (see p.80) immediately before the turn, and then set off in the new direction.

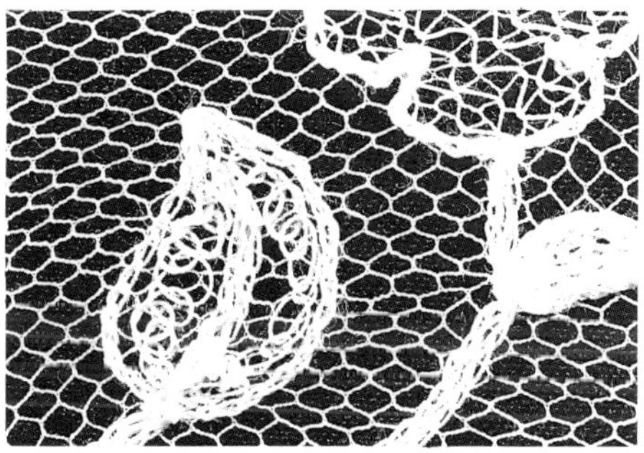

Above: loosely worked tambour stitch (known as fluderstitch).
Below: reverse side, back-stitch and linked spots.

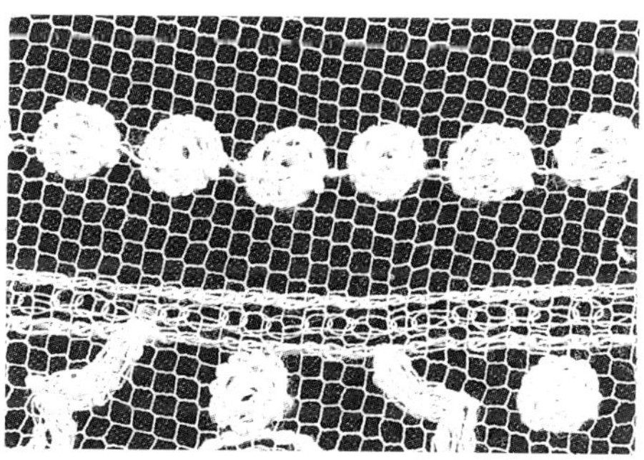

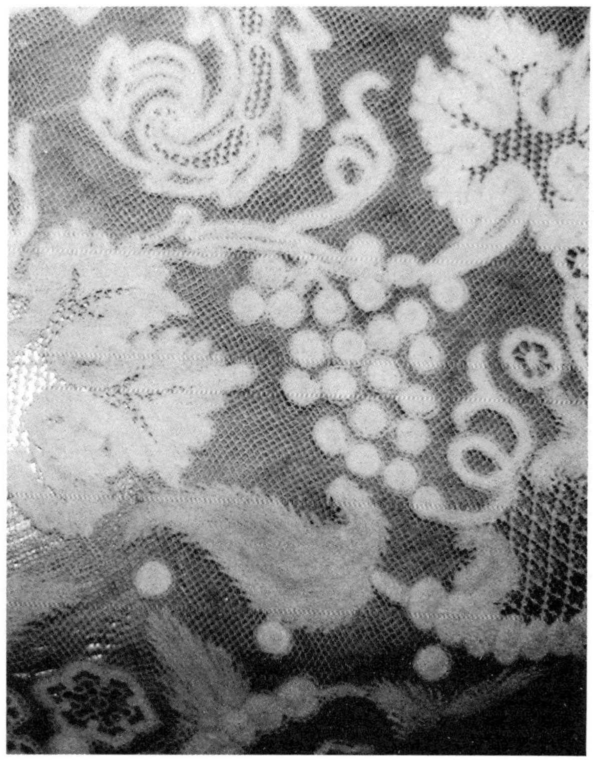

Above: a richly tambboured veil from St Clare's Convent, Kenmare. Some of the heavily tamboured leaves have spiky borders (see p.95). Around the grapes, thin fluderstitch forms a simulated bloom, and halos the vine leaves. A dramatic selection of run stitches fills the caskets.

Spots may be similarly terminated, but if they are situated close together, carry the thread forward beneath the net to the next position, then insert the hook through the net from above and continue stitching.

Complex tambour-stitch groupings are rare in Limerick, though found elsewhere (see ch.5), and much of the beauty of the lace rests on the design. On the square net it was important to work the 'long' or the 'short' way of the meshes, but some of this variety is lost in the round nets where all the sides are the same length. The orientation of the cross-sides is still important, as it is in run laces.

Dilation of the mesh holes also appears. The sides may be held back by tambour overcasting, or by needle overcasting followed by a ring of tambour-stitching.

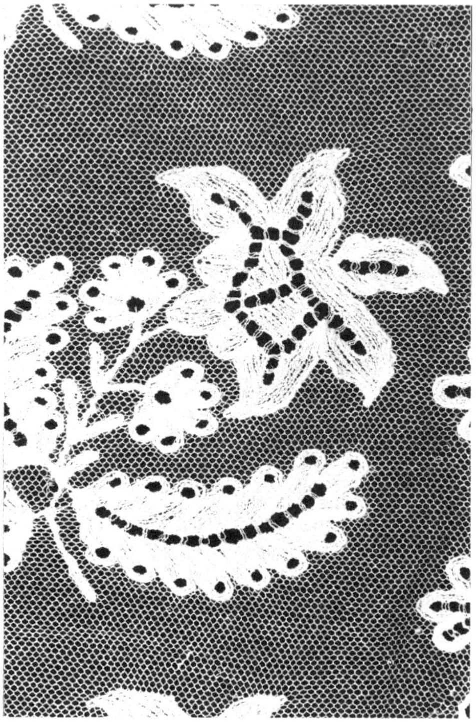

Details of a baby gown. Right: patterns of enlarged holes form the only decorations within the tambouring. Note how the parallel arrangement of the lines of stitches fits with the design.

Below: note the run outline and the border of machine picots, tamboured on.

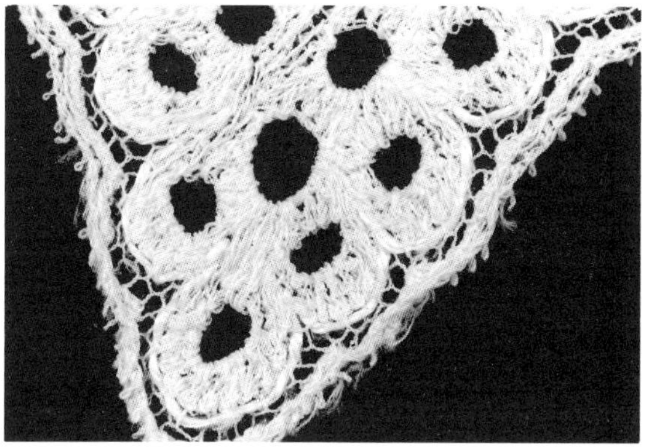

Stitch names as ever tend to be endearingly local. They are not used here. We suggest that readers insert their own names if they wish.

RUN DECORATION

Although *Needlecraft* states that 'in old times run and tambour work was occasionally combined in one piece of lace, but the fashion did not appear to take' [100], the editors were not infallible, and a great many Limerick-tambour laces do have a varying number of run fillings, some of them being in fact almost a 50% blend of the two techniques.

A small selection of such laces is described on pp.97-115. Most commonly the outline and the solid areas are tamboured, and the caskets run. The position of the stitches is indicated on the accompanying stitch maps.

A most unusual reversal of this solid tambour/decorative run is found in the shawl collar with long ties, already described (p.57-61); also in a black motoring veil embroidered in a vermicilli manner possibly by machine. It dates from the beginning of the century and shows several strands of thick silk run in to make a heavy outline while the inside is lightly shaded with slim tambour work.

The geographical origin of machine pieces is always difficult to determine since the technical aspects scarcely vary and fashion-conscious designs may override national barriers.

The baby gown, thought to be from Limerick, is decorated in very dense plain tambour work with no fancy stitches apart from enlarged holes. The motifs have a run outline.

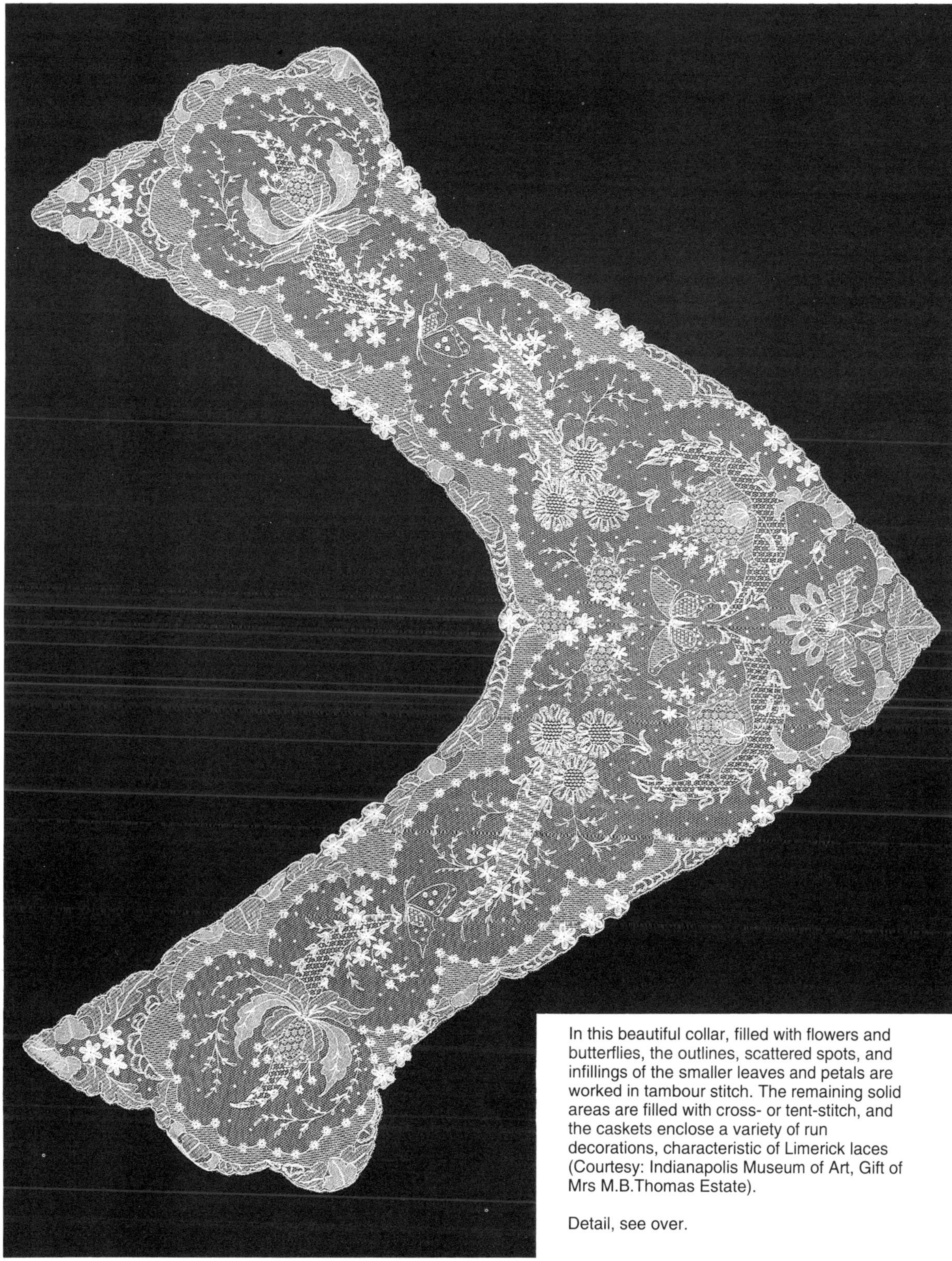

In this beautiful collar, filled with flowers and butterflies, the outlines, scattered spots, and infillings of the smaller leaves and petals are worked in tambour stitch. The remaining solid areas are filled with cross- or tent-stitch, and the caskets enclose a variety of run decorations, characteristic of Limerick laces (Courtesy: Indianapolis Museum of Art, Gift of Mrs M.B.Thomas Estate).

Detail, see over.

COLLAR WITH LILLE BOBBIN LACE BORDER (Alternative areas of origin are Coggeshall or Lierre)

The brief instructions for each stitch are given with the cross-sides in horizontal alignment.

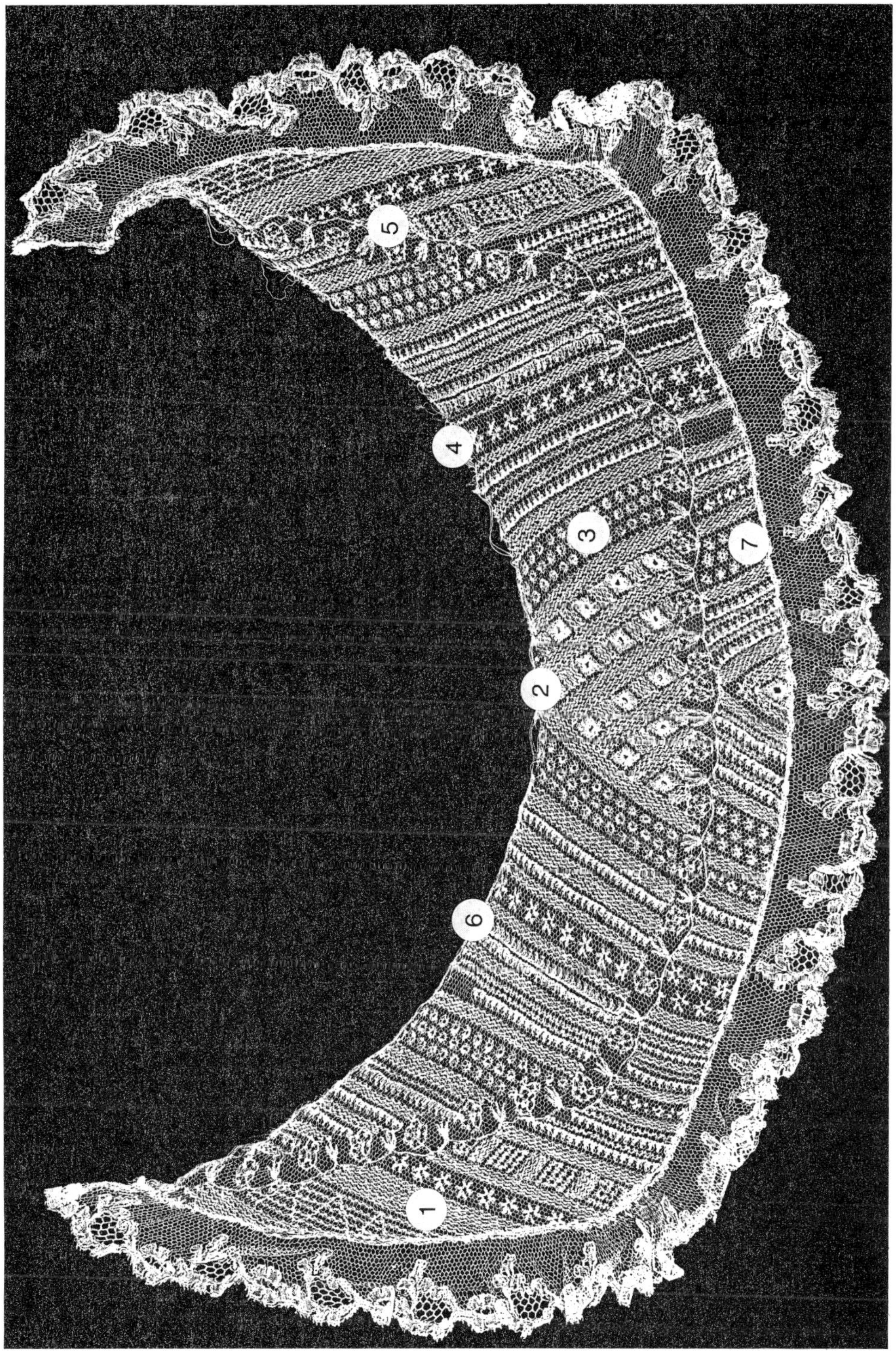

Stitch 1. Zigzag chain-stitch

Work diagonally across the net. Each stitch covers one hole and is placed at an angle to the previous one to form a zigzag line.

Cut the thread at the end of every row and begin again so that all the rows start from the same side. If a small frame is used and turned between rows, the work can be continuous, but the stitches will point in the opposite direction from one row to the next.

Stitch 2. Diamonds from a central hole

These spreading diamonds cover 5 holes at their widest point and are 10 rows tall. The twist-sides of the adjoining meshes stick out like spokes, six from each slanting side. There are 24 stitches per diamond, passing from the same hole to each of the surrounding meshes.

Enlarge the central hole, and attach the thread.

From the first stitch, pull loop (a) through and stretch it forwards until it reaches hole (1).

Bring up loop (b) through the tip of (a), then stretch it carefully back to the centre, tightening loop (a) as you go, until it is reduced to a small chain-stitch.

Keeping (b) on the hook, bring up loop (c) through the same hole as (a), stretch it forwards to hole (2), pick up the thread to make loop (d), and draw it carefully back to the centre.

Continue in this way until all 24 stitches are completed. Two small chain-stitches will appear at the base of each ray.

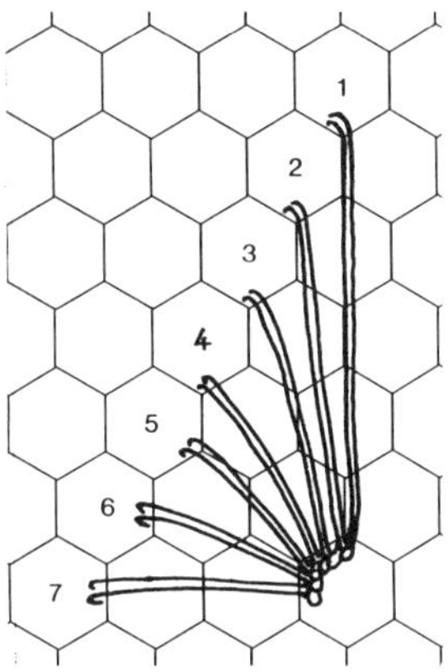

Right: stitch 1.

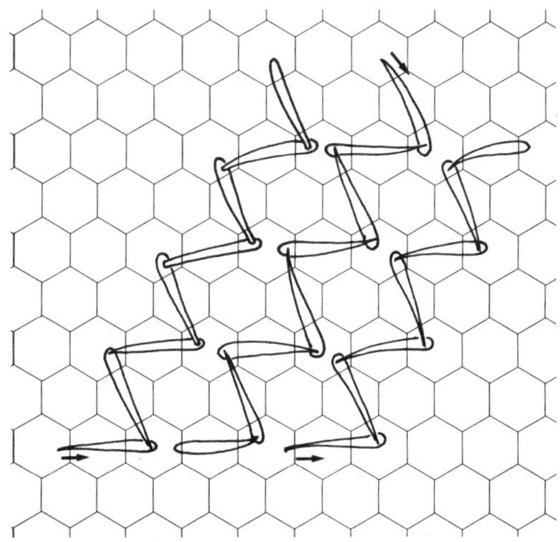

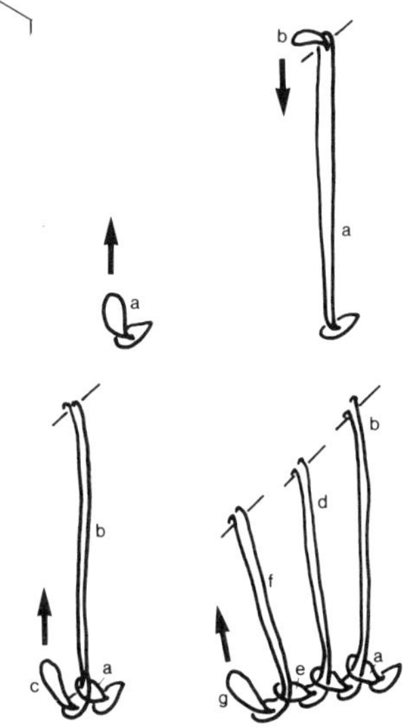

Stitch 2. Working the chain-stitch diamonds.

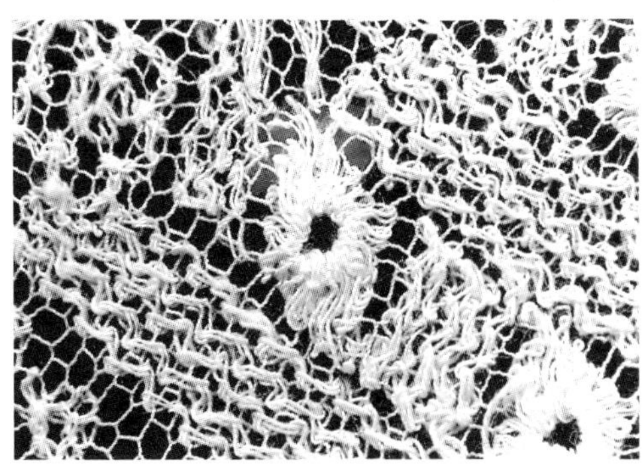

Stitch 3. Hollow circles

Work 6 tambour stitches in a circle around a central hole, then carry the thread forward diagonally beneath the net so that the finished circles are spaced one hole apart. Note how the stitches are displaced inwards when arranged in a ring.

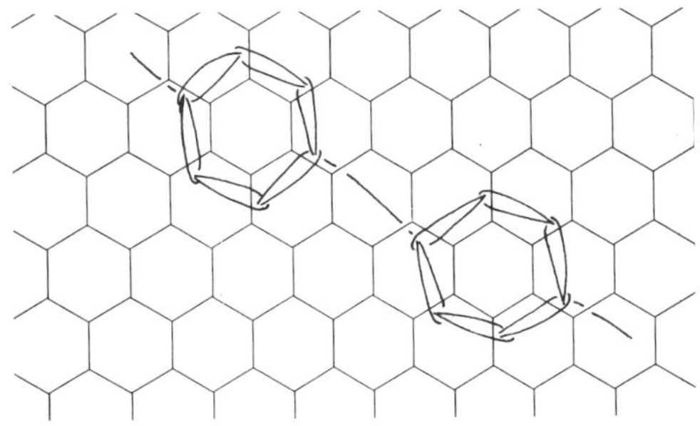

Left: St.3. The photograph also shows st.1, 2 and 5.`

Above L: st.4.

Stitch 4. Six-armed starfish

It is not clear how this stitch is made.

Stitch 5. Flower

The 6 petals are arranged about a central hole, and each tamboured petal encloses a single hole. Construct a ring of stitches in the middle, then pass outwards to the overlapping petals, following the direction of the arrows and stitches. Each flower covers seven meshes in all.

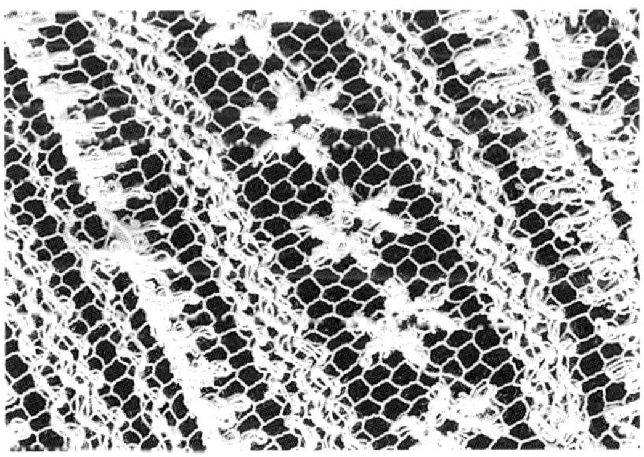

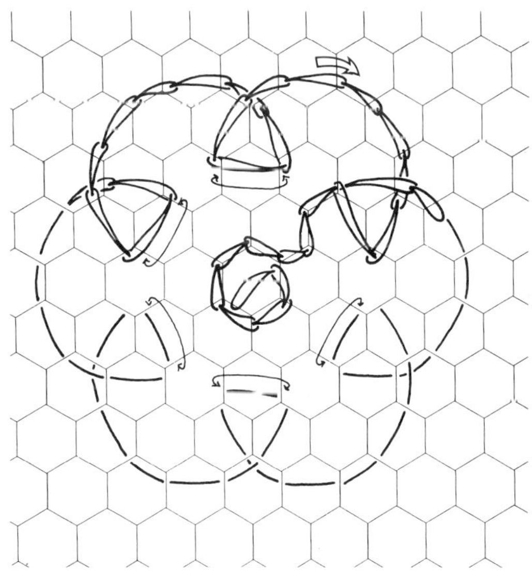

Above: st.5.
Left: a simple tambour outline to practise.

Stitch 6. Tambour fern-stitch

Work diagonally across the net. Using the same technique as in stitch 2, arrange the stitches in groups of 3, one tall flanked by two shorter, separating the triads by single stitches only one mesh high.
The returning threads can be seen clearly passing down the reverse side of the holes. The chain-stitches at the base emphasize the slanting course of the stitches.

Stitch 7. Rosettes

These too are worked in a similar manner to stitch 2, the chain-stitches forming a tight ring in the centre, while the loops form the six extensions.
As with the other stitches in this collar, the rosettes are worked diagonally and linked together by the feed thread passing beneath the net. Rosettes of consecutive rows lie horizontally to each other.

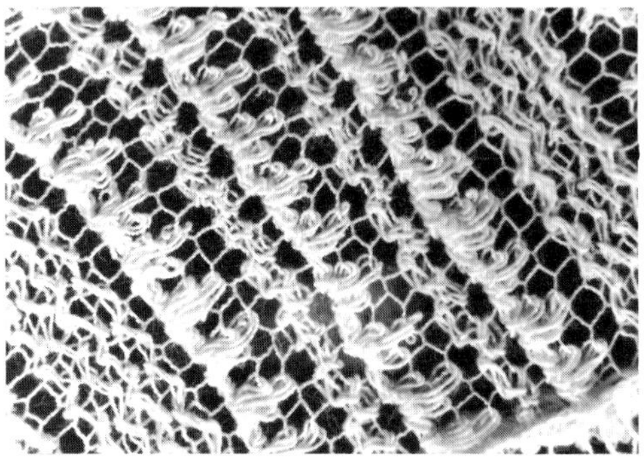

St.6.

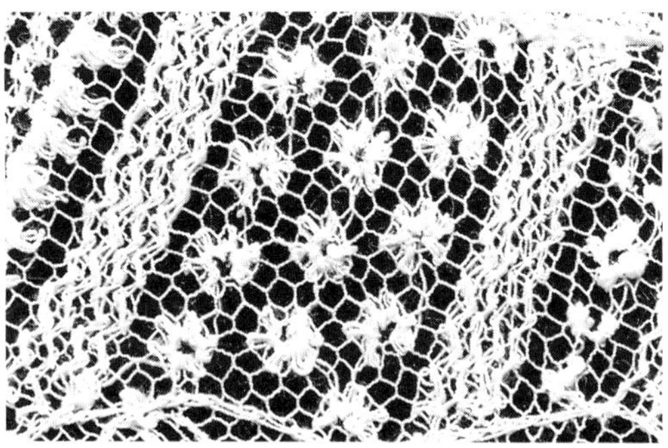

Left: st.7.

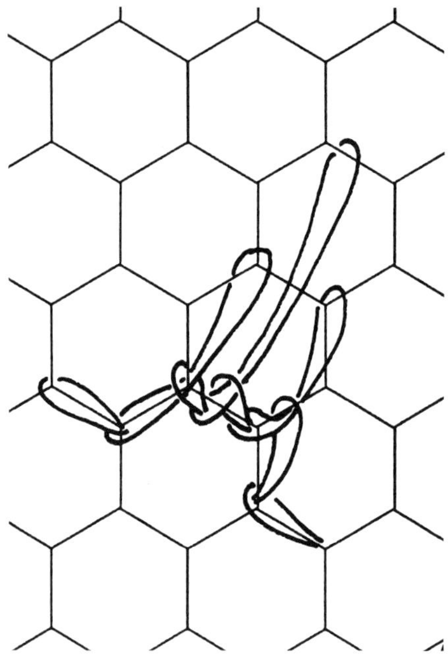

Left, and opposite, details of the large shawl.

LARGE TRIANGULAR SHAWL

This beautiful sophisticated design, very professionally executed, may indicate a Brussels origin. The swirling flowers, buds, leaves and tendrils, worked entirely in tambour in two thicknesses of thread, give an impression of the utmost complexity. In fact, the only added decoration is a scattering of tiny spots through the net, and lines of larger spots in formal alignment or spraying outwards from the stems.

Stitch 1. Small spots

As each spot is completed, make a knot, remove the hook, carry the thread beneath the net to the position of the next spot, and begin again.

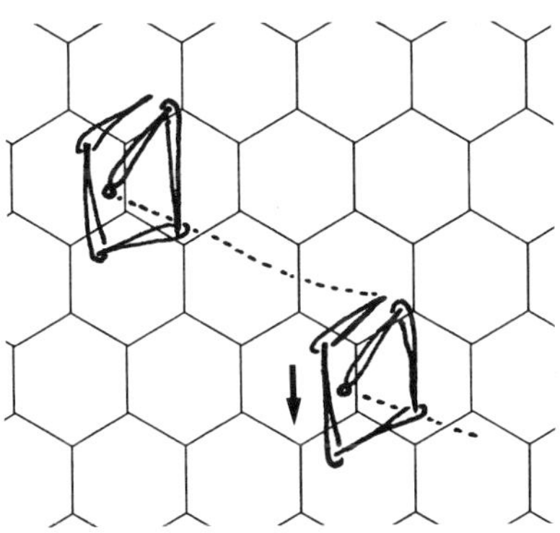

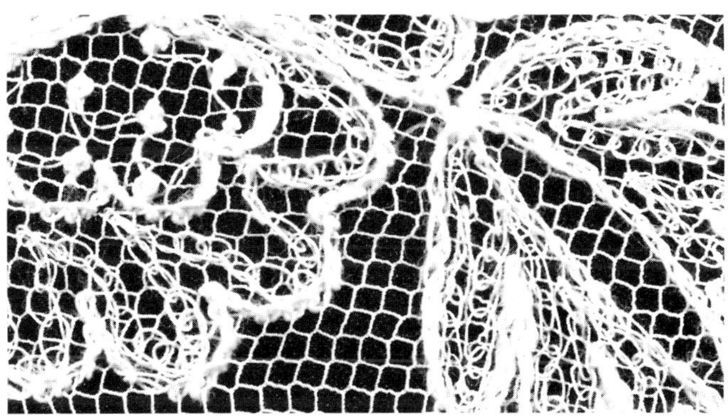

Further detail of the shawl.

Stitch 2. Large spots

These cover 4 meshes in every direction. Work as shown, and link the spots together as for stitch 1.

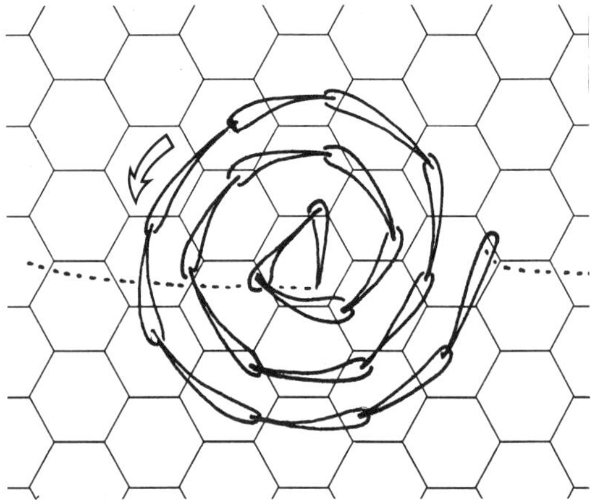

Embroidered machine net in fashion, April 1801. On the right, black silk patent net forms a collar and trim for the pink tunic worn over a white muslin gown. (Courtesy: Chertsey Museum).

A bonnet of rice straw decorated with rosebuds and embroidered blonde lace (see p.168).

Pattern and finished embroidery, a pheasant in a bower of blossom, Beveren-Waas, bonnet type (d), see p.176. (Courtesy: Trees van Ierssel-Ruygers).

On the right, the run outline completed; on the left a portion of the finished piece, bonnet type (c). (Courtesy: Trees van Ierssel-Ruygers).

Run stitches from an Italian sampler (Courtesy: Dieter Maier).

PEACOCK VEIL

This handkerchief-sized piece is centred with net instead of a woven fabric, and may be a veil to cover the face of a sleeping baby. It is worked entirely in tambour stitch.

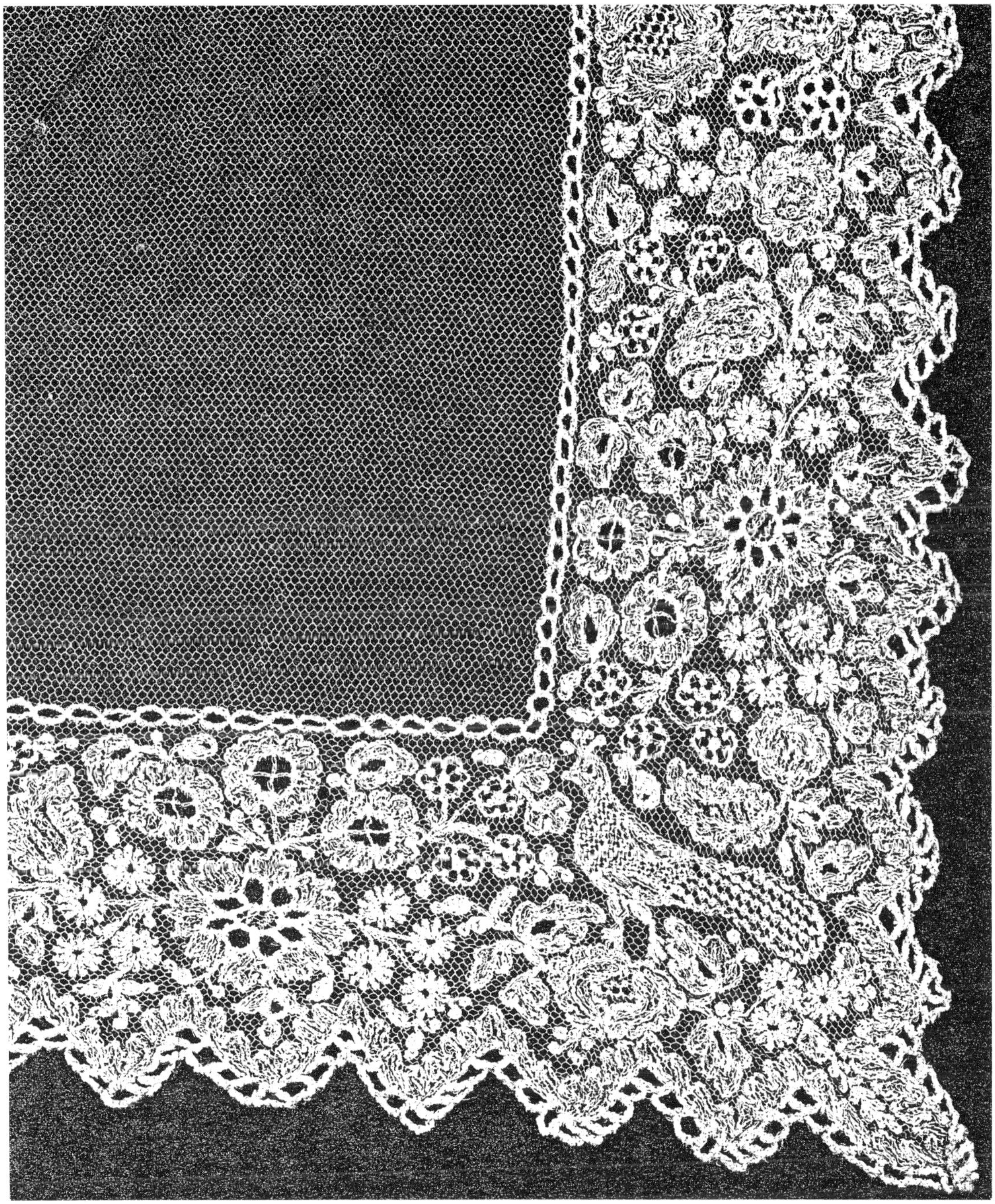

Stitch 1. Leaf vein

Row 1: Take sinuous lines of tambour stitch diagonally within the leaf, following the arrows.
Row 2: Repeat row 1 to form a mirror image. Each unit of the chain will enclose a single hole.

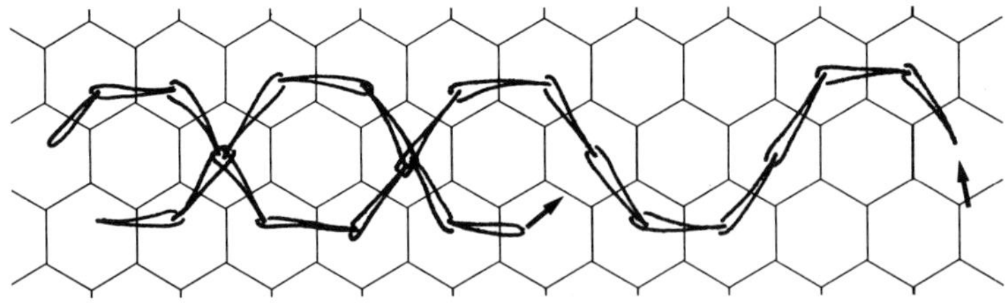

St.1. Here, worked horizontally.

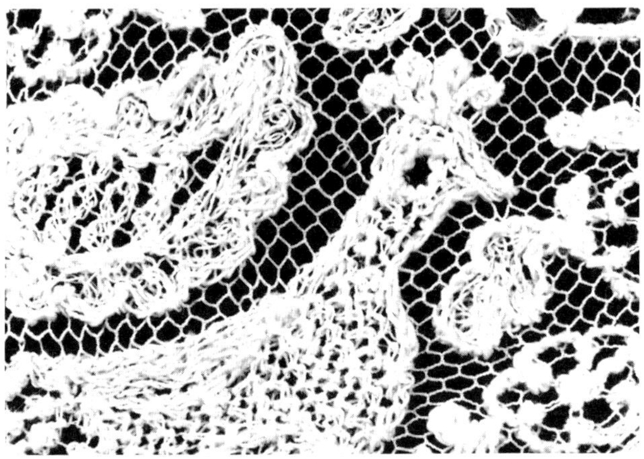

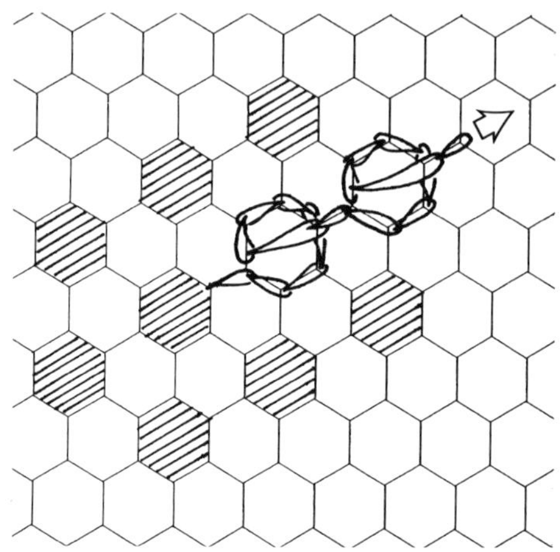

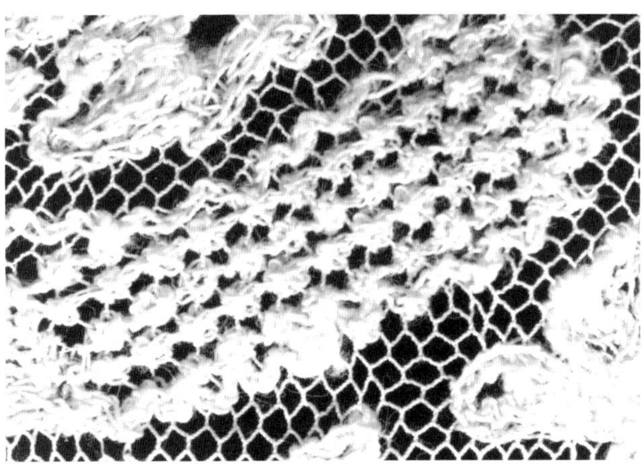

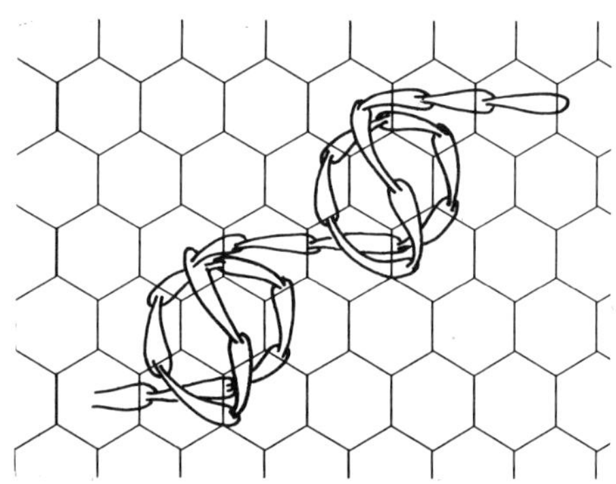

Above L: st.1. Below L: st.3. St.3 and 3A (on right).

Stitch 2. Zigzag bands

The peacock's body is filled with zigzag lines of stitches (see stitch 1, p.86). Narrow bands alternate with wider ones where the stitches cross two meshes at a time.

Stitch 3 and 3A. Spots

Spots fill the peacock's tail, like eyes. They are linked across the surface by tambour stitches, and the rows are worked LR RL in a diagonal direction. Two forms are possible, a simpler and a more complex.

Stitch 4. Flower (see stitch 5, p.87).

Stitch 5. Solid-petalled flower

It is not clear how this stitch is worked.

St.5.

The openwork border of the veil.

YOUNG WOMAN'S BONNET, 1st HALF 19th CENTURY

The bonnet is worked entirely in tambour, on a round net. As with the large triangular shawl, the complexity of appearance depends on the intricacy of the design rather than on the use of numerous fancy stitches. It may be from Lierre.

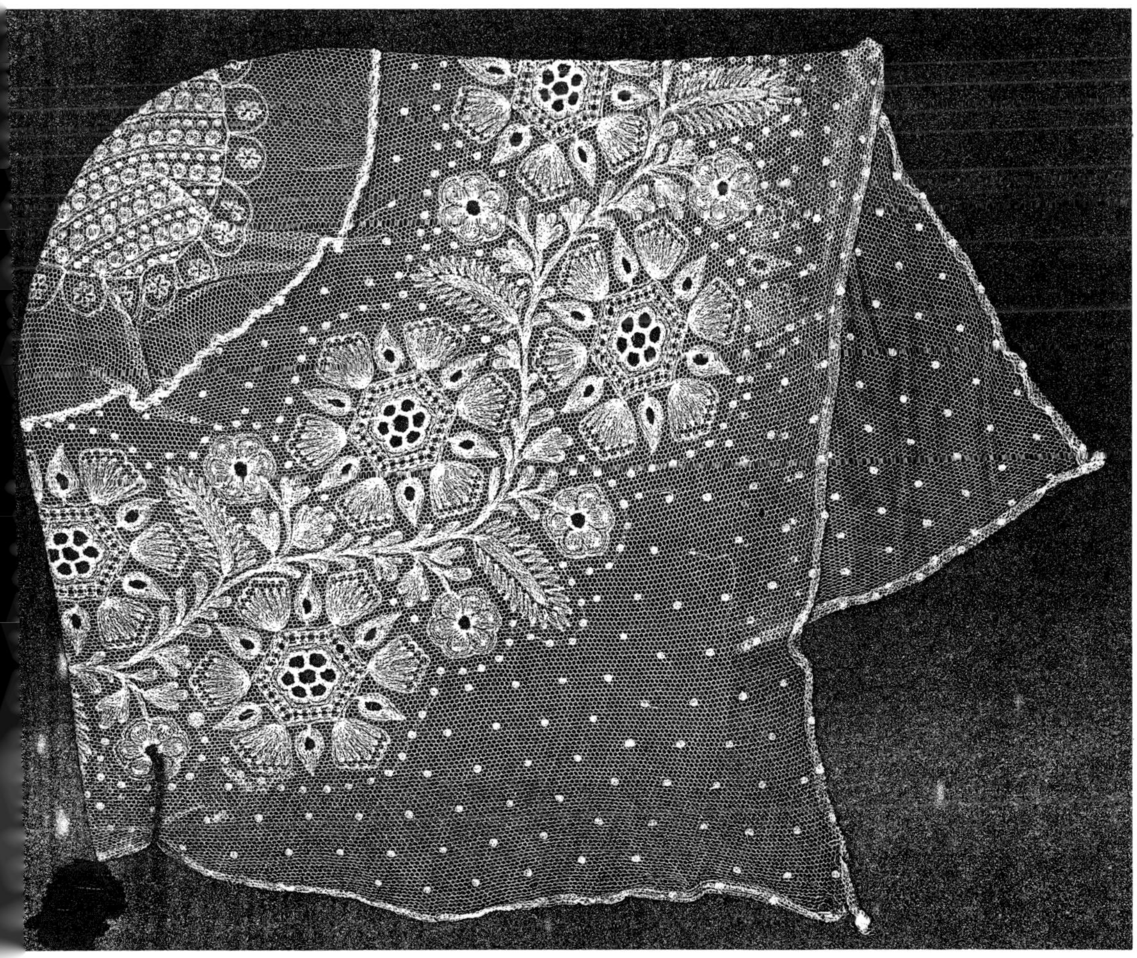

Stitch 1. Small spots

These form delicate lines, like suspended stars linked together and strung along the borders of the decorative band which surrounds the bonnet.
Work as stitch 1, p.89.

Stitch 2. Larger spots

These are similar in appearance to stitch 2, p.90, but the horizonatlly-arranged spots are joined by tambour stitches over the surface, instead of by threads carried beneath the net.

Stitch 3. Small flowers

Follow the arrows shown in the line drawing. Finish off each flower independently.

Stitch 4. Honeycomb border

Using a stiletto, slightly enlarge the holes marked (o). Follow the course of the arrows.
Work back in the reverse direction to complete the honeycomb.

Stitch 5. Enlarged holes

In the flower centres, dilate 7 mesh holes, using a stiletto. Tightly overcast the crushed walls using a needle, then tambour across the overcasting using a small zigzag stitch.

St.1.

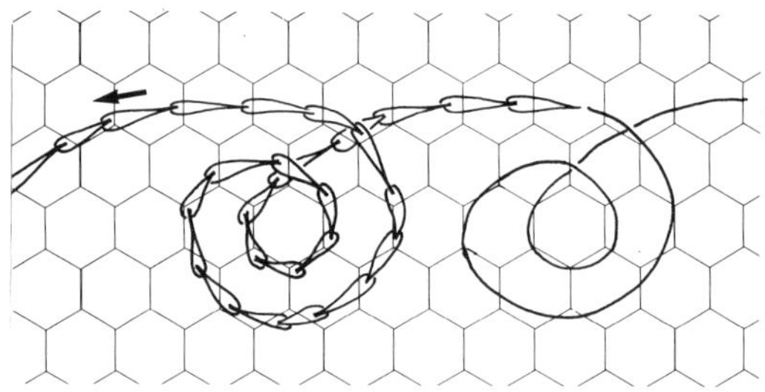

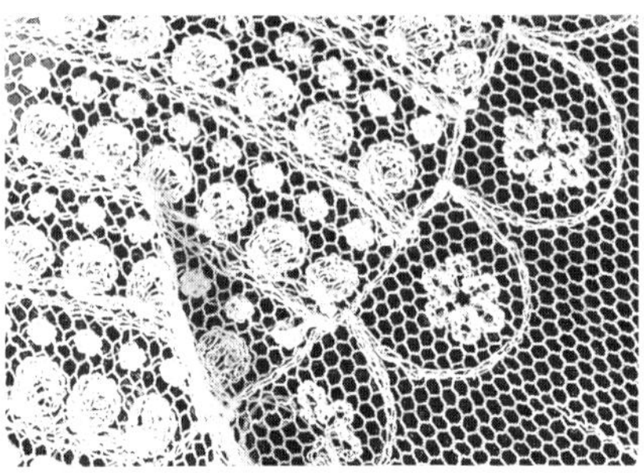

St.2.

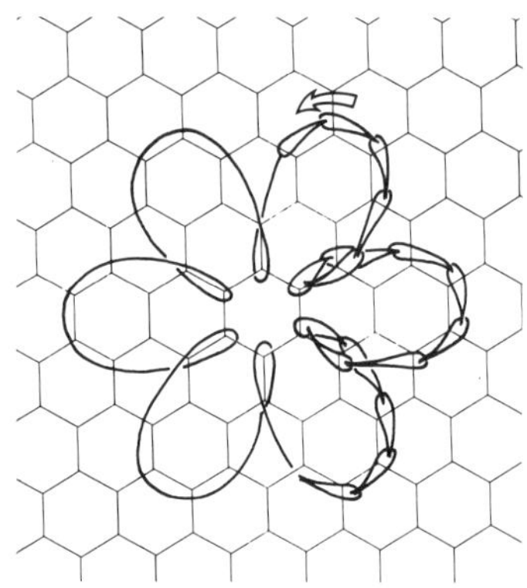

St.3.

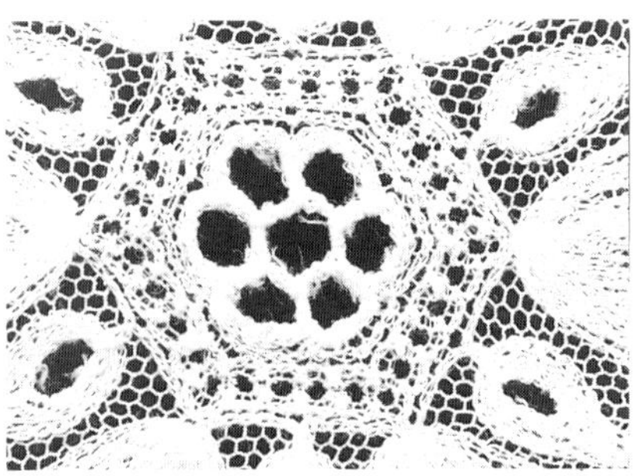

R: st.4. L: st.4 and 5.

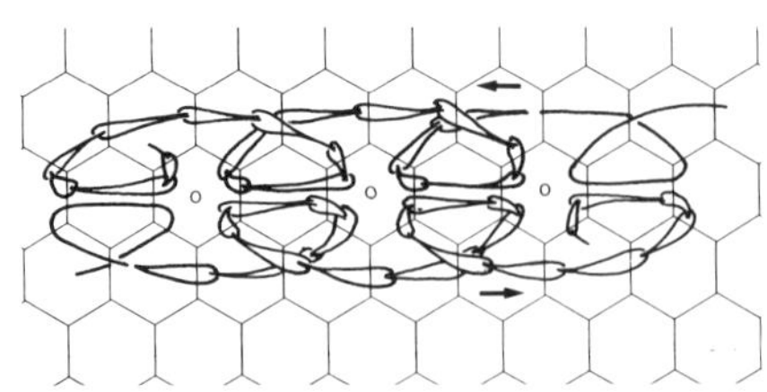

SPIKY COLLAR

This all-tambour collar has several unusual features: picots bristle from the solid design areas like thorny spines and the casket stitches pull apart the mesh sides binding them together like slender thongs. The bizarre effect is augmented by huge holes, their central net cut away, which crown the flowers; while the tambour stiches along the petals of the solid areas are so congested that the original outline is sometimes hidden.

This collar has been included for its unique qualities, though it is almost impossible to transfer its distorting thread movements from the original diamond-mesh net to the round isodiametric form.

The embroidery thread is slightly thicker than the mesh sides.

Picots

Work as the rays of stitch 2, p.86), but make all the spikes equal in length.

Details of spikes and holes.

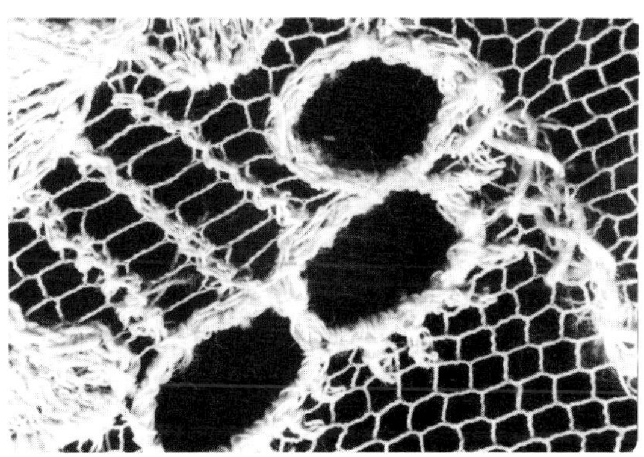

Stitch 1. Tightened zigzag

This is structurally similar to stitch 1, p.86, but the zigzagging tambour stitches are tightened until they lie almost straight, stretching the twist sides between them into ladder-like struts. Use a bodkin to assist the stretching.

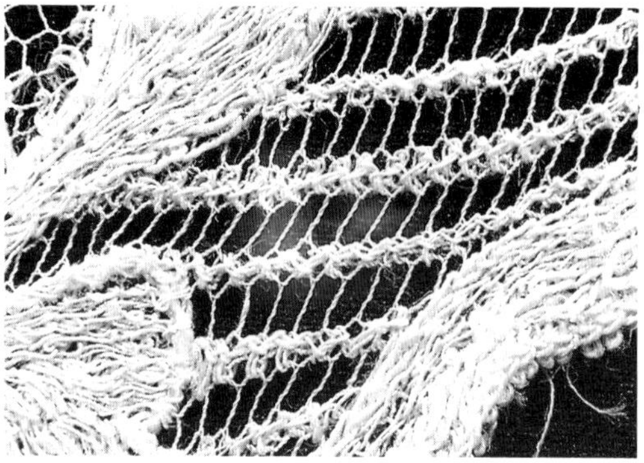

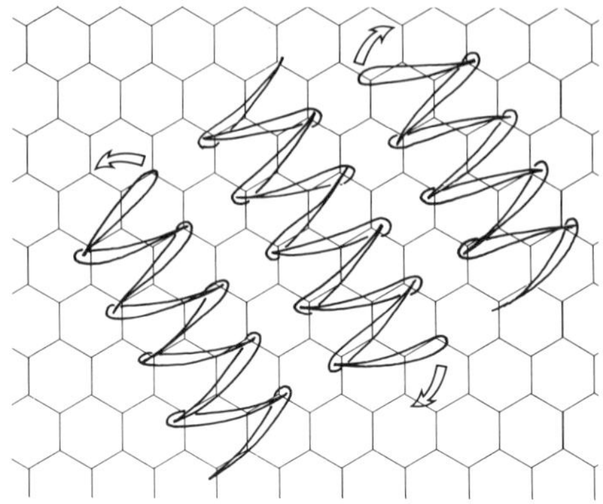

Stitch 2. Binding spots

These are essentially similar to stitch 3, p.92 , but each stitch is spread over several meshes so that when it is tightened extensive gaps appear around the spots.

Round nets are insufficiently flexible to allow such wide mesh displacements.

The collar is surrounded by a border of brief machine picots chain-stitched to the net.

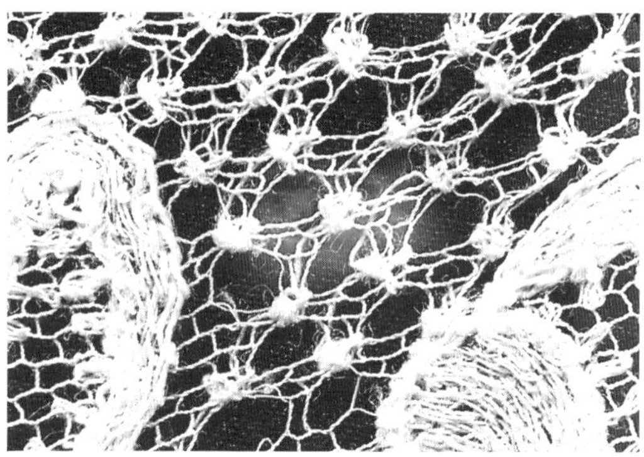

LONG STOLE

This artistically designed stole is worked in tambour on a round net ground. Tambour fills the solid areas and provides decorative elements such as spiral spots (stitch 2, p.90) and slightly enlarged holes. There are 3 run fillings.
The cross-sides are aligned horizontally.

Stitch 1. Diagonal cobweb-stitch (see stitch 1, p.32).

Stitch 2. Diagonal cross-stitch (see stitch 3, p.33).
The casket is diagonally striped with rows of cross-stitch separated by lines of empty holes.

Opposite below left: corner and border motif of a long stole.

Stitch 3. Diagonal figure-of-eight stitch

Note the path of the embroidery thread between one 8 and the next, and the reversal of the cross-over direction in alternate stitches.

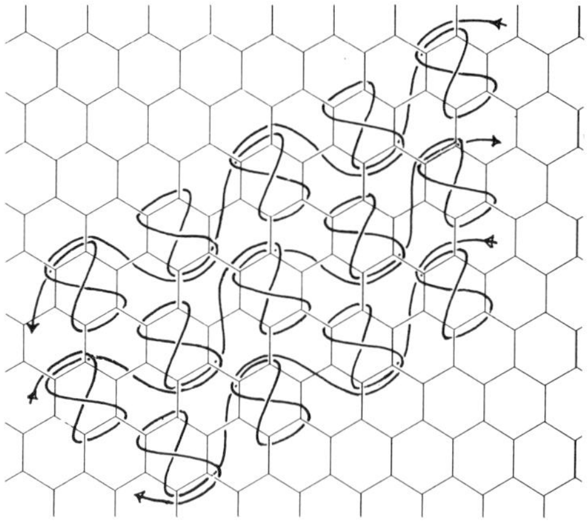

Within these reflexed sepals a small tamboured spot covers the central hole.

Details of the stole working.

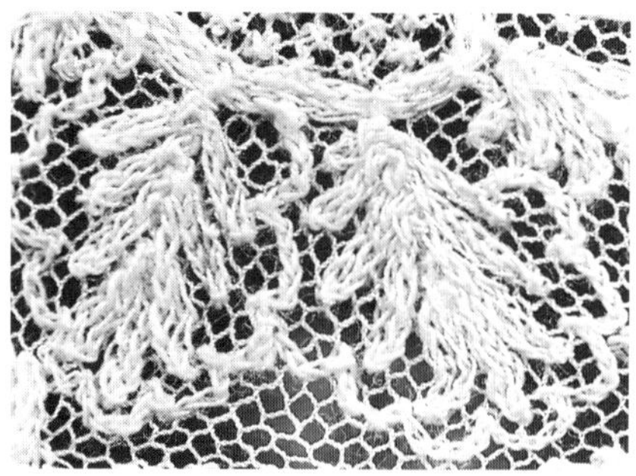

HARP AND IVY LEAF COLLAR

This is a fairly even mix of run and tambour stitches on a 4-twist diamond net.

The clustered forget-me-knots are heavily tamboured, so that they stand out sharply, while the harp and strings are worked more lightly. There are no fancy tambour stitches.

The shamrocks and ivy leaves have a run outline of fine double thread, and run caskets.

The solid design areas are filled with mixed cross-stitch and tent-stitch, producing a variegated effect.

For each casket stitch the cross-sides are aligned horizontally.

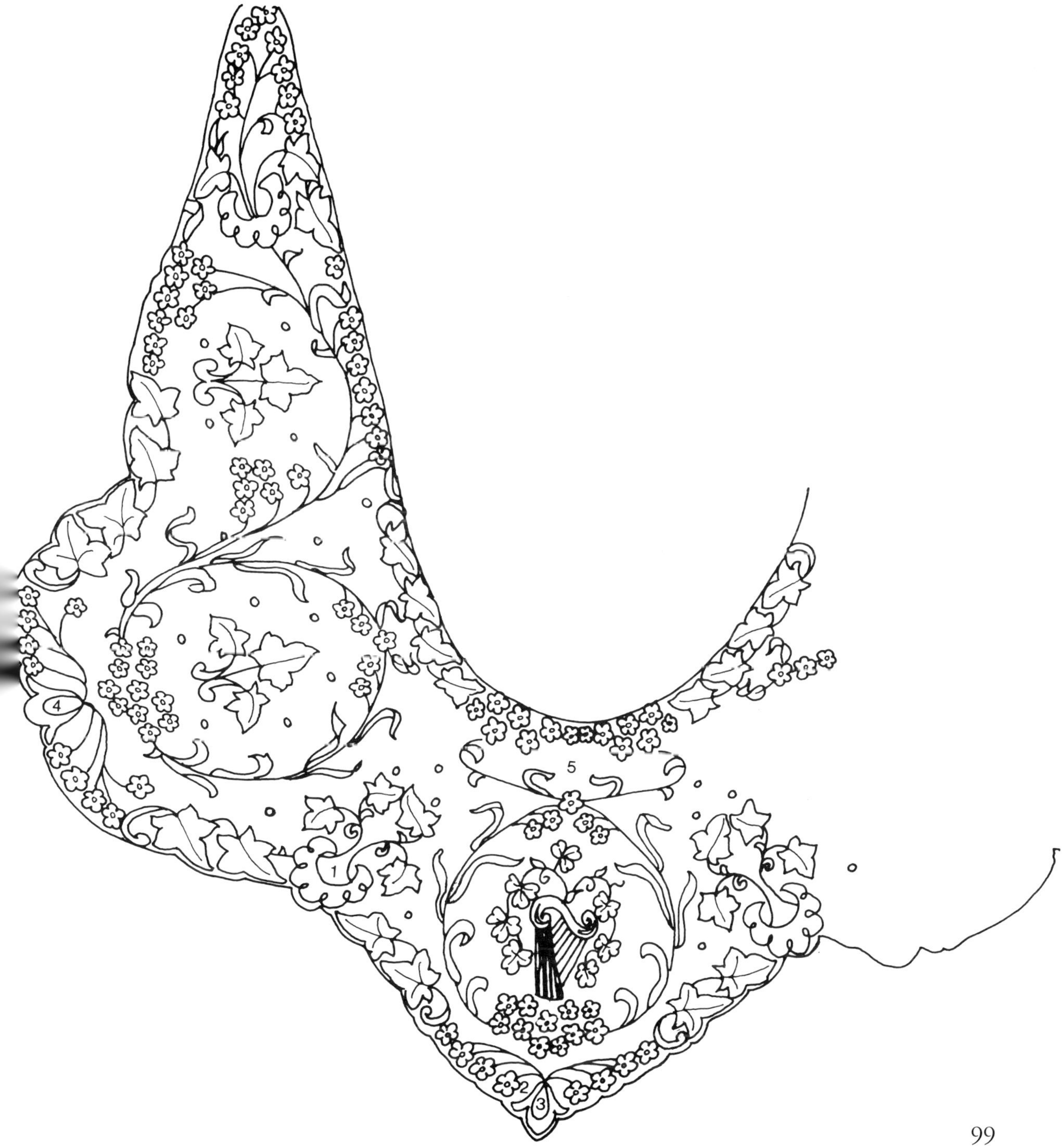

Stitch 1. Empty cross-stitch lattice (see stitch 1, p.38, but omit the satin-stitch spots).

Stitch 2. Horizontal cross-stitch
The casket is horizontally striped with rows of cross-stitch separated by lines of empty holes. The direction of the cross-overs is the same in every row (see p.28).

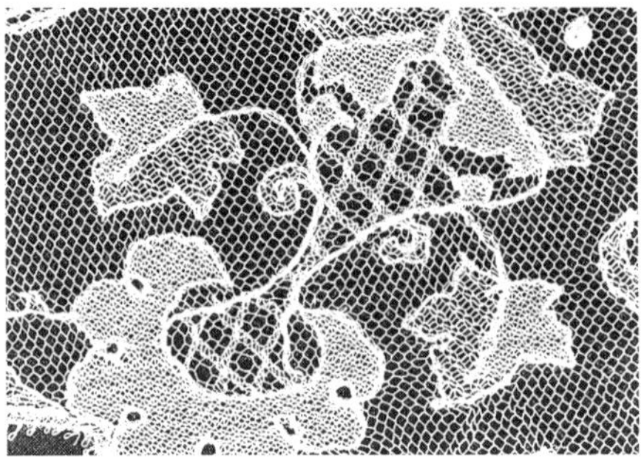

St.1.

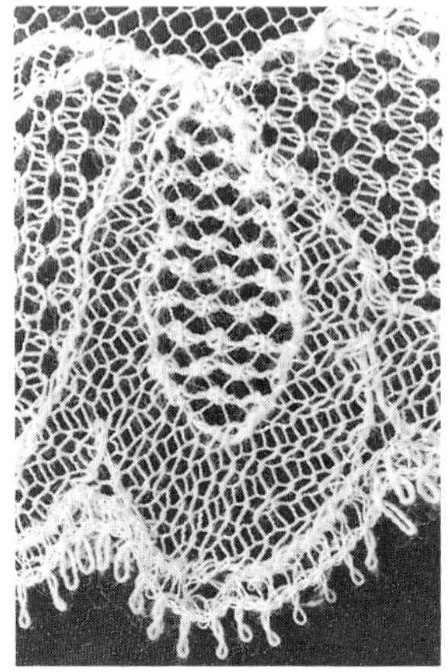

St.4.

Stitch 3. Cobweb-stitch (see stitch 2, p.33).

Stitch 4. Figure-of-eight stitch in vertical rows. (see stitch 10, p.36).
The threads of the stitches are slightly tightened so that the surrounding twist-sides ray outwards from them.

Stitch 5. Giant satin-stitch zigzags
In the original collar, each stitch covers 5 meshes, and each diamond encloses 9 holes, but the smaller size shown in stitch 5, p.34, fits better into a round net.

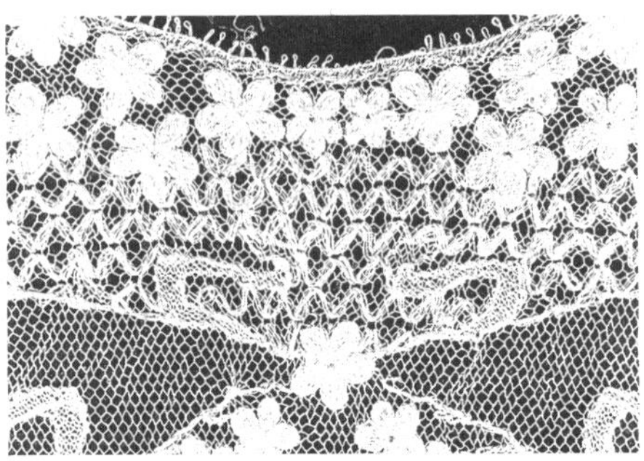

St.5.

The close tambouring of the flowers.

LARGE SHAWL COLLAR

This very large shawl collar measures 100 inches (250 mm) along its upper border.
The outline is tamboured on a 4-twist net.
The solid design areas are sometimes filled with tambour stitches, sometimes with cross-stitch or tent-stitch. The same embroidery thread, just slightly thicker than the mesh sides, is used throughout, but the closer proximity of the stitches in the tamboured areas, compared with the run, provides a strong contrast. There are 8 different run stitches in the caskets, but none of tambour.
For each stitch, the cross-sides are aligned horizontally.
Refer to the stitch map for the numbered positions.

A central theme of the design adapted to form the corner of a veil.

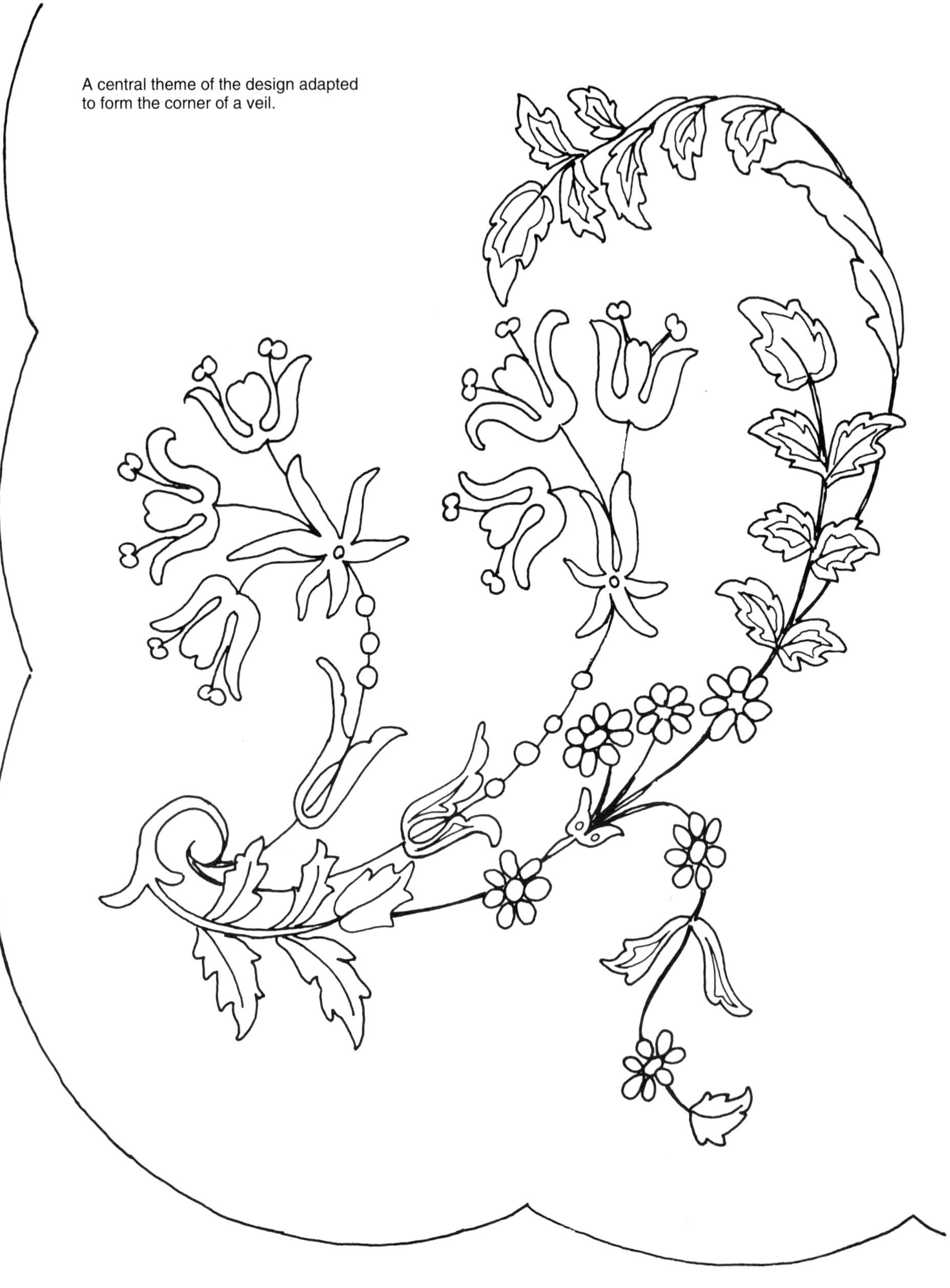

A stitch map of a detail of the shawl arm, and an inner motif.

A stitch map of the shawl.

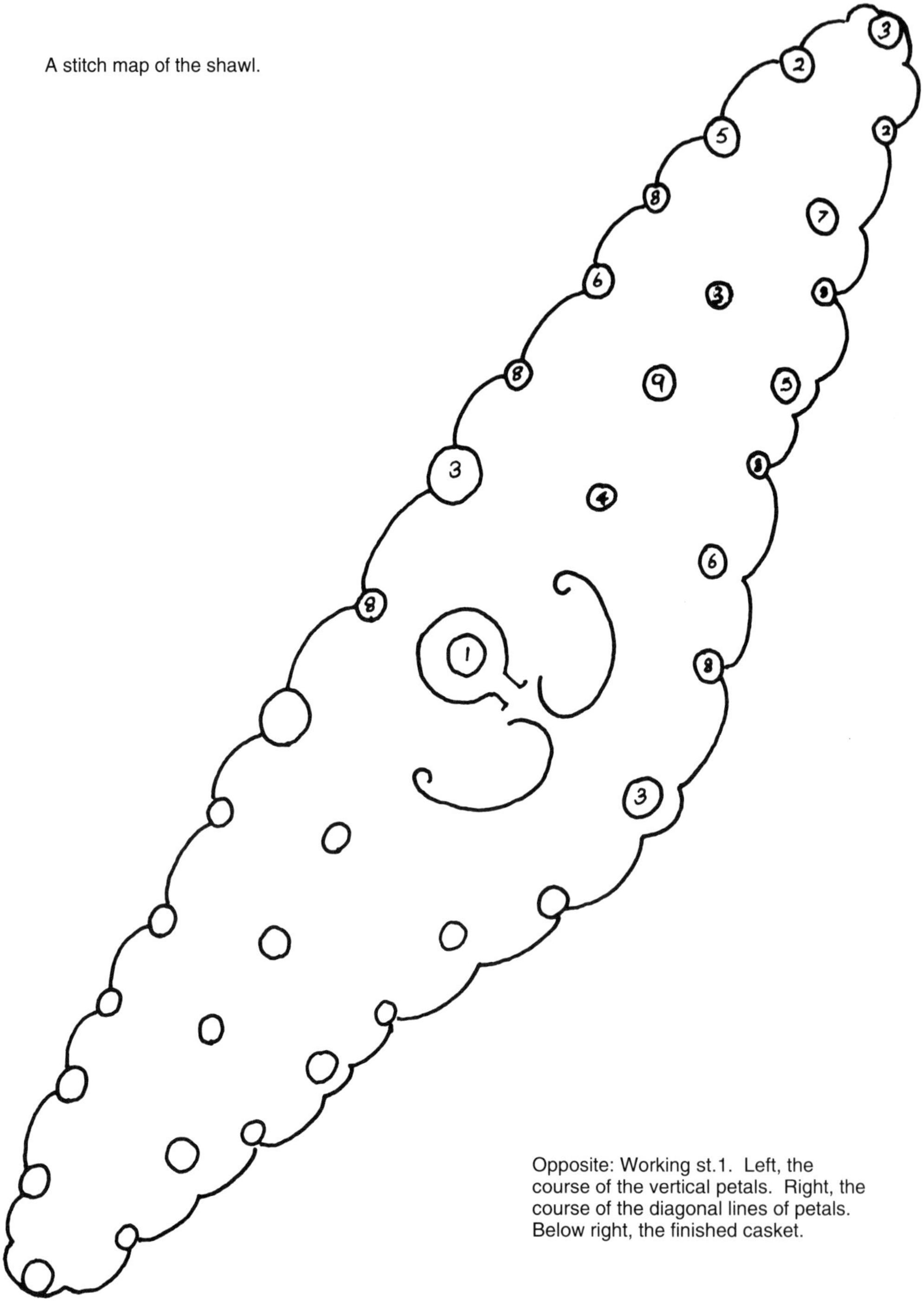

Opposite: Working st.1. Left, the course of the vertical petals. Right, the course of the diagonal lines of petals. Below right, the finished casket.

Stitch 1. Kaleidoscopic petals

The final effect, of 6-petalled flowers merging and overlapping with each other on all sides, is bewildering, but on examination the casket is found to contain only simple diamond-shaped areas of graduated running-stitch worked both vertically and along opposing diagonals.

Vertical petals Run the thread downwards so that it spirals around first one, then two, then three meshes, before narrowing again to two, then one. Continue in this way, referring to the drawing, until you reach the far side of the casket.

Space the next vertical row of petals 5 holes away, so that the widest part of the second row fits against the narrowest part of the first.

Repeat until the casket is filled.

Diagonal petals Work rows similar to the vertical ones, filling the casket first in a top-L/bottom-R direction, then from top-R/bottom-L.

The finished appearance is shown in the photograph. Except at the narrow link-points between one petal and the next, each lozenge is bounded by a line of holes.

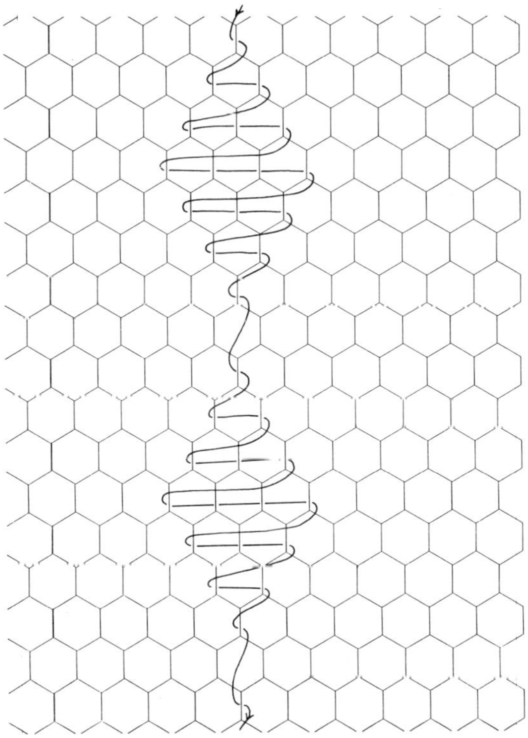

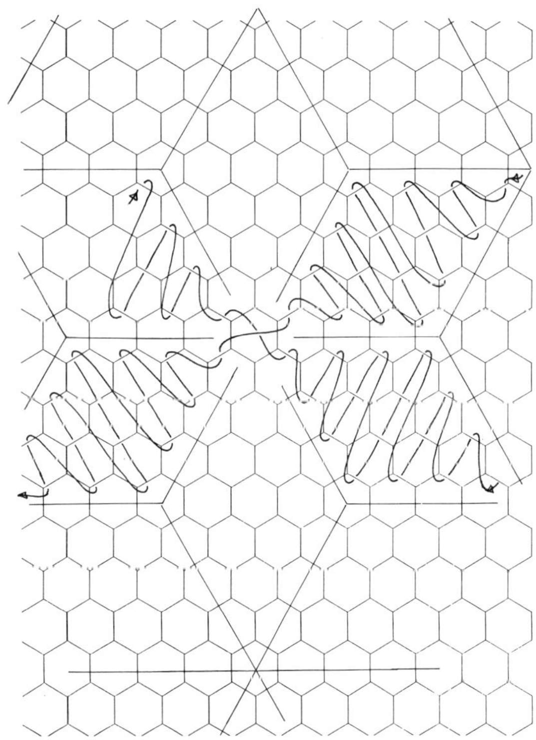

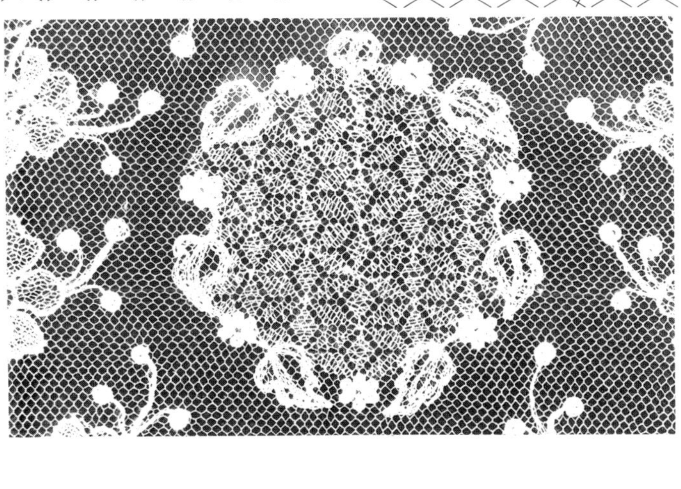

Stitch 2. Satin-stitch knobs

The stitches are worked in two diagonal sets:

Set 1. Top R to bottom L: Row 1. * Work tight satin-stitches 2.5 times around opposing twist-sides, pulling on them to enlarge the holes on either side. Pass beneath the next hole.
Repeat from * to the end of the row.
Leave a line of holes.

Row 2. Work back again in a similar manner.

Repeat these two rows until the casket is filled.

Set 2. Top L to bottom R: Row 1. Work satin-stitches tightly around the remaining twist-sides of the same meshes so that they encircle the stitches of Set 1, producing a bulky spot. Carry the thread across the back of the next hole, and repeat to the end of the row.
Leave a line of holes.

Row 2. Work back in a similar manner.

Continue until the casket is filled. The uppermost stitch will lie in the same direction in every knob.
In a round net, the crossing of the opposing threads is no longer at right angles.

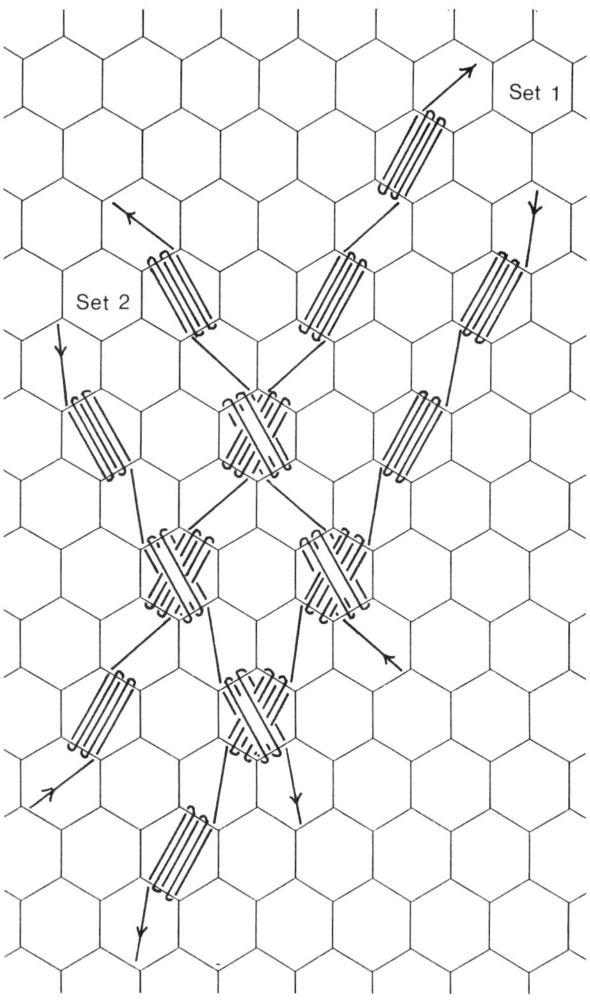

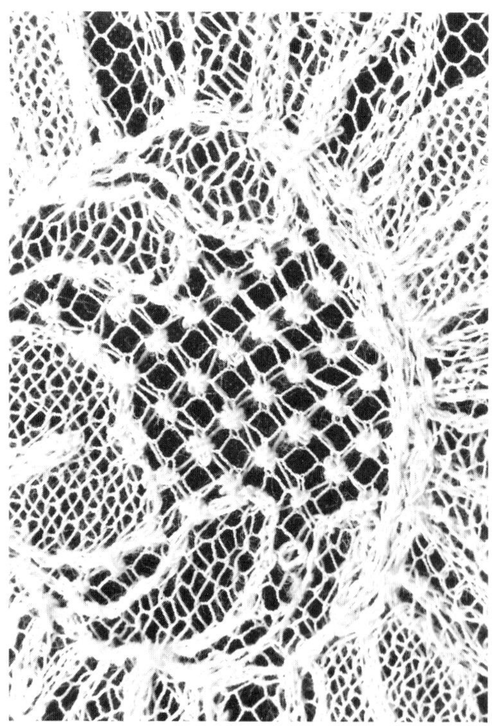

Stitch 3. Lattice of crossed figure-of-eight stitch (see stitch 10, p.36 and stitch 3, p.50).

The crossed figure-of-eight stitches are worked in vertical and diagonal rows so that, as they cross, the casket takes on the appearance of a diamond-paned window.

Set 1. Work crossed figure-of-eight stitch in a vertical row downwards. Leave a line of empty holes. Then turn the lace round and return in the opposite direction, creating a mirror-image of row 1.

Continue until the casket is filled.

Set 2. Work in the same way as set 1, but running the stitches diagonally from top R to bottom L. At the points of intersection with set 1, pass the thread beneath the stitches already worked, then continue.

Each unit of the lattice encloses 7 holes.

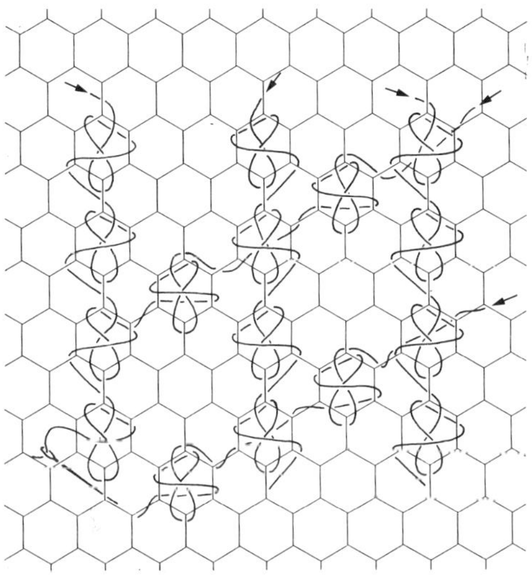

St.3.

St.5.

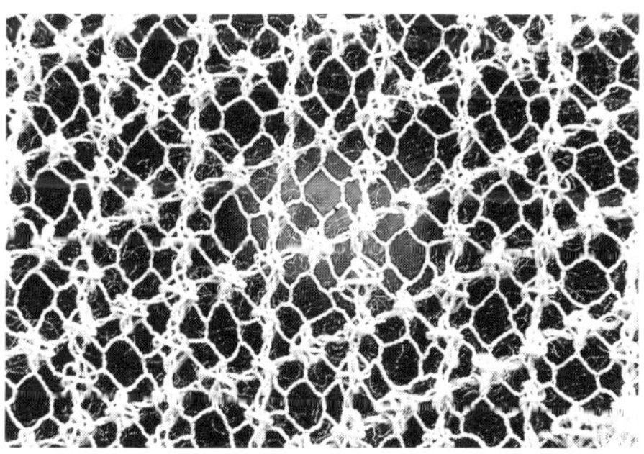

St.3.

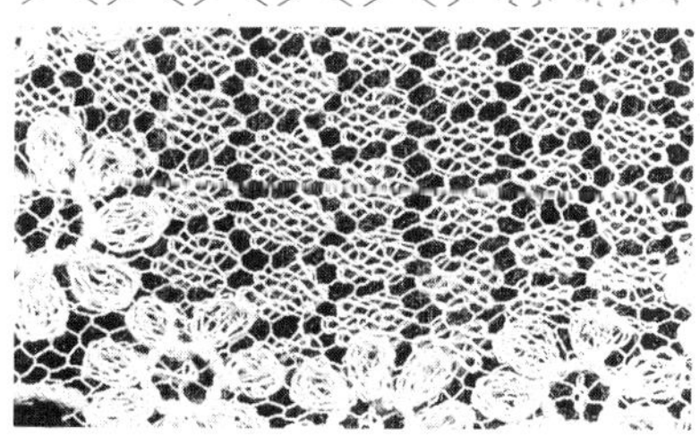

St.5.

Stitch 4. Horizontal cobweb stitch (see stitch 2, p.33).

Stitch 5. Cross-stitch clusters

Row 1: Work 2 cross-stitches in a horizontal direction, following the arrows.

Row 2: Work 3 cross-stitches.

Row 3: Work 2 cross-stitches.

Repeat rows 1-3 vertically up and down the casket, leaving a line of holes between the sets.

Alternate the position of the clusters in adjacent sets.

Apart from this misalignment, the sets are mirror-images of each other.

Stitch 6. Cross-stitch clusters and ten-point stars

Set 1. Work a single set of cross-stitch clusters, as for stitch 5.
Set 2. Next to set 1, work a vertical line of half-stars, positioning them as shown. Complete the other half of the stars on the return row.
(refer to stitch 6, p.60).

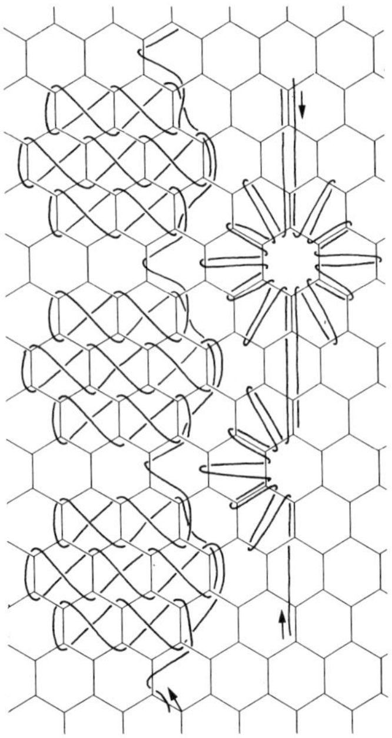

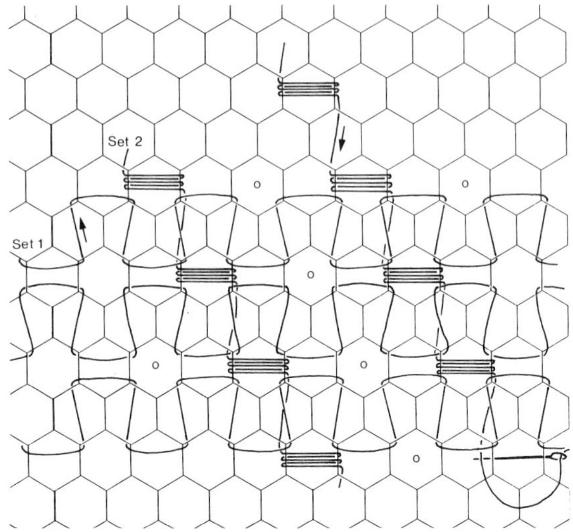

St.6.

St.7.

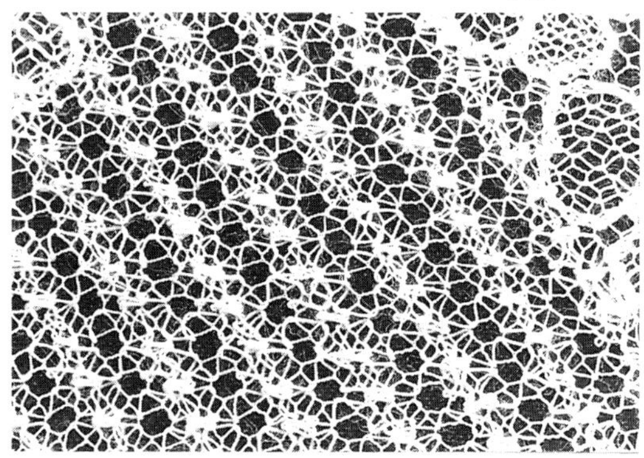

Stitch 7. Cobwebs and satin-stitch ties

Set 1. Fill the casket with horizontal cobweb-stitch, worked in every row. Keep the straight sides opposite each other, as shown. Note that the straight thread goes over the cross-sides of the meshes in one row and under them in the next. Slightly enlarge the holes (o) between the stitches as you work.
Set 2. Fill the casket with diagonal rows of satin-stitch spots, leaving three lines of holes between the rows. Pass the thread 2.5 times around the cross-sides which lie between the cobweb rows, alternating with the enlarged holes. Take the thread from one spot to the next along the back of the net.
The embroidery thread is the same thickness in sets 1 and 2, but the extra closeness of the stitches in set 2 makes them look heavier.

Stitch 8. Buttonhole-stitch ovals

The prettily interlocked ovals are worked as two rows, LR and RL, but the lace is turned between rows so that the embroidery proceeds constantly in a LR direction.

Follow the course shown in the drawing, noting the over-under movements of the thread.

Alternate the ovals in successive rows so that they fit closely between each other.

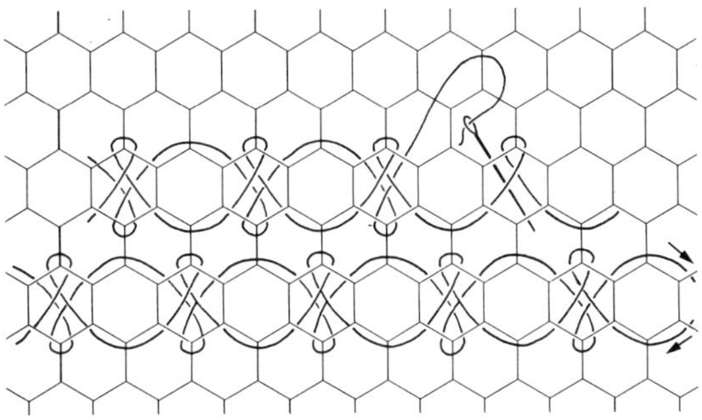

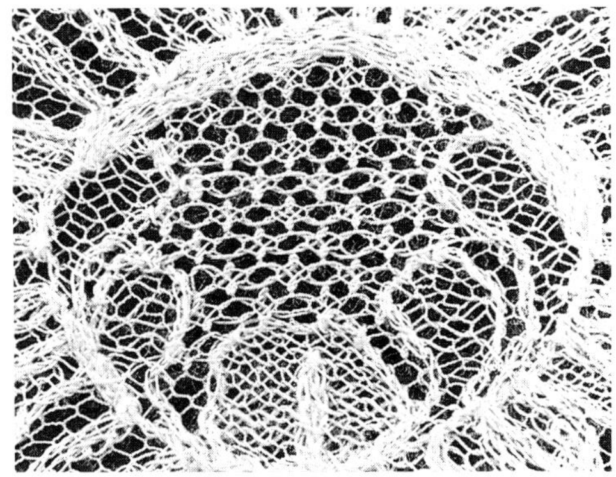

Above: st.8.

Below: st.9.

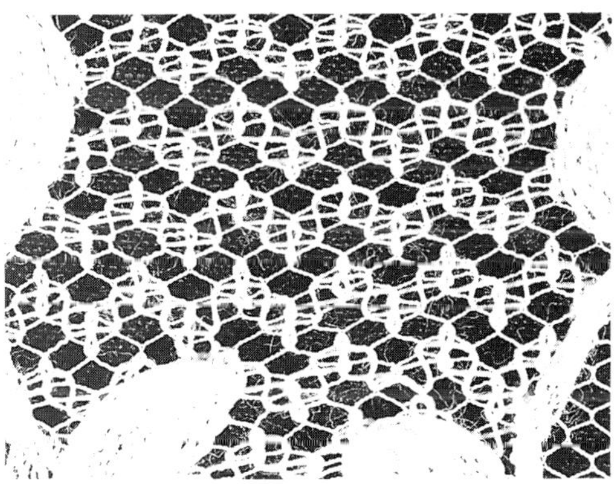

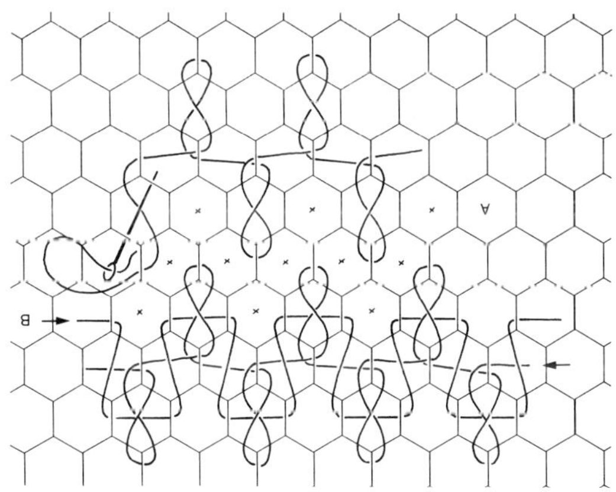

Stitch 9. Superimposed figure-of-eight and cobweb-stitch arranged as chevrons

Set 1. Work figure-of-eight stitches horizontally, taking alternate stitches one row down (see A in the line drawing).

Continue until the casket is filled, repeating the ups and downs in successive rows, giving the appearance of a zigzag line of holes (x) between them.

Set 2. Work horizontal cobweb-stitch around the figure-of-eight stitches, arranging the straight sides to lie below them and the slanting threads above (see B).

(compare stitch 3, p.47).

Stitch 10. Crossed figure-of-eight stitch

Worked vertically, this stitch fills the centres of the small tamboured flowers which cluster along the border of the collar.

Stitch map of the small veil with roses.

SMALL VEIL WITH ROSES

The outline and some of the veins are tamboured, but both the solid areas and the caskets are filled with run work.

The holes of the net are round.

The tambour thread is slightly thicker than the mesh sides. The run thread is sometimes slightly finer, but this varies in different stitches, depending on the effect required.

The solid design areas have light tent-stitch fillings and also denser areas of darning-stitch in which the first set of threads is heavier than the second (see working, p.27).

The delicate fragmented nature of the design and the wind-blown scattering of the run fillings in the leaves and petals, makes the lace look more complicated than in fact it is.

For each stitch, the cross-sides are aligned horizontally.

Stitch 1. Spoked wheels

The wheels are worked in vertical rows, each as a series of six interlocking triangles, tightened to stretch the holes around them.
Row 1: Using a stiletto, slightly enlarge the holes (o) to establish the parameters of the first wheel.
Beginning at A, follow the arrows and the numbered segments in the order 1 to 6.
The outer border of each segment crosses the face side, while the inner threads pass beneath the net.
When the first wheel is complete, take the thread to position B.
Lock it once around the cross-sides, then pass it beneath the net for three holes to position 1a. This is the starting point for the next wheel.
Continue in this way across the casket.
Row 2: Repeat row 1 so that the wheels lie close to, and alternating with, those already made.

Stitch 2. Crossed figure-of-eight stitch (see stitch 10, p.36).

Rows of crossed figure-of-eight stitch are worked vertically up and down the casket, alternating with each other in adjacent rows.

Stitch 3. Cross-stitch lattice with satin-stitch spots

Set 1. Working as stitch 3, p.107, construct a lattice of crossed figure-of-eight stitch. Each lozenge encloses a central hole surrounded by six others.
Set 2. Over the central mesh of each lozenge, work 5 or 6 satin-stitches around the twist-sides, using a thread the same thickness as the tambour outline. Crowd the stitches together to produce a large raised spot.
Finish off the spots individually by taking the thread through to the back. Sew in at once, or leave hanging to be finished off later.

Stitch 4. Cross-stitch and fans

Rows 1 - 4: Work cross-stitch diagonally across the casket along two rows of meshes.
Row 5: In the next row, following the thread course shown in the drawing, work a series of fans, each with 5 sticks.
Bind the intermediate meshes tightly together with satin-stitch encirclements.
Leave a line of holes.
Repeat these five rows until the casket is filled.

Stitch 5. Cross-stitch and satin-stitch spots

Rows 1 - 2: Work satin-stitch spots diagonally, so that they alternate over two rows.
Rows 3 - 6: Work two diagonal rows of cross-stitch as in stitch 4.
Rows 7 - 8: Repeat rows 1 - 2.
Leave a line of holes.
Repeat this sequence until the casket is filled.

Stitch 6. Interlocking tent-stitch

Follow the instructions for stitch 1, p.44, but use a single fine thread instead of a thicker double one.

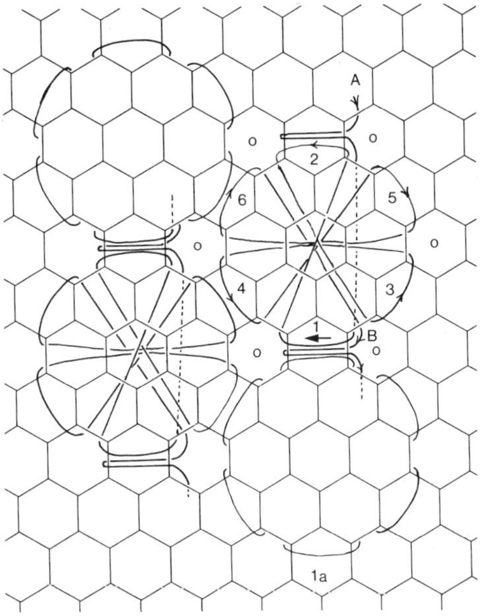

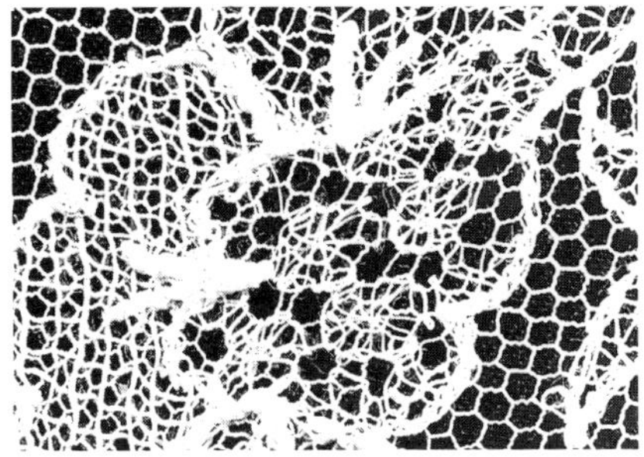

St.1.

Right: St.3.

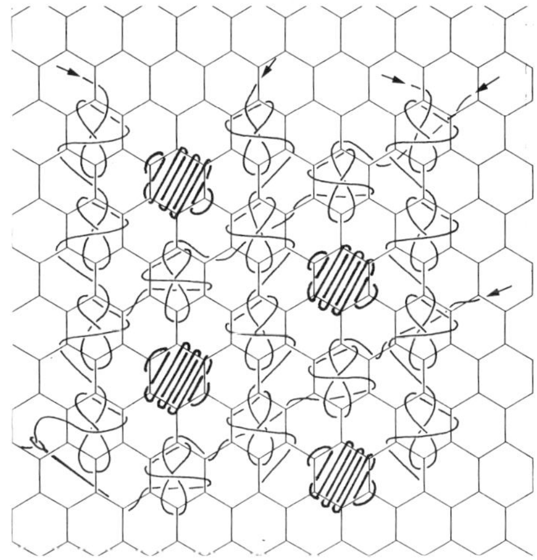

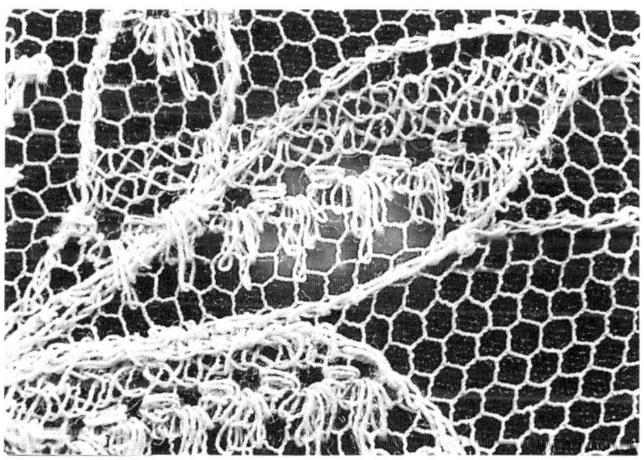

Below: St.5.

St.4.

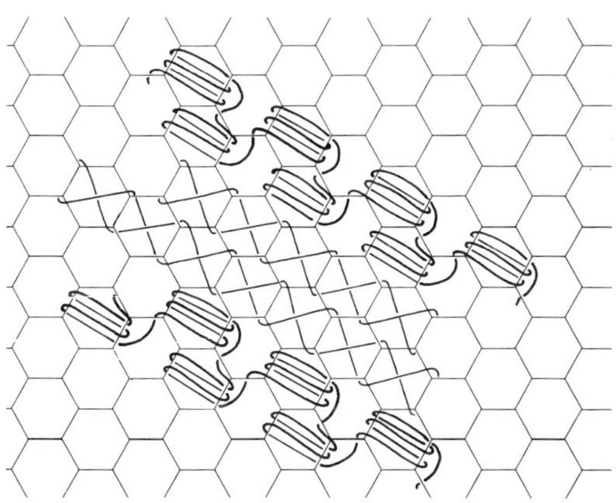

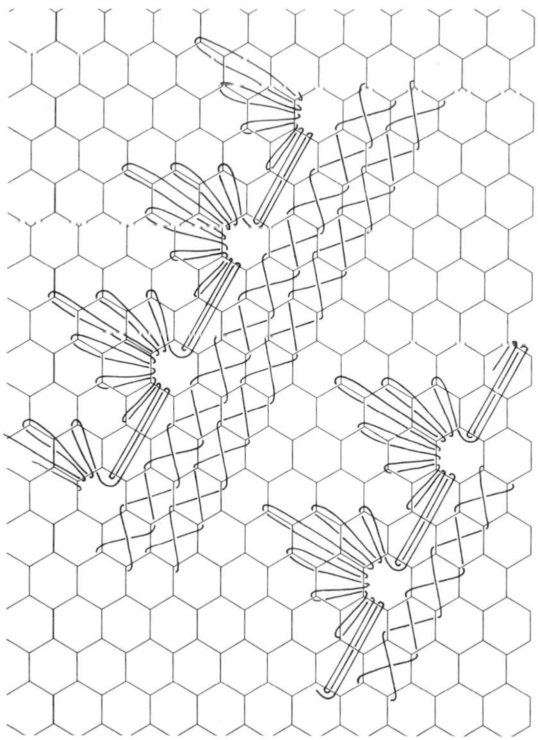

BONNET VEIL WITH PALMATE LEAVES

This striking design is simple to work.

The narrow tambour outline encloses solid areas filled with darning-stitch.

There is almost no additional decoration. A single casket filling is provided by **figure-of-eight stitch** within the hearts of the 7-lobed leaves. Although the appearance is of vertical rows, the stitches are linked diagonally, and the figures-of-eight tightened, to extend the rayed effect of the surrounding mesh sides.

Stitch map of the bonnet veil.

Figure-of-eight stitch from the leaf centre.

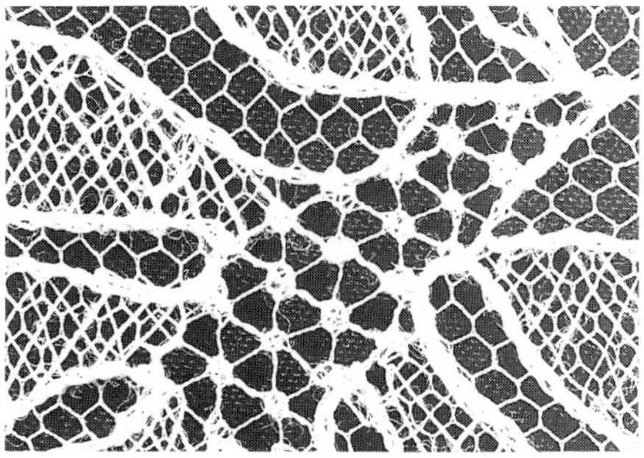

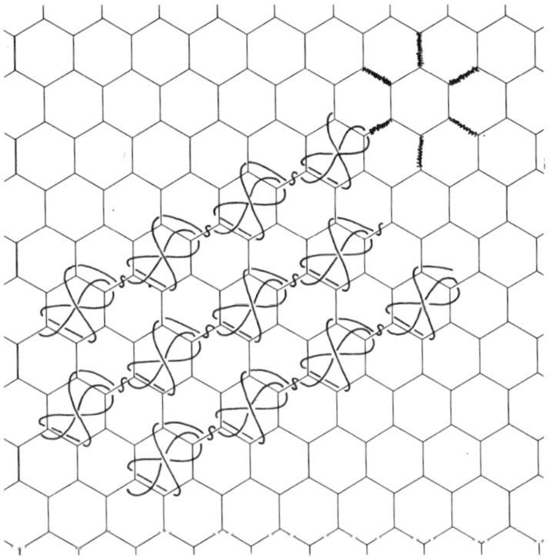

Finishing touches When the embroidery is complete, the net will need to be cut to the required shape or size. The raw edges will then be exposed to fraying and loss of shape. A protective treatment of the border is most easily carried out while the work is still in the frame. As with the run laces (p.19), this can be done by:

(a) attaching a machine-picot border, using overcasting or tambour stitches;
(b) tambouring along the shape of the border and then overcasting the tambour with fine thread;
(c) stitching on a narrow border of lightweight bobbin lace.
(d) overcasting or buttonholing with thick thread. This was used only for heavy furnishing laces.

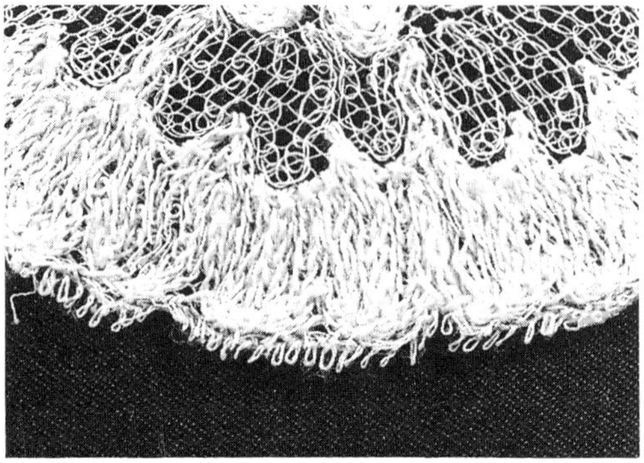

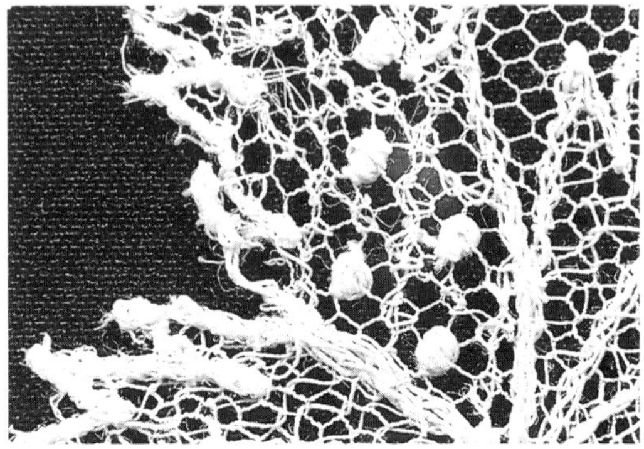

Above: (a). Below: plain tamboured border.

Above: (b). Below: (c).

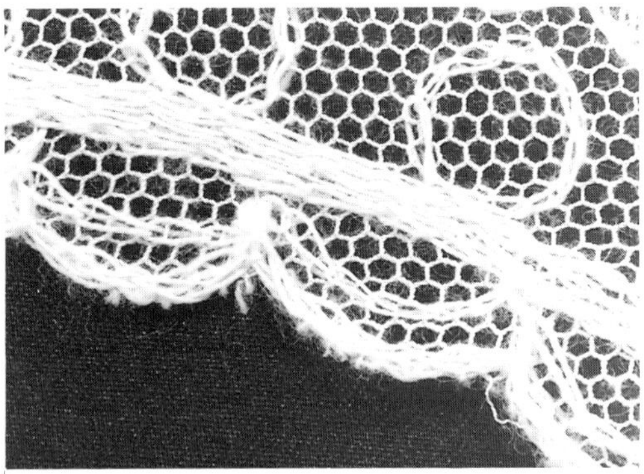

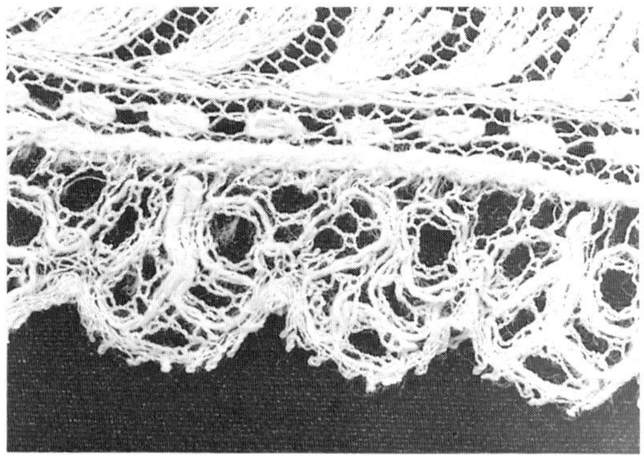

TAMBOURED BEADS AND SEQUINS

The decoration of machine nets by strands or clusters of glittering sequins, crystal, pearls, jet and steel beads was a natural extension of cotton-thread embroidery. There is no record of the technique being practised in Limerick, but nets with attached artifacts were certainly haute couture from the 1890s to the 1930s [101].

To work: String all the beads you are likely to need for one motif on to the free end of the feed thread, and hold beneath the net. As each tambour stitch is made, release one bead to lie against the lower surface, pull the next bit of thread through to the upper surface and make a chain stitch. Hold up the next bead, and repeat.

If you find you have too many, or too few, beads, simply knot the thread, cut, remove or add beads, and rejoin.

When the work is complete, the upper surface with the tambour stitching will become the reverse side and the lower surface with the beads the face side.

A similar technique can be used for bugles (long cylindrical glass beads) and for sequins.

The Cornely machine could also stitch on beads and sequins, though in s slightly more complex manner. Both sewing thread and beads were fed from above, from rotating reels (universal feed).

Instead of the single hooked needle, as in hand-work, the up and down movements of a straight eye-pointed needle above the net were coordinated with the horizontal movements of a hook below the net. This caught the thread from the needle and pulled it forward in the form of a loop. The next descent of the needle passed through this loop, then its thread was again pulled forwards, ready for the next descent. Thus chain-stitches were produced on the underside, and backstitch on the upper.

At the same time, the string of beads was laid over the surface of the net, adjusted so that each backstitch would catch the string down to the net, allowing the beads, one at a time, to be attached.

Tambouring in brilliant silk or metal threads, often on vividly dyed nets, was found in England, France and Belgium in the late 19th century (see chapter 5).

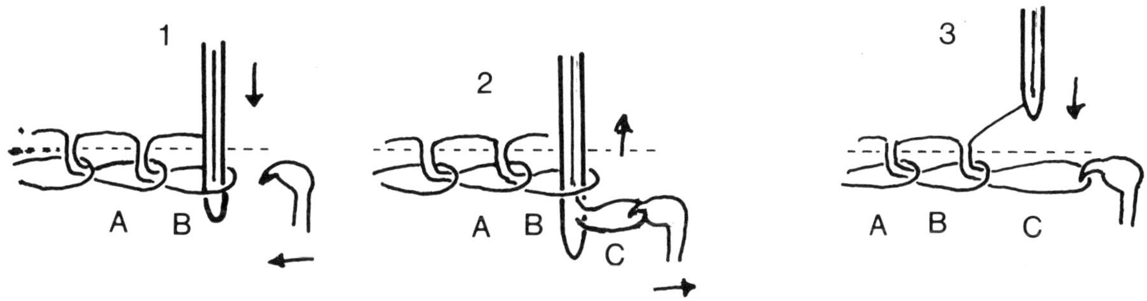

The chain-stitch action of the Cornely.
1. The hook on the right waits to catch the thread from the needle as it descends.
2. The thread is hooked and pulled into a loop. As the needle ascends, the net is moved, placing the threaded needle tip over the newly formed loop.
3. The needle descends again, ready to repeat the sequence.

Note: the freedom from that constant rotation of the thread associated with hand work, reduces the likelihood of the twist becoming unbalanced.

OTHER TAMBOURED LACES

Coggeshall in Essex, Lunéville in France, Lierre in Belgium and Hamilton in Scotland, as well as north-eastern Switzerland were all important centres for the commercial manufacture of tambour embroidery on net. In chapter 5, some attempt has been made to investigate their distinctive features, but the extreme scarcity of reliably provenanced pieces makes it impossible to draw any truly dogmatic conclusions.

Genuine Irish pieces and patterns may appear in museum collections anywhere in the world, even when those collections are specifically devoted to the embroidered machine nets of the local industry.

Harps and shamrocks occur, and are regarded as noteworthy elements of Irish design, but there was nothing to prevent their being copied elsewhere. Some Irish manufacturers attached seals to their laces in an attempt to establish the authenticity of their origin, but it is unlikely that any of these survive to help us now [102].

Heavily spotted blizzard effects and an almost over-rich formality are often associated with Belgium.

Clusters of wild flowers such as sheep's parsley, bluebells, primroses and ivy leaves, filling the net with graceful sprays, are characteristic of Coggeshall, but found also - more sparingly used and more distantly spaced - in Limerick laces.

Apart from this, differences in the number and type of decorative stitches, in the method of outlining and the treatment of the solid design areas, as well as documented patterns, provide some pointers to the date and place of origin of individual pieces.

Commercial production was, as always, directed by the flimflams of fashion, and any successful centre was likely to be avidly copied by others, aiming to horn in on the profits of a rich but uncertain market.

Below: based on a design from Carisbrooke Castle, I.O.W.

Below: based on a design of 1852.

The five designs (pp.118-120) can be worked in plain tambour stitch, or with some decorative tambour caskets, or some run caskets.

Based on a Vienna design, c1913.

4. INSTRUCTIONS FOR MAKING LIMERICK CUT CAMBRIC

Definitions A woven fabric (initially cambric, muslin or lawn) is cut into shapes according to the pattern, to form the solid areas of the design.

Appliqué Before cutting, the fabric is attached all round the pattern outline by means of a couched cord to a layer of net which lies beneath it.

Guipure The pattern is again outlined by a couched cord, but there is no net and the attachment is between cord and fabric only. Before cutting, the motifs are linked together by buttonhole-stitched bars passing over the surface from one to the next without penetrating the fabric.

Appliqué and guipure may appear together in the same lace.

In appliqué, the caskets for the decorative stitches are created by cutting away the surface layer of cloth to expose the net. On this the fancy stitches are worked.

In guipures, cutting away the cloth does not expose any net since there is none there. An empty hole is left, which is then filled in with needlelace and needleweaving stitches. Additional decoration is provided by twisting the outline cord along the outer border of the lace into a series of circular loops known as TWIRLS. Very occasionally these also appear around or within the motifs.

POPS, or buttonholed circles, are traditionally found in the appliqué form.

The historical relationship between the industries of Limerick and those first established on the Bath and Shirley estates in County Monaghan, some 170 miles to the north, is unclear. Both began around 1820, the first reputedly inspired by a Brussels bobbin lace appliqué onto hand-made net; the second by Italian - Caulfield and Saward say Indian - embroidered appliqué [103]. There are no universally agreed features to distinguish the north and south industries - and of course, as with all Irish laces, there were quite a number of scattered subsidiary centres of manufacture.

Some examples show a strong Brussels influence but, considering the nonchalence with which Brussels designs were 'borrowed', and the enormous prestige of all Brussels laces in the market place throughout the 19th and early 20th centuries, this is scarcely remarkable. 'Brussels lace, that magnet of attraction to ladies, is contributed in great abundance and beauty, by many manufacturers of the Belgian capital' [104].

A mixed appliqué and guipure lace, designed by Alice Jacobs at the Metropolitan School of Art, Dublin. Purchased by the Department of Agriculture and Technical Instruction at the Royal Dublin Society summer show, 1901. (Courtesy: National Museum, Dublin).

Materials

NET Sufficient two-twist bobbinet is needed to leave an adequate margin around the design, to be trimmed back later.

WOVEN FABRIC A cotton, or cotton-and-nylon organza, should be translucent enough to allow the pattern to be seen through it, but not so transparent that, in the finished piece, the net below is immediately apparent, or the impact of the design will be weakened.

Heavier nets will need a denser overlying fabric to give sufficient solidity of contrast.

Both net and fabric should be pre-shrunk. If this is not already done, soak both in hot water, spread them out, blot away excess moisture, leave until almost dry, then iron gently to avoid pulling out of shape, and use spray starch for a crisp finish, if required.

PATTERNS AND SUPPORTS For small pieces a card can be used, and for large pieces a frame with rollers. In the past, instead of a frame, even quite big pieces of cambric and net were tacked to several layers of sheeting or unbleached calico underlaid by brown or grey paper, all soft enough to be rolled up so that only the piece of the work currently being embroidered was exposed.

The pattern may either be placed beneath the net, in which case it must be sharp enough to be seen through both net and cloth, or it may be stamped, transferred, drawn or pounced directly onto the cloth, using some colouring matter that will wash out later. William Briggs, who invented the hot-iron transfer method in 1874, also supplied 'kits' for making appliqué work under the name 'Carrickmacross' in the early years of the 20th century.

In the first instance, a colour such as blue, green or burnt yellow for the pattern makes it easier for the white threads to be seen, and reduces the strain on the eyes. Only a single repeat need be drawn. When it has been completed, move the pattern along, tack it in place, and begin again. However, when the pattern is marked on the top layer, it has to be drawn in its entirety.

Photocopying the pattern on to strongish paper saves time, energy and thought. Nevertheless, practise in drawing or tracing patterns is an excellent preparation for creating designs.

As in the other embroidered machine nets, the pattern fixes the shape to be followed by the worked outline. Caskets should be indicated on a stitch map, and the selected stitches should be numbered.

NEEDLES For tacking the layers together - larger sharps;
For couching stitches to hold down the cord - finer sharps, such as no.10.
For the casket stitches - blunt-tipped needles, or sharps used eye-first, so that the threads of the net are not snagged.

THREADS Coloured sylkos or similar for the temporary tacking stitches which simply hold the layers together so that they cannot slip. Colour prevents their being confused with the permanent white stitches when they are later removed.

Crochet cottons 12, 20, 30 or 40 for the outline cord. Choose the thickness according to the size of the article, and the effect you wish to achieve. The thicker the cord, the more raised up from the surface it will appear.

Fine threads which will not tear the material, for the couching stitches, but they must also be strong.

Cotton thread counts of 50, 60, 80 or 100 for the casket stitches, to be compatible not only with the outline cord but with the mesh-size of the net.

An ordinary THIMBLE may be helpful when making the tacking stitches, to push the needle through the various layers.

The coiling of twirls.

SCISSORS with one tip blunted are needed to cut away the unwanted cambric when the design areas have been defined.

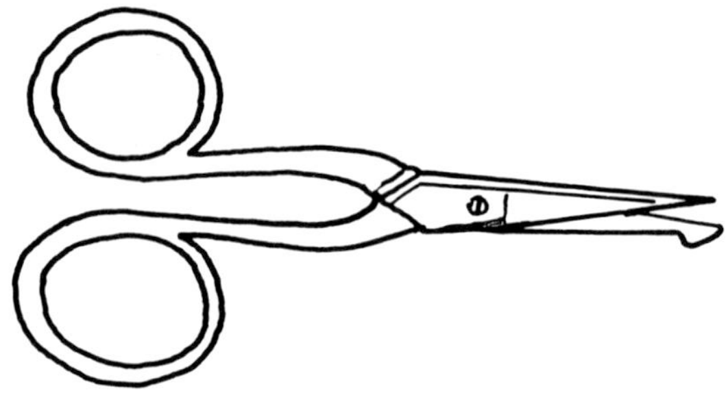

Setting up

Appliqué Arrange the layers - backing, pattern, net and organza - one above the other, with the organza on top, cut slightly larger than the total size required.

Run a coloured thread through the cross-sides of a row of meshes, and another at right angles to it. Align the net so that the guide threads are straight-on to the pattern.

Pull a single warp and weft out of the organza, clear of the pattern area, and arrange the gaps thus formed parallel to the coloured threads, warp against warp. In this way diagonal pull, and pouching of the net and fabric as the work proceeds, will be avoided.

Using rust-proof insect pins, fasten the layers together, then attach them by close tacking stitches also in coloured thread, passing close to the design motifs but never across them. The holes made by the needle and pins will then be cut away with the excess fabric. Begin tacking in the middle of the piece and work outwards, smoothing away any irregularities as you go, to avoid wrinkles.

If necessary, the raw edges can be protected by folding the backing over the edge of the organza and net and tacking it in place like a protective frame.

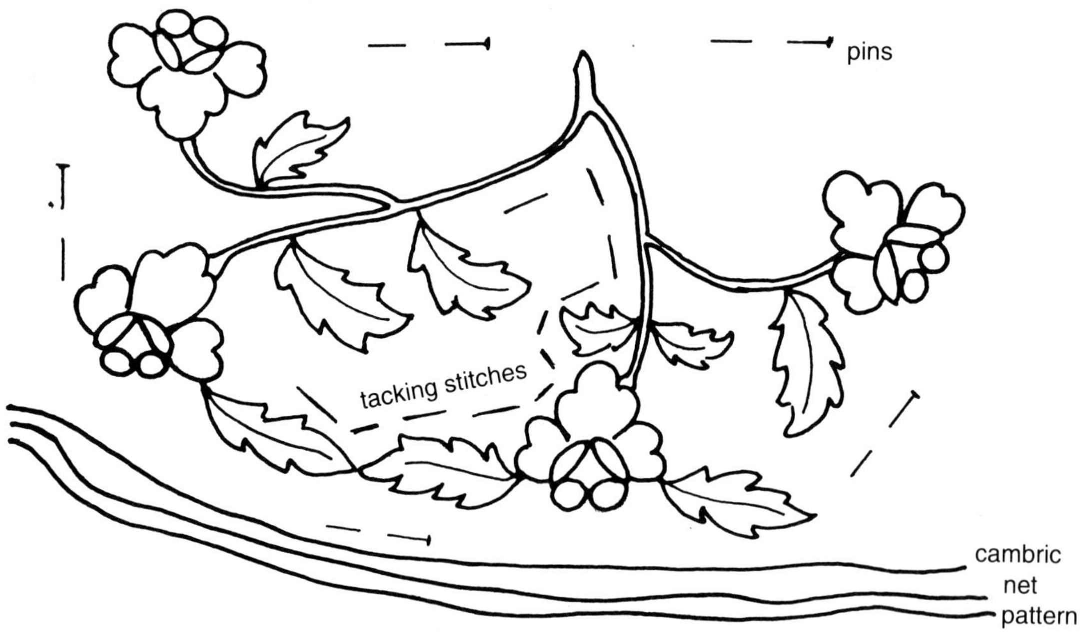

OUTLINING

The outline cord can be used as a continuous length. Tie a loose knot several inches from its end. Cut approximately 2ft lengths of the couching cord, thread the smaller sharp needle, and work two or three close buttonhole stitches over the cord to hold it down. Then begin to make couching stitches over it. The knot is simply to prevent the cord from slipping forwards under these.

Alternatively the outline cord can be pulled through from the back. In the antique laces it is hard to tell which technique was preferred.

Hold the cord so that it is taut but not over-stretched. Keep the couching stitches small, firm, straight across the cord rather than oblique, and close enough to each other to prevent both fraying and the pulling away of the cloth from the cord.

The couching stitches must pass only through the organza and net, never through the pattern or other underlying layers since they are to be preserved in the finished piece while everything below the net will be removed, and any 'permanent' stitches passing through them will be broken.

Work the couching stitches with the needle pointing inwards towards the motif. If worked in the reverse direction, the stitches would pull outwards against the fabric, loosening the warps and wefts instead of compacting them, and increase the chance of the outline pulling completely away.

The outline fixes the shape of the design, so proceed carefully, making the curves smooth and angles sharp.

Where the cord turns back on itself, and runs double, as between the petals, take the couching stitches over the double cord.

As you complete the outline, overlap the two ends, and let the couching stitches bind them together. Cut away the knot left at the beginning.

Guipure The motifs will be most sharply clear when the outline of each is completed independently, rather than letting the cord run on from one to the next in a maze-like fashion. The couching stitches simply bind the cord to the organza.

VEINS The cord is used not only to define the motif's outline, but also its caskets, and sometimes to form the veins, delineating a central core within which fabric and net were cut away, and the space bridged by bars. Brenan complained that the workers cut away the midrib area to such an extent that 'a mere fringe of cambric was left, and the leaves presented the appearance of having been eaten away' [105]. However, used discreetly, the effect can be pleasing.

Thinner veins can be worked in without cutting, simply as strands passing over the leaf surface. Fix the thread at the leaf base, carry it up almost to the tip, pass it lightly through the material, hold it there with a backstitch, and return to the base, spiralling around the first thread and catching it down to the organza (and net) at short intervals.

TWIRLS

As you couch the outline cord along the outer border, twist it into a loop. Make two or three overcasting stitches to hold the loop form. Continue couching along the outline, then make another twirl. The twirls can be spaced evenly, or as best fits the design.

The direction of twisting of the twirls must depend on the thread. There is considerable risk of the cord overtwisting, and unless this is corrected the loops will spring up from the lace, refusing to lie flat. Correct the tendency as soon as it appears by turning the reel in a counteracting direction.

The twirls are usually all the same size, but this is not essential as long as any differences are are not the result of negligence but part of the plan.

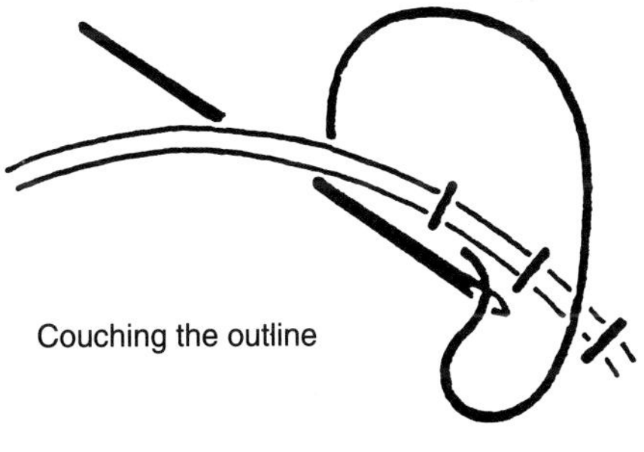

Couching the outline

Twirls, a simple vein, and the doubled cord clamped by a single row of couching stitches.

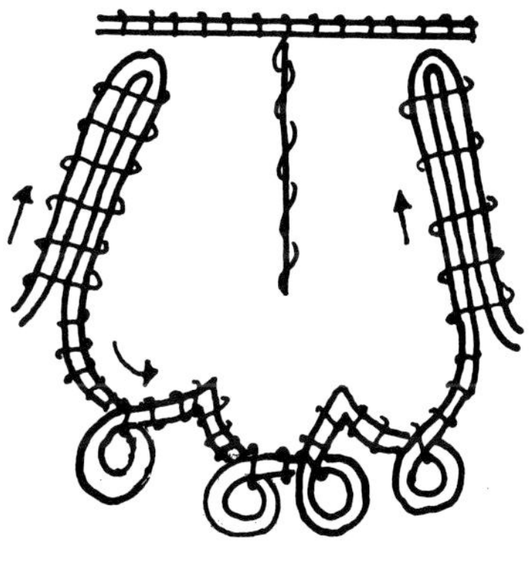

POPS

These decorative circles are described on p.138.

A more dramatic three-dimensional effect can be achieved by inserting the tip of a stiletto into the selected holes, winding the thread around it, then buttonholing over the padding, with the needle pointing outwards from the centre.

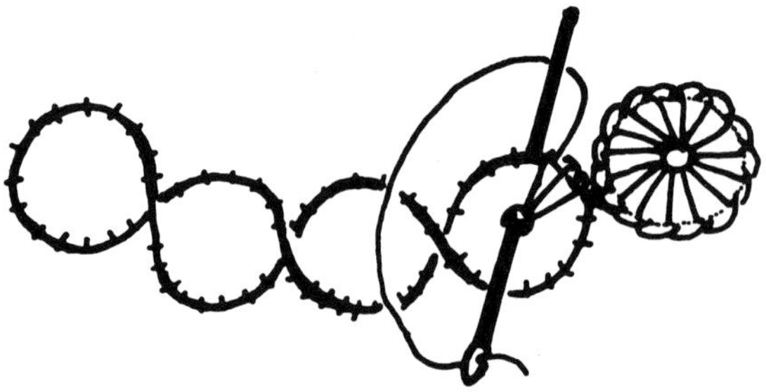

The spiralling couched cord creates the outline, and radial buttonholing converts it into a row of pops.

CUTTING AWAY THE SUPERFLUOUS FABRIC

Appliqué As soon as the outline has been fully completed, the organza not involved in the design can be removed, though to avoid undue strain on the cord and its stitches it may be better to cut it away piece by piece when the net needs to be exposed for the addition of pops or of filling stitches.

To cut away the organza over the caskets, hold the lace face side up. Begin by raising the cloth away from the net with the tip of a needle, then make a cut just big enough to allow the rounded tip of the scissor blade to enter. With enormous care cut around just inside the cord, but without actually touching the couching stitches, and taking great care not to damage the net beneath.

Guipure Removal of the organza over the caskets will reveal empty holes. Do not remove any further excess fabric until all the bars are complete.

The pattern below the net, or organza, can now be removed - but if left it will continue to provide support, as well as an easy-to-refer-to stitch map of the intended fillings, so there is some argument for leaving it in place.

INFILLING

Use the fine thread and begin each casket with a small knot or tight buttonhole stitch tucked beneath the outline cord. Alternatively run the thread between the muslin and net for a short distance, bring it to the front and make a small tight buttonhole stitch.

At the end of each row, whip the thread two or three times around the cord down to the mesh row where the next line of stitching will begin.

Some of the casket stitches of cut cambric appliqué are also found in Limerick run and tambour laces.

Dense satin-stitch florets in a casket filled with cross-stitch.

COLLAR WITH SATIN-STITCH FLOWER HEADS

This unusual collar shows a fabric of the most delicate muslin attached to a fine square-meshed net totally covered with areas of decorative run stitches which glide neatly and unobtrusively between the leaves and stems of the motifs so that, in looking at the lace, the design is not befogged by the decoration but stands out with total clarity, starkly highlighted by its clustered groups of satin-stitch flowers.

The collar has been cut from the net with the cross-sides aligned vertically, like warps.

The embroidery thread is only marginally thicker than the mesh sides, and the resultant delicacy of the large casket areas enables them to blend modestly into the background.

The casket stitches are relatively simple and include cross-stitch and tent-stitch.

The collar shows no pops or twirls and no picot border. The motifs are outlined with an outer and inner cord to secure the loosely woven fabric.

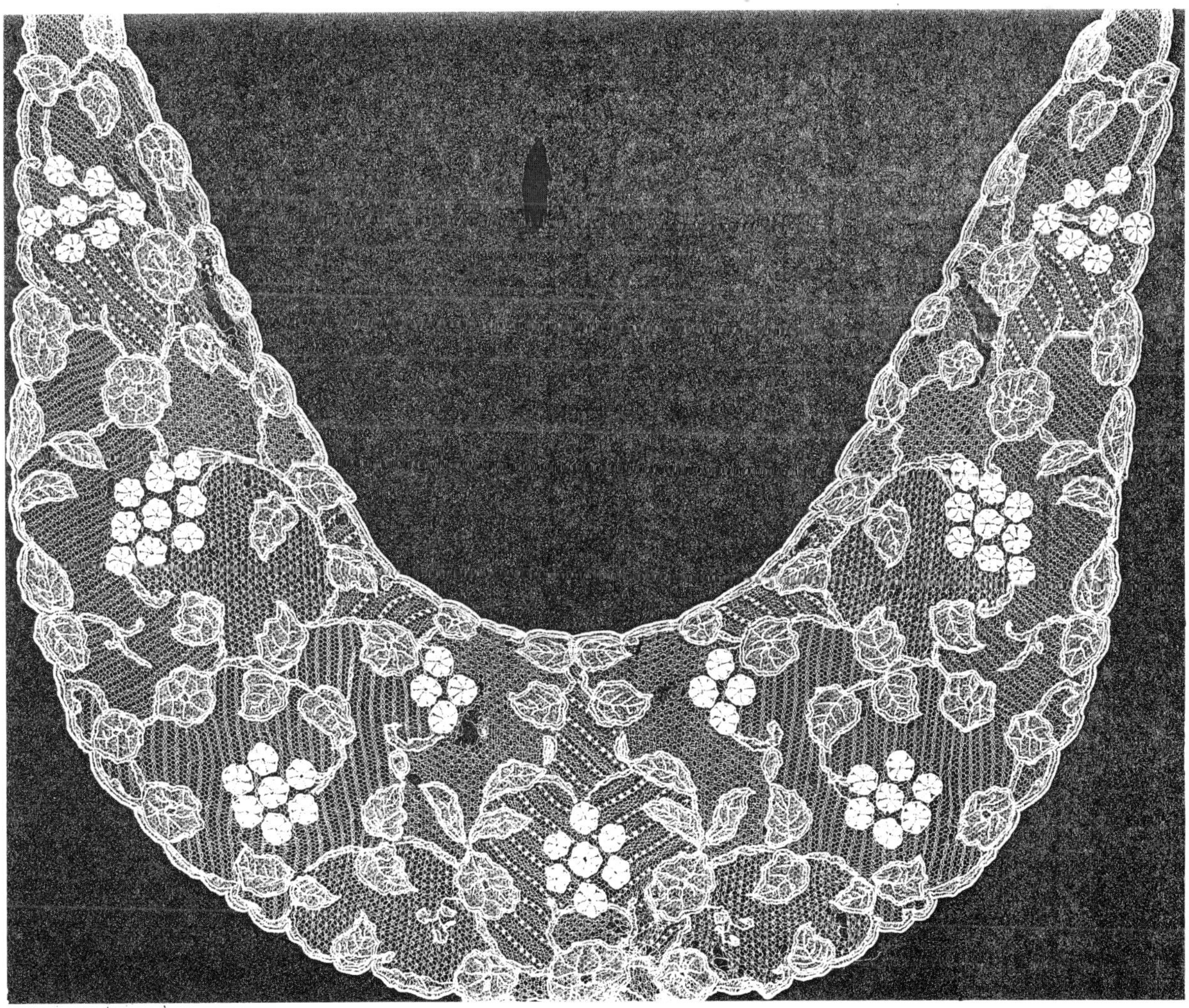

Stitch map of a portion of the collar.

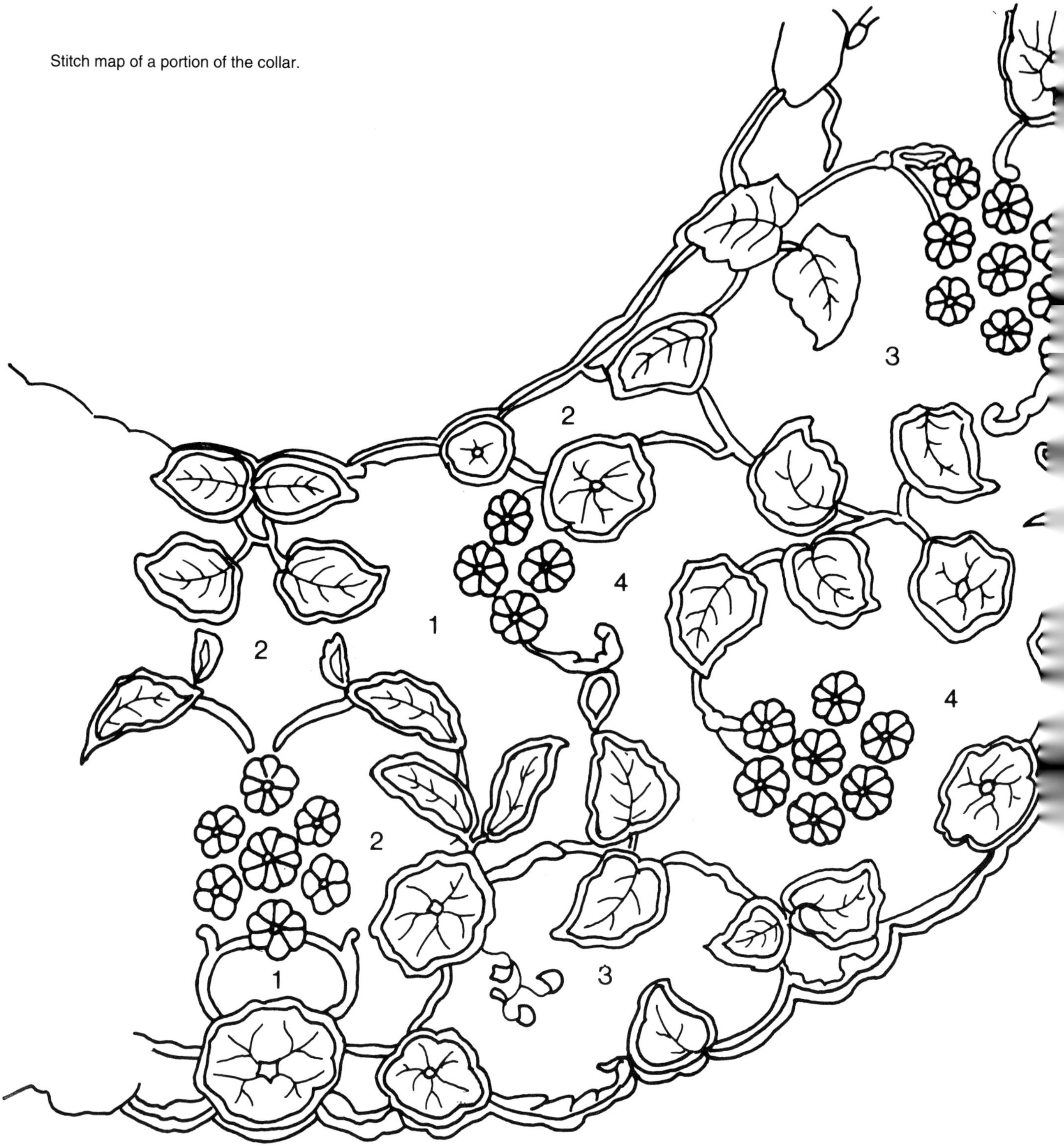

Stitch 1. Cobweb-stitch (see st.2, p.33).

Stitch 2. Satin-stitch and tent-stitch

* Rows 1 and 2: Work tent-stitch diagonally across the casket, LR and RL.
Leave a line of holes.
Row 3: Work 3 or 4 satin-stitches over the twist-sides of alternate holes, linking the spots together by whipping the thread along the tops of the meshes.
Leave a clear line of holes.
Repeat from * until the casket is filled.

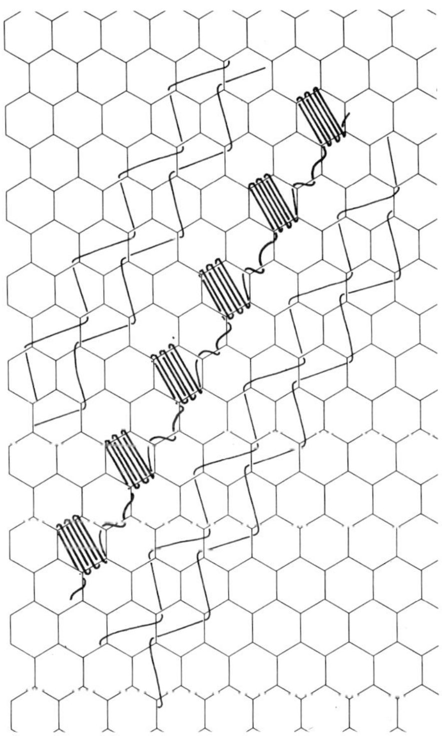

Stitch 2.

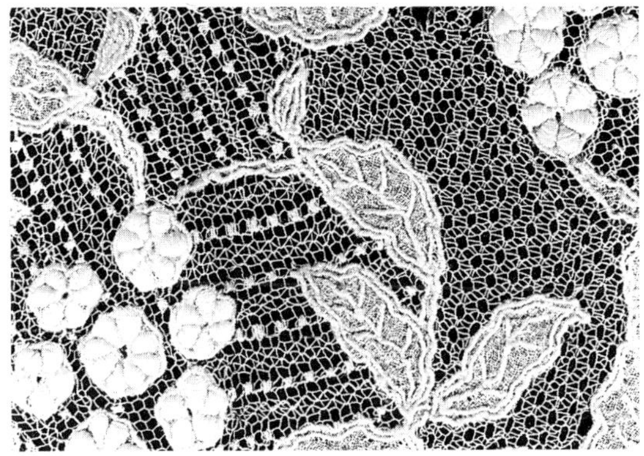

Stitch 3.

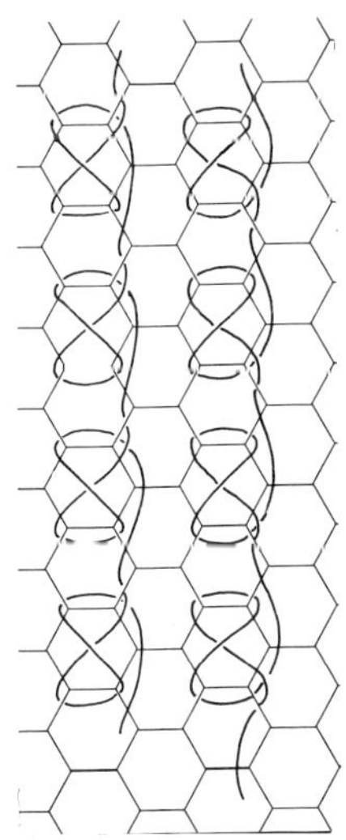

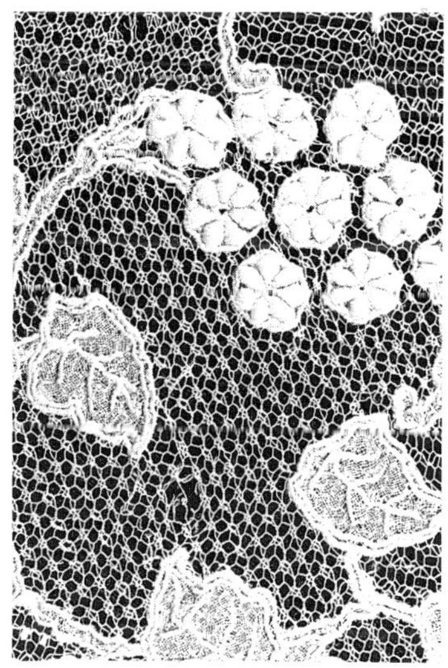

Stitch 3. Figure-of-eight stitch

Work figure-of-eight stitch over alternate holes, passing along the cross-sides and linking the stitches together as shown.
Turn the work between rows. The direction of the thread-crossings will appear reversed.

Stitch 4. Cross-stitch

Work rows of cross-stitch along the cross-sides as shown, leaving two lines of holes between each row.

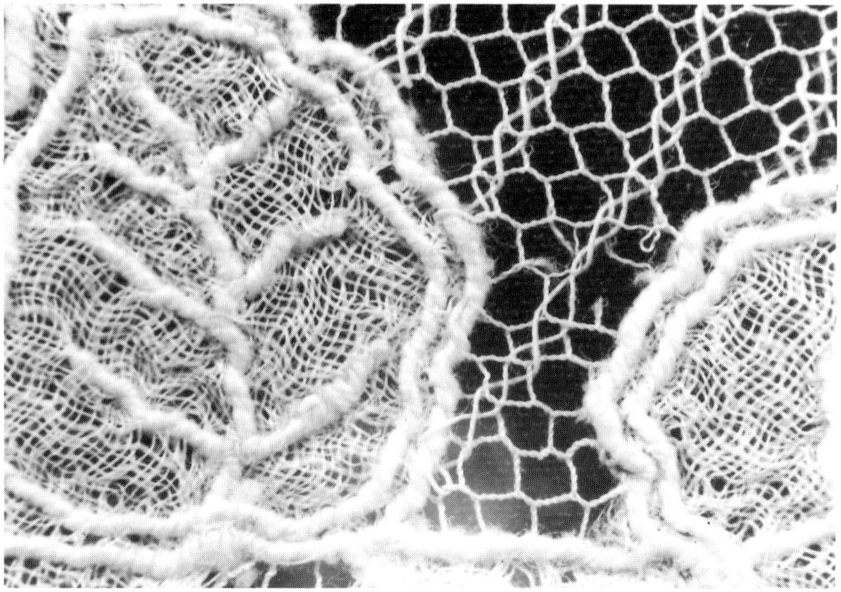

Detail of the couched cord. Note the texture of the muslin, and the branching vein.

Stitch 5. Diagonal tent-stitch

Work diagonal rows of tent-stitch, reversing their direction in alternate rows to form mirror images.
Leave a line of holes between the rows.
Continue until the casket is filled.

Different casket stitches occur in the globular groups of feathers covering the body of the various swans. Further decorative stitches are found in the partially guipure borders (Courtesy Joslyn Baker).

THE SWANS

This drift of swans, taken from a set of table mate, resembles the swans of Bruges, conceivably inspired by nostalgic memories of the Belgian nuns at the Convent of the Good Shepherd.

The embroidery thread is noticeably thicker than the mesh sides.

The couching stitches holding down the outline have been spaced too far apart and the fabric has started to come away from the net.

There is a certain deviation between the mats, some of the stitches being worked from the face and some from the reverse side of the lace. They were perhaps a cooperative effort, and a not entirey professional one. In all, 19 decorative stitches are described and illustrated.

A single mat.

Stitch 1. Cobweb-stitch

Work as stitch 1, above (p.129), but do not enlarge the holes.

Stitch 2. Crossed figure-of-eight stitch, a variation

Work vertically up and down, turning the lace between rows so that the thread movements appear reversed. Follow the arrows and the direction of the needle shown in the line drawing.

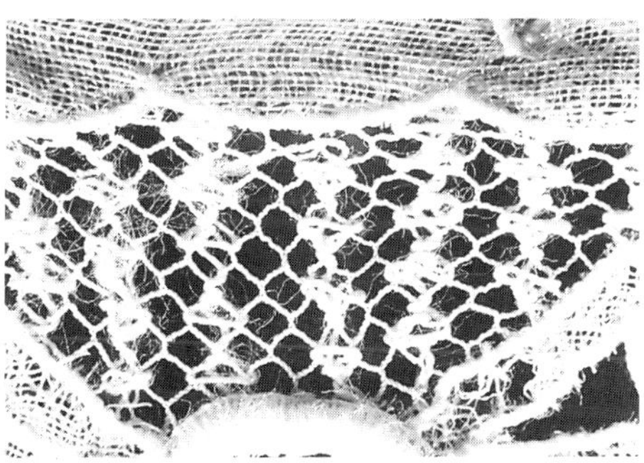

Stitch 2.

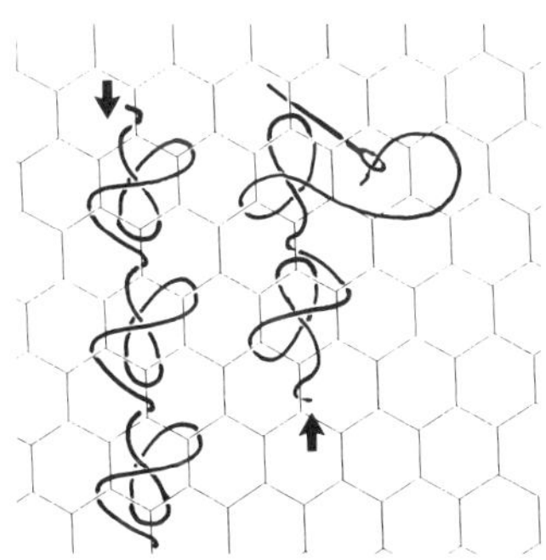

Stitch 3. Superimposed figure-of-eight and cobweb, with compressed cross-stitch stripes

Rows 1 and 2: Work a row of compressed cross-stitch (see cross-stitch B, p.28).
Leave a line of holes.
Row 3. Work cobweb-stitch with its horizontal sides passing above the net surface.
Row 4: Superimpose a row of figure-of-eight stitches over the cobweb-stitch, passing the linking threads across the reverse side of the meshes.
The zigzags alternate in the manner of chevrons.
The working of this stitch is not entirely consistent. In some caskets it appears identical to stitch 3, p.47, with the horizontal strokes of the cobweb-stitches passing over instead of under the figures-of-eight.
Also, some of the repeats of rows 1 and 2 have the points facing towards each other instead of alternating, so that the effect is of diamond shapes, not chevrons.

Stitch 4. Superimposed figure-of eight and cobweb, with tent-stitch stripes

* Rows 1 - 3: Work 3 rows of tent-stitch, reversing the net left-right between rows so that the direction of the strokes is mirror-imaged.
Leave a line of holes.
Rows 4 and 5: Work figure-of-eight stitches over the cobweb stitches as shown in stitch 3, above. The points of the zigzags face towards each other.
Leave a line of holes.
Repeat from * until the casket is filled.

Stitch 5. Six-point stars

This is worked in the same way as stitch 5, p.51.

Stitch 6. Tent-stitch trellis with needlewoven rosettes

Construct a trellis of tent-stitch, taking the diagonal rows first from bottom-L to top-R and spacing the rows 5 holes apart.
Repeat in a top-L to bottom-R direction, at each intersection passing the needle beneath the stitch already made.
The units enclose 25 holes in all.
From the central hole of each lozenge, work 6 blocks of needleweaving, like the petals of a flower. Leave the thread hanging at the back, to be sewn in later. Alternatively, whip the thread along the mesh sides to the position of the next rosette.

Stitch 7. Tent-stitch trellis with needlewoven crosses

Construct a tent-stitch trellis, 5 holes wide, as in stitch 6, above.
Needleweave a cross in the centre of each lozenge.

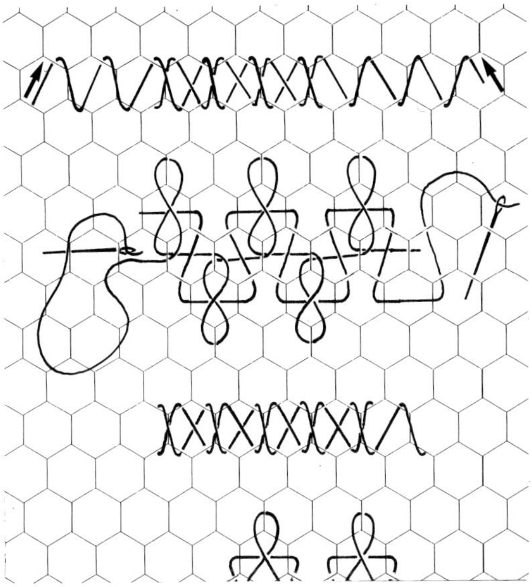

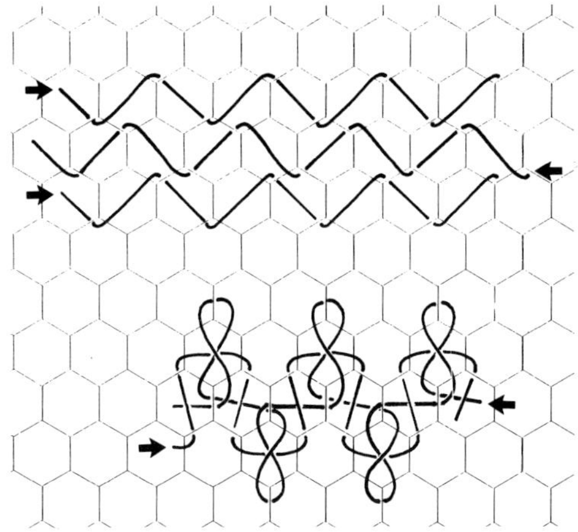

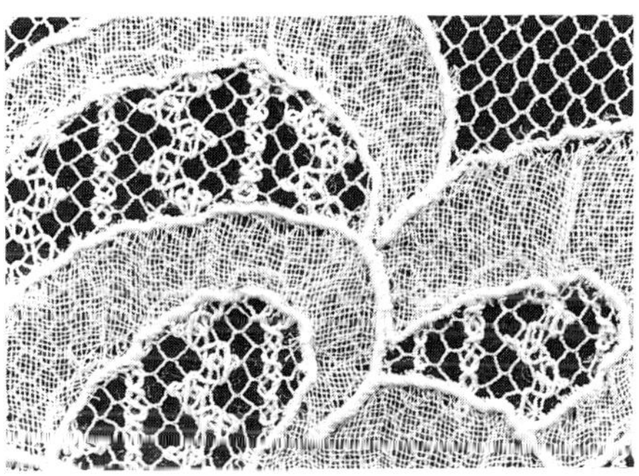

St.3.

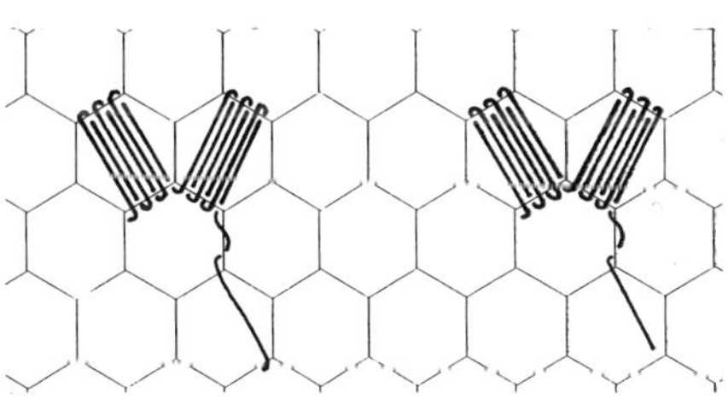

Left: satin-stitch. Right: needleweaving.

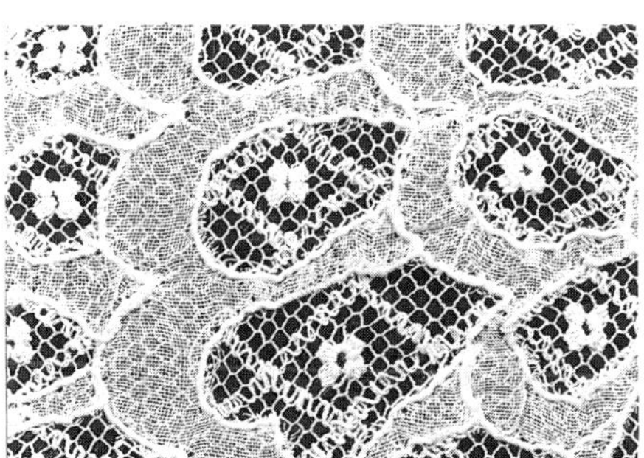

Stitch 6.

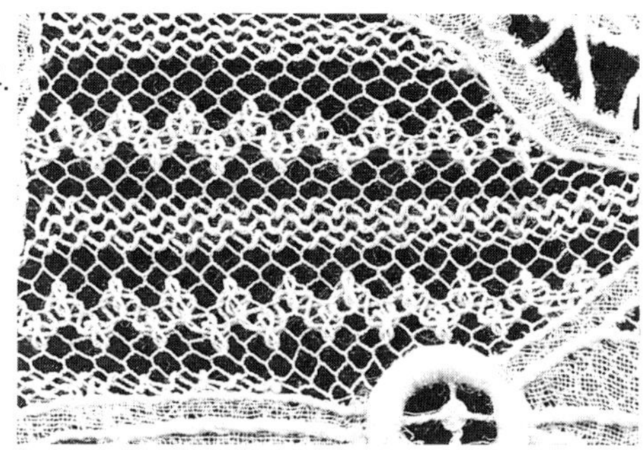

St.4.

Stitch 8. Tent-stitch with reinforced cobweb-stitch

Set 1. * Work two rows of tent-stitch horizontally across the casket, reversing the slant of the threads in the second row.
Leave a line of holes.
Set 2. Work a row of cobweb-stitch. In the horizontal portion, which lies at the front of the net, take the thread twice around to form a narrow satin-stitch slash.
Repeat from * until the casket is filled.

Stitch 9. Cobweb banded with satin-stitch, alternating with tent-stitch stripes

This is an elaboration of stitch 8, above.
The cobweb-stitches are worked in paired rows, their horizontal sides facing each other.
Row 1: Work cobweb-stitch across the casket, taking the thread a second time around the upper and lower meshes which the stitch encloses.
Row 2: Turn the work round and return, again working in a LR direction. This time take the thread 1.5 times around the mesh sides distant from row 1, But only once around those adjacent to row 1.
Leave 2 lines of holes.
Rows 3 - 4: Work tent-stitch as shown.
Repeat these 4 rows until the space is filled.

Stitch 10. Superimposed cobweb and figure-of-eight stitch, with slanting satin-stitch stripes

(Note: the photograph is of the face side, but the stitches have actually been worked from the reverse side, following the removal of the pattern). The movements are described as worked.
Row 1: Work 3.5 satin-stitch slantwise across 2 vertically placed meshes, as shown, to form a spot. Link consecutive spots by coiling the thread around the intervening mesh side.
Leave a line of holes.
Row 2: Work a complete row of figure-of-eight stitches.
Row 3: Return along the same meshes, placing a cobweb-stitch over each figure-of-eight. The horizontal bars pass under the middle of the 8, and the slanting threads over the surface.
Leave a line of holes.
Repeat rows 1 to 3 until the casket is filled.

Stitch 11. Tent-stitch lattice enclosing quartets of circles

Set 1. Construct a tent-stitch lattice 5 holes in width, as in stitches 6 and 7, above, p.132.
Set 2. Run circles around the 4 middle holes of each lozenge, passing in and out of the mesh holes. Take the thread across the back of the net to the next lozenge, and repeat. Alternatively, leave the thread at the back of the work to be sewn in later.

Stitch 12. Lattice of crossed figure-of-eight stitch

Fill the casket with a lattice constructed as in stitch 3, p.112. Leave the centres empty.

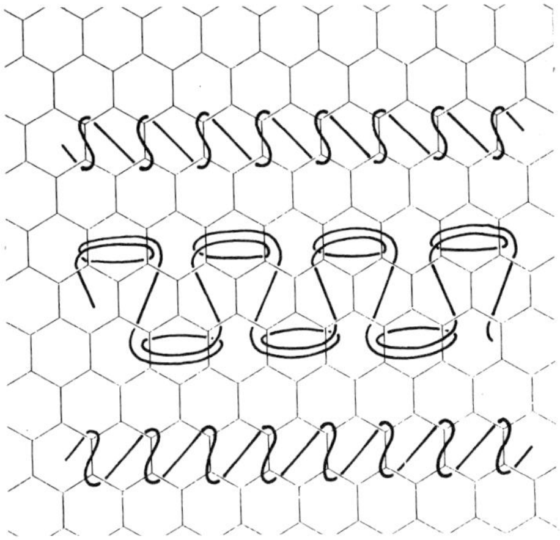

St.8.

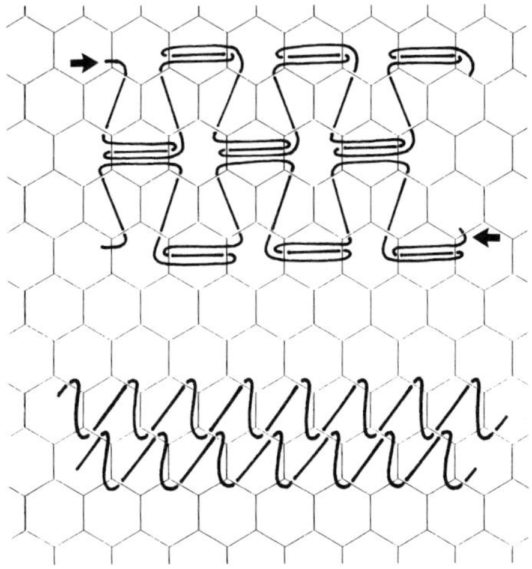

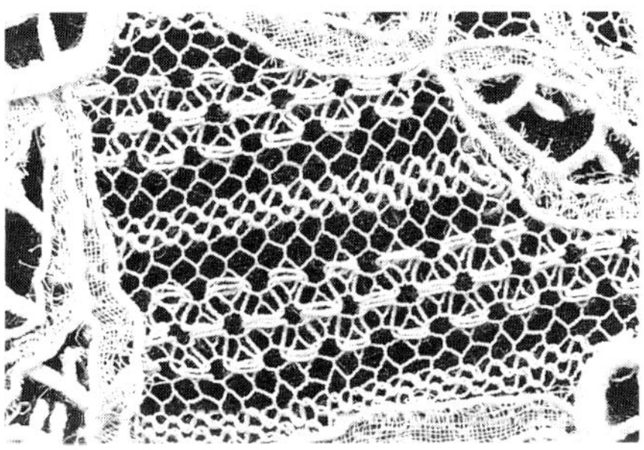

St.8.

Stitch 9.

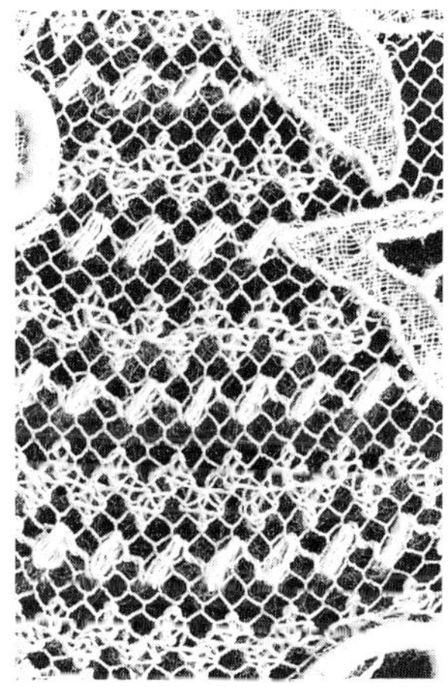

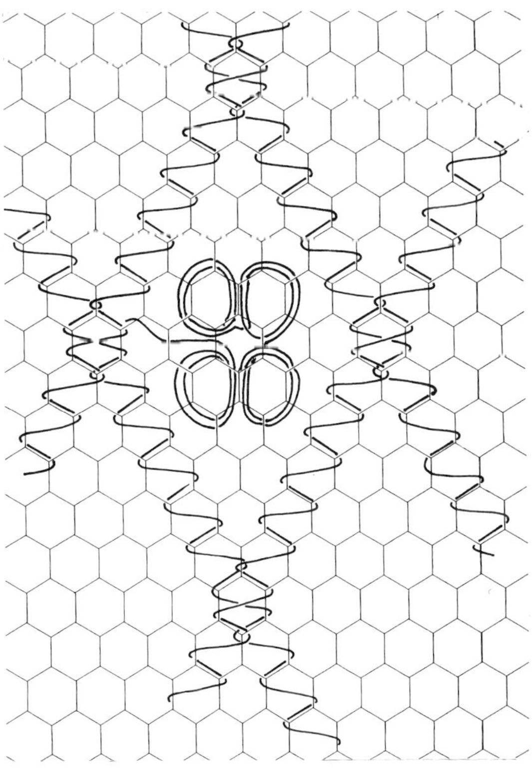

Stitch 11.

Stich 10.

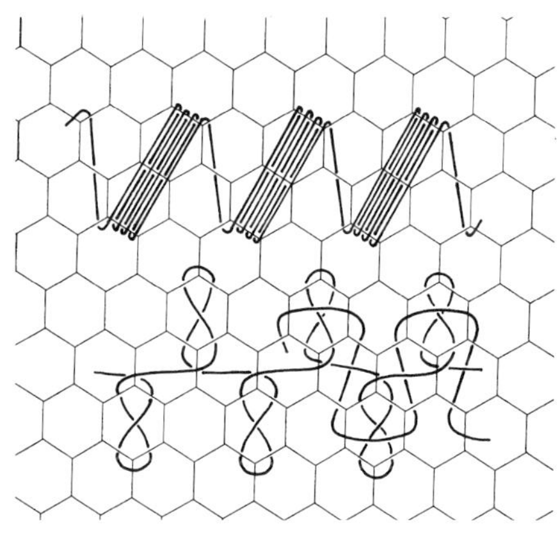

Stitch 13. Cross-stitch fangs (Compare stitch 5, p.107).

The fangs are constructed of cross-stitch, compacted as in stitch 3. Each fang consists of three horizontal rows of 3, 2 and 1 stitches, staggered by one hole at the left side of each row.
Sketch A shows the RL and LR rows.
Sketch B shows the sequence of working. Follow the arrows and note the over-under intersections of the thread.
Sketch C shows the position to be occupied by the following row.

Stitch 14. Cross-stitch fangs and satin-stitch spots

* Work the fangs as for stitch 13, above.
Leave a diagonal line of holes.
Work a double row of satin-stitch spots, taking the thread 3.5 times around each mesh. Link the spots with a single thread across the back of the net. (Compare stitch 5, p.113).
Leave a line of holes.
Repeat from * until the casket is filled.

Stitch 15. Stripes of tent-stitch, with satin-stitch spots

The rows are worked diagonally.
Row 1: Work a single row of tent-stitch, taking the thread under two cross-sides and over two twist-sides as shown.
Leave 2 lines of holes.
Row 2: Work a double row of satin-stitch spots in a zigzag manner, taking the thread 3 times around each pair of twist-sides.
Link the spots together by taking the thread across the back of the net, as indicated by the needle in the drawing.

Stitch 16. Compacted cross-stitch and satin-stitch spots

* Rows 1 and 2: Work a single row of compacted cross-stitch horizontally across the net.
Leave a line of holes.
Row 3: Work a row of satin-stitch spots, each stretched over 2 meshes, and linked by whipping the thread around the intervening mesh sides (see stitch 10, p.134).
Leave a line of holes.
Repeat from * until the casket is filled.

Stitch 17. Triplets of satin-stitch spots, with tent-stitch stripes

This is yet another variation of familiar stitches.
* Row 1: Work satin-stitch spots horizontatlly in groups of three, spacing the rows one hole apart and each time insetting the left side by one mesh.

Link the triplets together as shown, and continue right across the casket.
Leave a diagonal line of holes.
Row 2: Work a row of tent-stitch parallel to the satin-stitch triplets.
Leave a line of holes.
Repeat from * until the casket is filled.

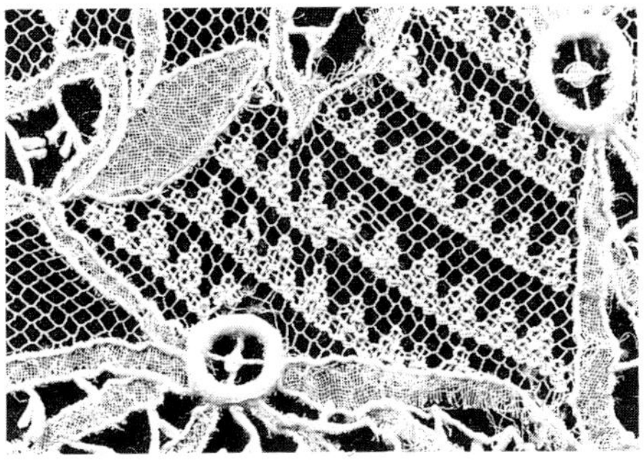

St.13.

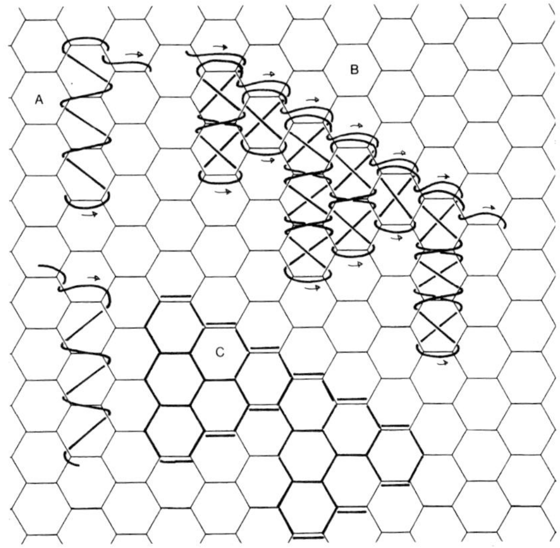

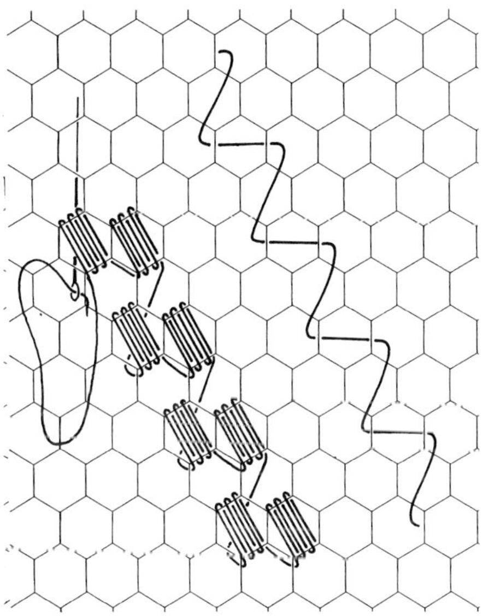

Stitch 15.

Stitch 14.

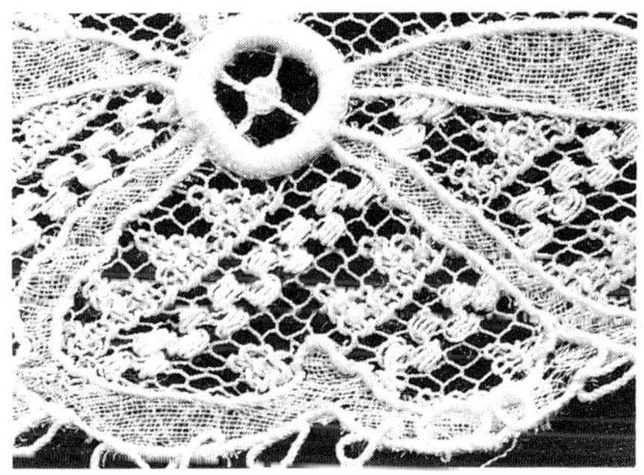

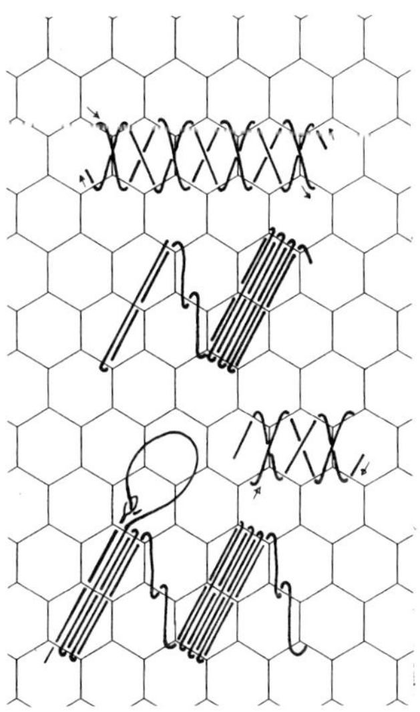

Stitch 16.

Stitch 17.

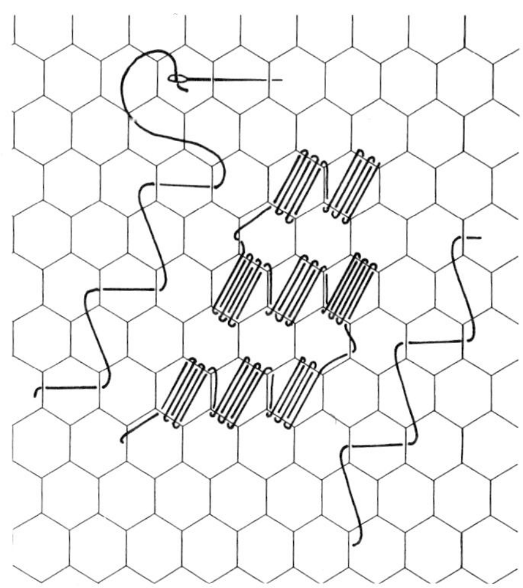

Stitch 18. A tent-stitch and satin-stitch sandwich

* Row 1: Work a diagonal row of close tent-stitches across the casket.
Row 2: In the next row of meshes, loop the thread 2.5 times around alternate holes to make the satin-stitch spots.
Row 3. Work another row of close tent-stitch, as shown. Note that the slope of the stitches is the same as in row 1.
Leave 2 lines of holes.
Repeat from * until the casket is filled.

Stitch 19. Paired fans and tent-stitch stripes

* Rows 1 - 3: Work 3 rows of tent-stitch. Turn the work between rows so that the over-unders of the thread movements appear reversed.
Leave a line of holes.
Row 4: Work a series of 5 stitches radiating from one side of a central hole. Carry the thread across to the next hole but one, and repeat.
Row 5: Work back again, making a mirror image of row 4. Leave a line of holes.
Repeat from * until the casket is filled.
(compare stitch 6, p.60, and stitch 4, p.112).

Pops

For the swan's eye, select the hole to be enlarged and pierce the fabric which lies over it. Using a stiletto or bodkin, push the mesh sides and the strands of cloth apart until the opening is clear.
Work a close ring of buttonhole stitches around the hole with the needle pointing outwards, so that the pop radiates across a ring of single meshes.
Pops scattered through the net can be worked either before or after the excess fabric has been cut away.

Bars

Running from the eye of the swan, down the side of the neck, is an open slash crossed by buttonholed bars. Use the outline cord to define the space, and work the bars while the net and fabric are still in place.
Attach a sewing thread securely to the outline cord at A. Carry it across to B, back to A, and back again, passing around the outline cord at the end of each trip. Finally, return to A, buttonholing over the three strands.
Whip the thread inconspicuously along the outline to position C, and repeat.
Take the greatest care that none of the bar stitches pick up the organza. They must pass over its surface, not through it.
In the drawing, the buttonhole stitches have been spaced apart to show them more clearly.
The separation of the upper and lower beak is indicated by a line of thread whipped over in the manner of a vein (p.125).

Picots

The flowers along the border of the mats have the net and organza cut away from their outer sepals. Within these spaces are bars, decorated by the addition of little thorns or picots. These, like the bars themselves are worked before the layers are cut away.
Buttonhole a short way along the bar. Take the thread outwards in a loop and pin its tip down to the organza. Return to the bar, secure the thread, take it out again to the pin. Return, buttonholing over the triple strand back to the bar, again taking care that none of the stitches pass through the layers below.
Repeat the picot formation as often as required.
Remove the pins so that the picots stand free.

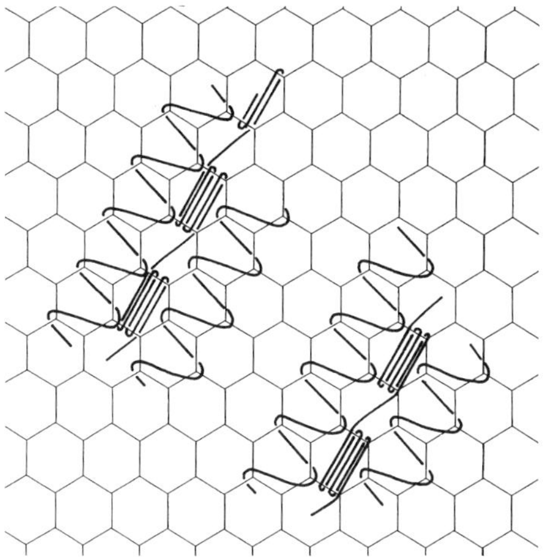

St.18.

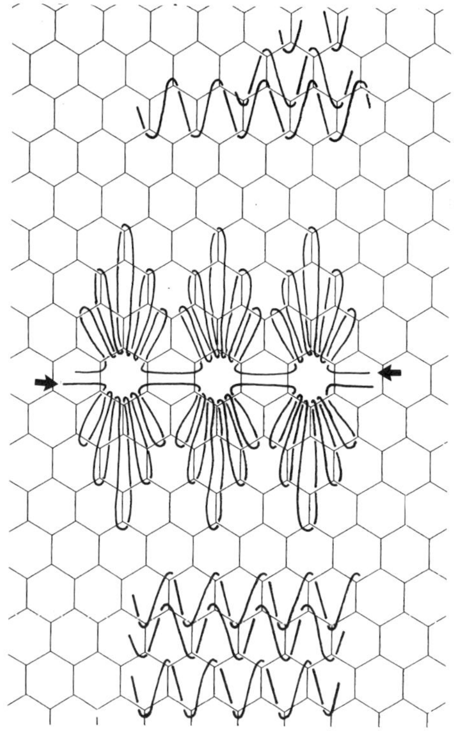

St.19.

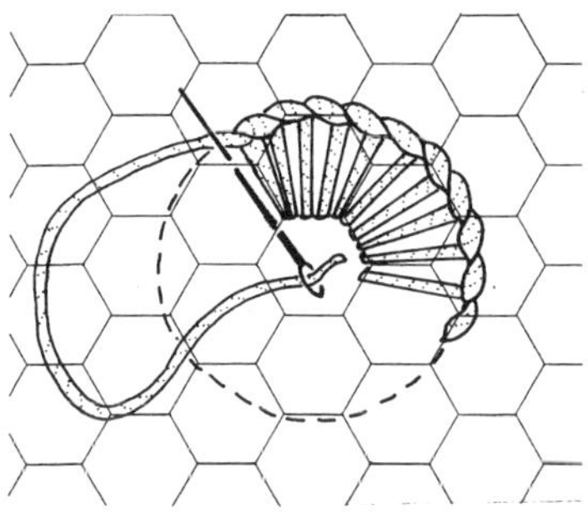

Making a pop.

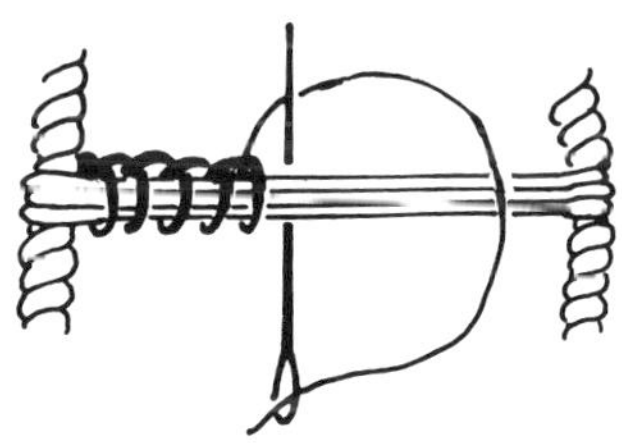

Making bars.

Stitch 19.

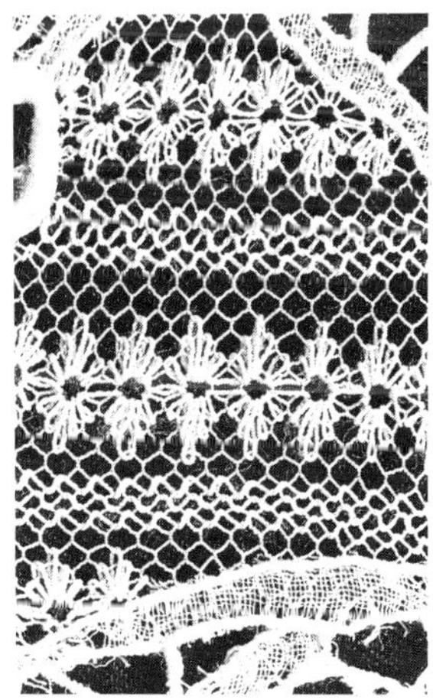

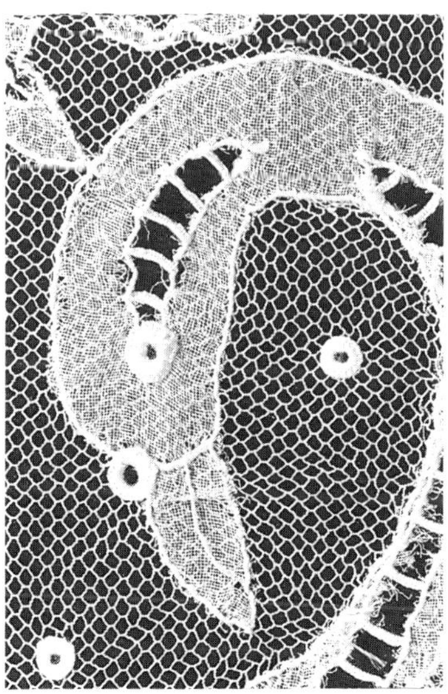

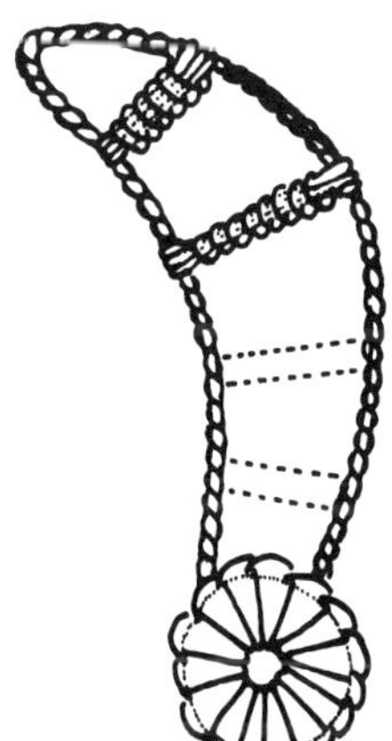

Picots.

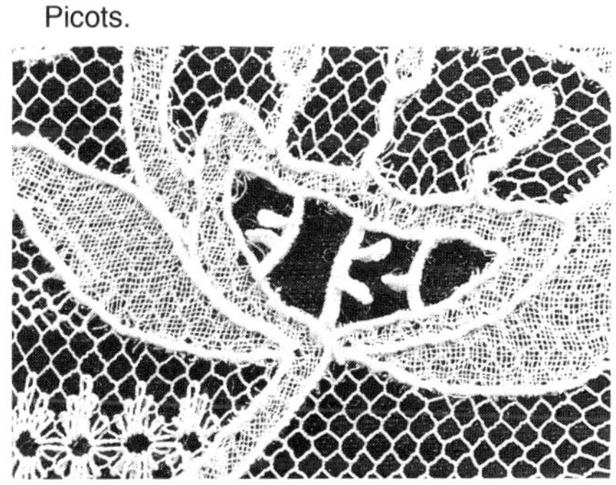

PLASTRON

This shape of collar, intended to cover the centre part of the bodice front, dates from the late 19th or first few years of the 20th century. The work is typical of the cambric appliqué commonly referred to as 'Carrickmacross'.

A neat line of twirls fringes the entire border, which has been closely buttonholed over.

With two exceptions (stitches 7 and 10), the caskets hold nothing new, but there are some ingenious rearrangements of stitches already known.

The cross-sides are aligned horizontally in each stitch.

The embroidery thread is marginally thicker than the mesh sides.

Stitch 1. Horizontal cobweb-stitch

Work as stitch 2, p.33.

Stitch 2. Six-point stars

Work as stitch 5, p.51.

Stitch 3. Diagonal cross-stitch with satin-stitch spots

The stitches are similar to stitch 4, p.64, but the cross-stitch rows are double, and the satin-stitch spots slant in the same direction in every row.

Stitch 4. Diagonal needleweaving

Take the needle over and under the twist-sides at each movement, instead of entirely over and entirely under as in satin-stitch (see p. 133). Weave in and out of the mesh as many times as necessary to fill the hole.

Pops, twirls, and the buttonholed outer border.

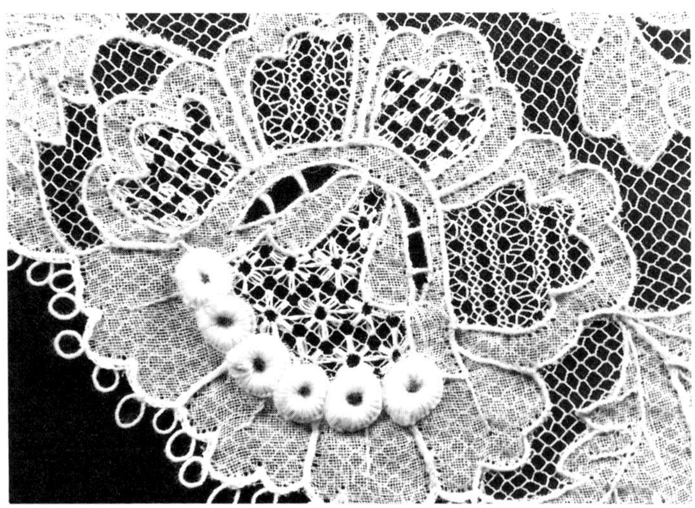

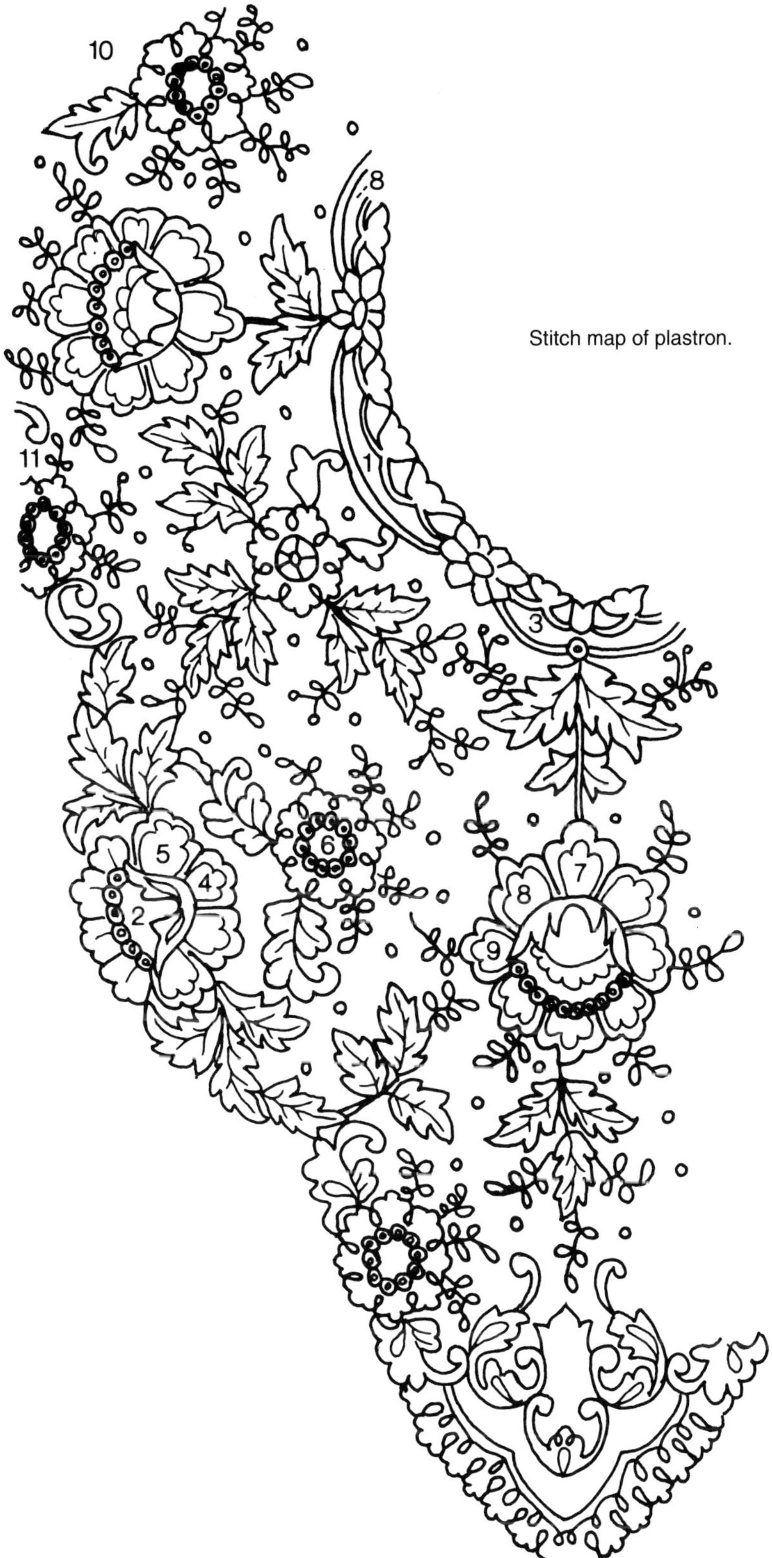

Stitch map of plastron.

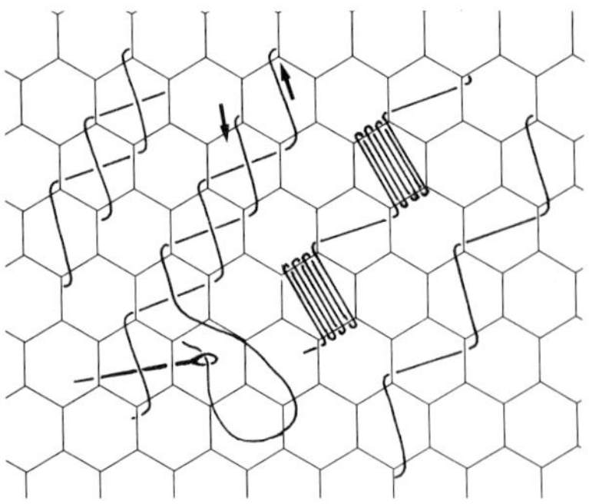

Stitch 3.

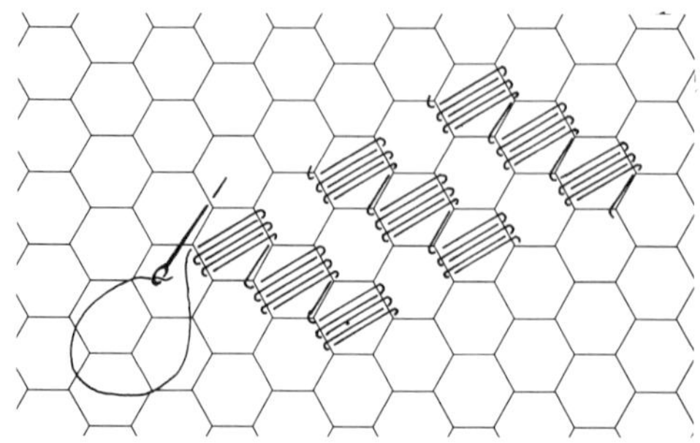

Stitch 4.

Stitch 5. Horizontal cross-stitch and cobweb-stitch
Work as shown.

Stitch 6. Diagonal satin-stitch spots
Note the course of the connecting thread. The slant of the stitches is the same in every row.

Stitch 7. Diagonal cross-stitch rows intersected by double tent-stitch
Set 1. Fill the casket with diagonal rows of cross-stitch spaced two lines of holes apart.
Set 2. Run double rows of tent-stitch diagonally across the casket, in the opposing direction, passing the thread beneath the stitches of set 1.

Stitch 8. Vertical rows of crossed figure-of-eight stitch
This is similar to stitch 10, p.36, but the slant of the crosses is the same in every row, an effect achieved by working all the rows in the same direction.

Stitch 9. Cross-stitch lattice enclosing four holes
Work as stitch 8, p.35.

Stitch 10. Half-rosettes with tent-stitch looping
Set 1. Follow the course of the short rows of tent-stitch, noting the over-under passage of the embroidery thread.
Set 2. In the next row, work a series of half-rosettes, winding satin-stitch pairs around each of three meshes, to form an arc. The middle pair of stitches takes in the turn-back loop at the end of the tent-stitch row.
In the drawing, the sets have been separated to make them clearer.

Stitch 11. Tent-stitch looping with satin-stitch spots
Set 1. Work as stitch 10, above.
Set 2. In the next row, work a column of satin-stitch spots.
Repeat sets 1 and 2 until the casket is filled.
No clear lines of holes are left between the rows.

Pops

These are used very effectively to add emphasis to the floral forms.
In this example the pops appear in series rather than in isolation, and the outline cord is used to support them. Take it sinuously along one side, then back along the other, couching it down in the normal way. Each circle will have one central hole. Pierce this and work as usual (p.138).

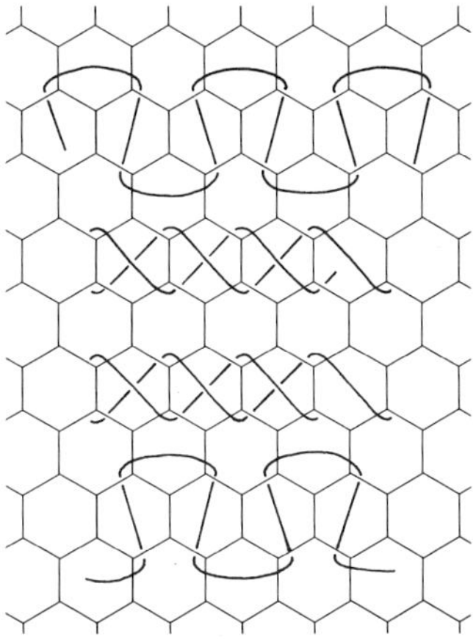

Stitch 5.

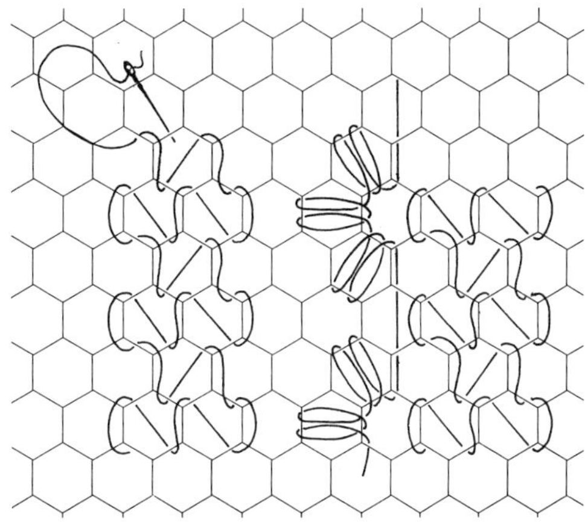

Stitch 10.

Stitch 6.

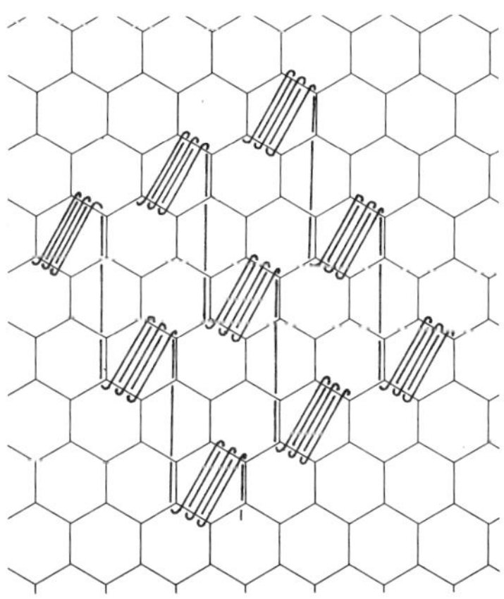

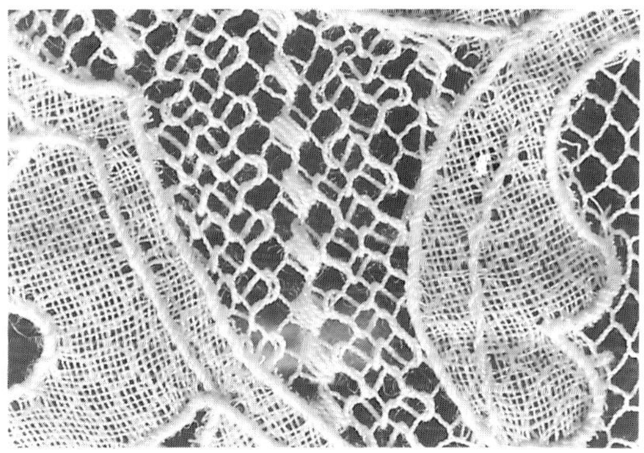

Stitch 11.

Stitch 7.

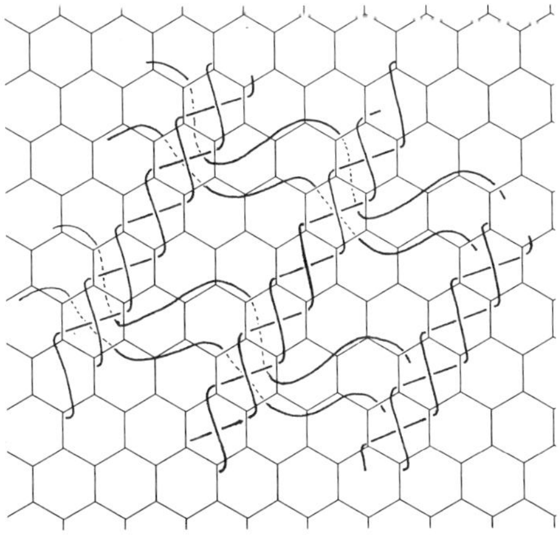

Raised pops, and border.

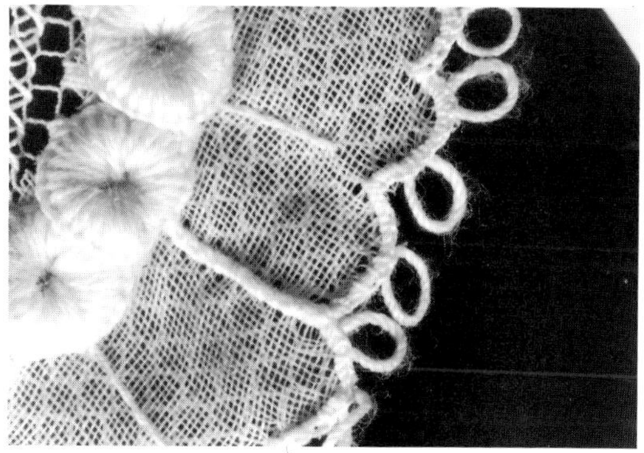

CUT CAMBRIC GUIPURE

Couch the outline as already described (p.125).

In the absence of any supporting background of net, bars or ties are constructed to hold the motifs together. They are formed exactly as described for the neck of the swan (p.138). Picots or *thorns* are often added.

Solid or open circles of cloth are scattered around and between the motifs like fat spiders hanging in their webs.

The outer border is usually decorated with picots rather than twirls.

Since there is no net, none of the run fillings described for cut cambric appliqué can be used. Instead, the decorative caskets are filled with a medley of needleweaving and needle lace stitches, the latter consisting of the basic detached buttonhole, and twisted buttonhole, stitches. These are abbreviated in the instructions to dbs and tbs.

Between rows, the embroidery thread is whipped around the outline cord where it is lost among the couching stitches, the thread of both being identical.

All the casket stitches are worked downwards, with the needle pointing towards the worker, a typical 19th-century practise. The organza covering the caskets must first be cut away to expose the empty space (p.126).

FLOUNCE The rather uninspired design is made up of short repeats.

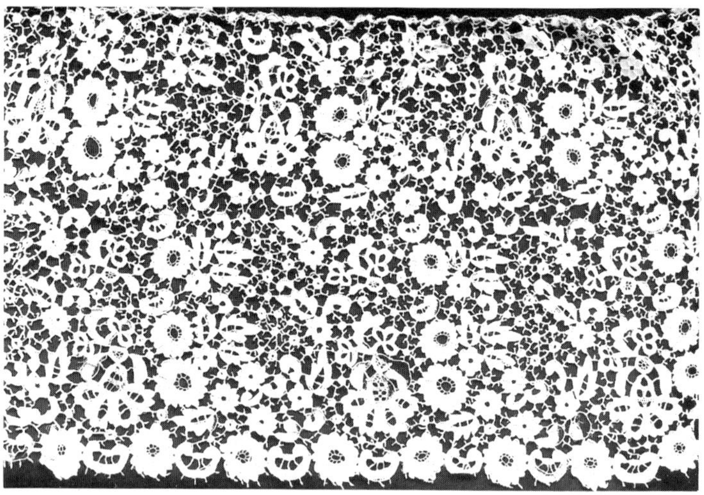

Circlets of cloth filling the blanker spaces.

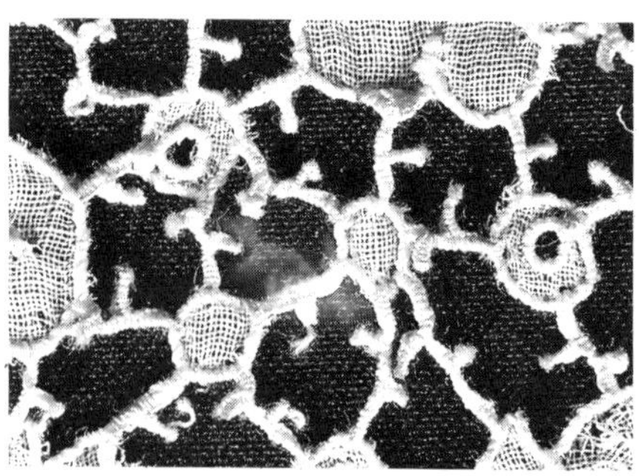

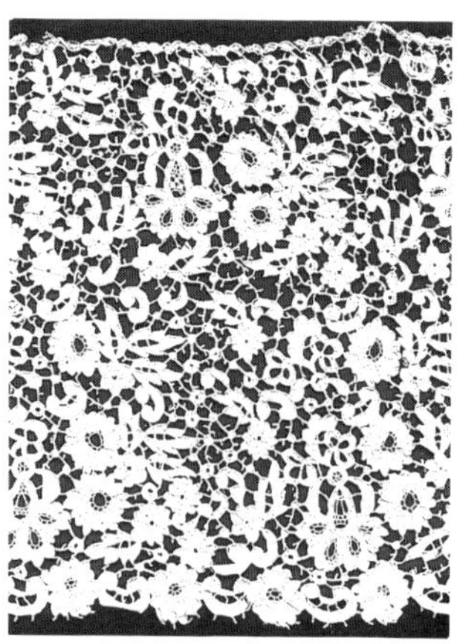

A picot border replaces the twirls.

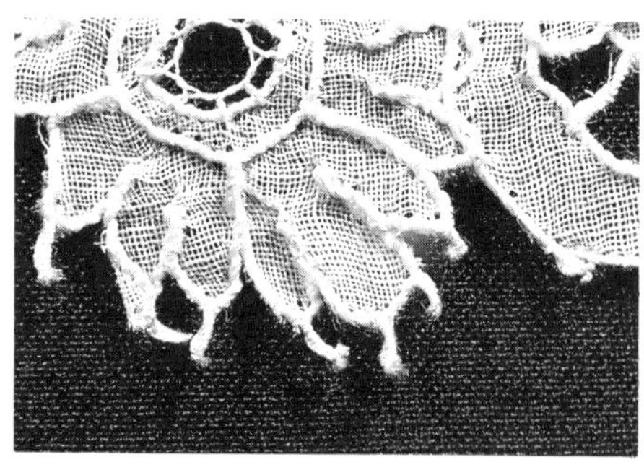

Left: a single repeat.

Stitch 1.

Preliminary row: Starting at the right-hand side, stretch a strand of thread across the top of the casket to the left-hand side.
Row 1: Work 3 detached buttonhole stitches (also known as single Brussels stitch) over this strand.
Row 2: Whip the thread back over the loops of row 1, from R to L. This is referred to as a whipped return.
Row 3: Work 4 dbs, one into each whipped loop of row 1.
Row 4: Whip the thread back around the loops of row 3, to the L-hand side.
Row 5: Take the thread down, over the outline cord, to the position indicated in the drawing. Carry it across to the R-hand side and fix it to the outline.
Row 6: Work a whipped return from R to L, taking the thread up to the second and fourth loops of row 3 in a needleweaving manner.
Row 7: Work 3 dbs so that they alternate with the needleweaving stitches of row 6.
Row 8: Whip the thread back from R to L over their loops.

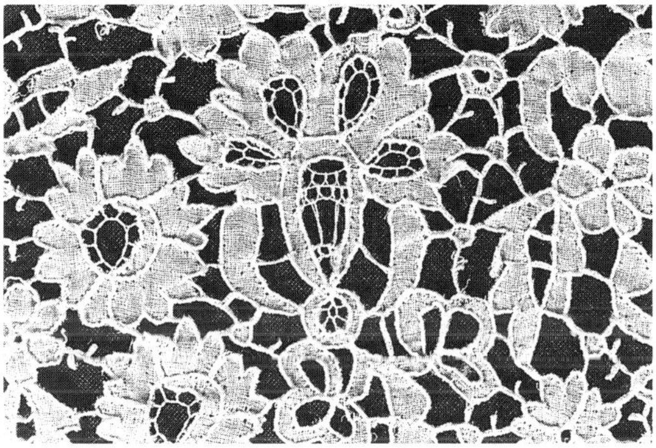

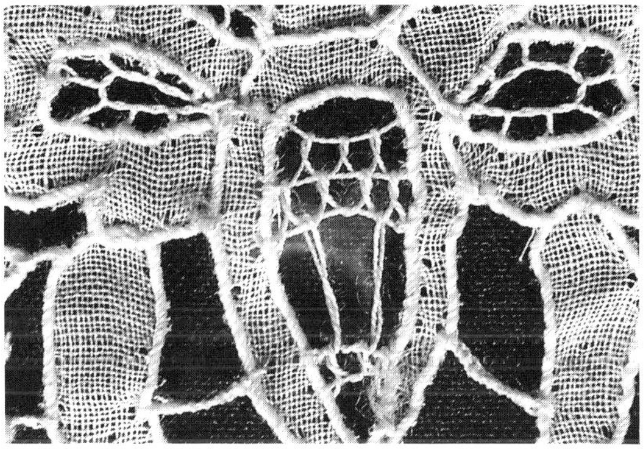

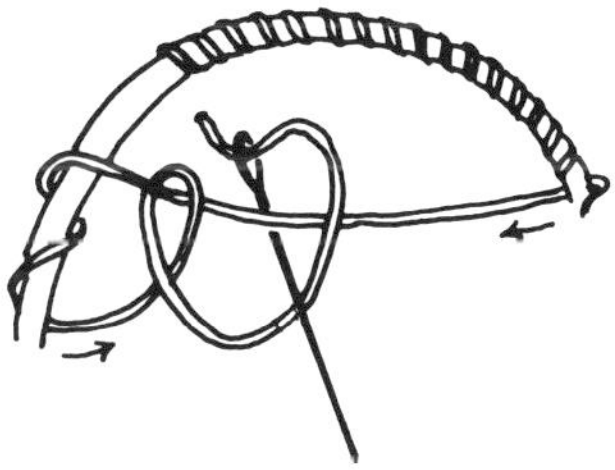

The preliminary row.

Rows 1 and 2.

Rows 3-7.

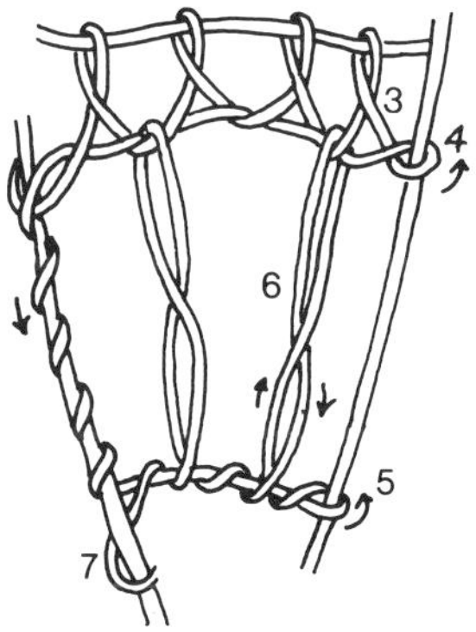

Stitch 2.

Row 1: LR. Work 6 dbs over the couched outline cord.
Row 2: Whip the thread back over each loop of row 1.
Row 3: LR. Work 6 larger dbs, taking one over each loop of row 1.
Row 4: Whip the thread back over the loops of row 3.
Row 5: Carry the thread down the L-hand side to position A, then across to B.
Row 6: RL. Whip back, taking the thread up to the loops of row 3, as shown, to form 3 pairs of long needleweaving stitches.
Row 7: Work 3 dbs to alternate with the bases of the needlewoven pairs.
Row 8: Whip back over the loops of row 7.
Rows 9 - 12: Continue in dbs until the casket is filled.
The working is sufficiently similar to stitch 1 that no drawing is needed.

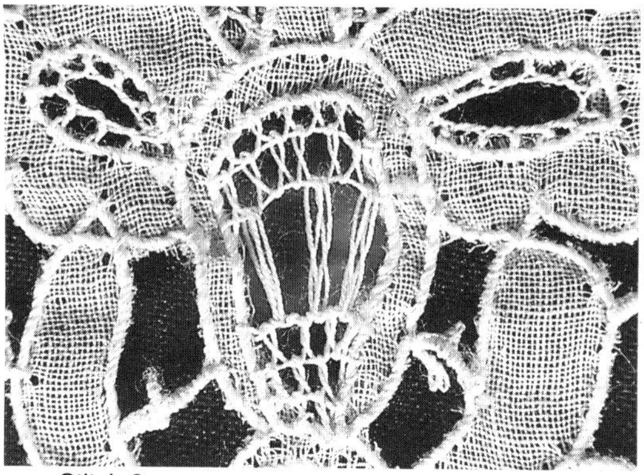

Stitch 2.

Stitch 3.

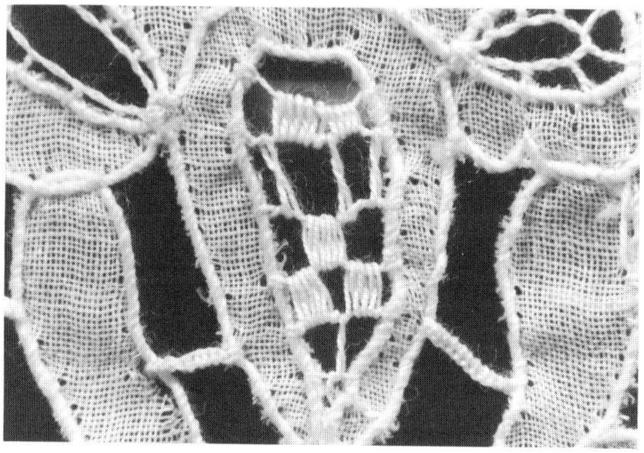

Stitch 3.

Row 1: Lay a strand from L to R.
Row 2: Whip the thread a short distance down the R-hand outline cord. Take a strand across to the L-hand side and secure it to the outline.
Row 3: LR. Work 8 short needleweaving stitches between strands 1 and 2.
Row 4: Lay a strand from R to L.
Row 5: Whip the thread back from L to R, working 3 long needleweaving stitches placed to the left, centre and right of the 8 stitches of row 3. (The original lace is slightly irregular in this respect).
Row 6: Lay a strand from R to L.
Row 7: Whip the thread back from L to R, working a block of 6 short needleweaving stitches, placing 3 on either side of the central stitches of row 5.
Row 8: Lay a strand from R to L.
Row 9: Whip the thread back from L to R, working 4 short needleweaving stitches on either side of the block in row 7.
Row 10: Line a strand from R to L across the narrow base.
Row 11: Whip the thread back from L to R, working a single long needleweaving stitch between the two blocks of row 9.

Stitch 4.

Row 1: LR. Work 6 dbs over the top outline cord of the casket.
Row 2: Whip the thread back over the loops of row 1, to the L-hand side.
Row 3: LR. Work 7 dbs, adding a half-loop at the R hand-side.
Row 4: Whip back over the loops of row 3.
Row 5: Take the thread down the L-hand outline cord, then carry it straight across and fix it to the R-hand side.
Row 6: Whip the thread back to the L side, at the same time needleweaving up to the 8 loops of row 4, in two wedge-shaped fans. Bind the lower ends of each group together with a knot.
Row 7: LR. Work 6 dbs in three pairs, alternating with the two wedges of row 6.
Row 8: Whip the thread back to the L-hand side.
Row 9: LR. Work 5 dbs into the whipped loops of row 7, adding a half-loop at each end.
Row 10: Whip the thread back to the L-hand side.
Row 11: Take the thread down the L outline cord, then carry it straight across and fix it to the R-hand cord.
Row 12: Whip the thread back, and at the middle point make a fan-shaped wedge of 5 needleweaving stitches, passing their top ends through the loops of row 9, and binding their lower ends together with a knot.
Row 13: LR. Work 2 dbs, one on either side of the wedge of row 12.
The drawing has been slightly simplified to show the needlewoven wedges, tied at the base, more clearly.

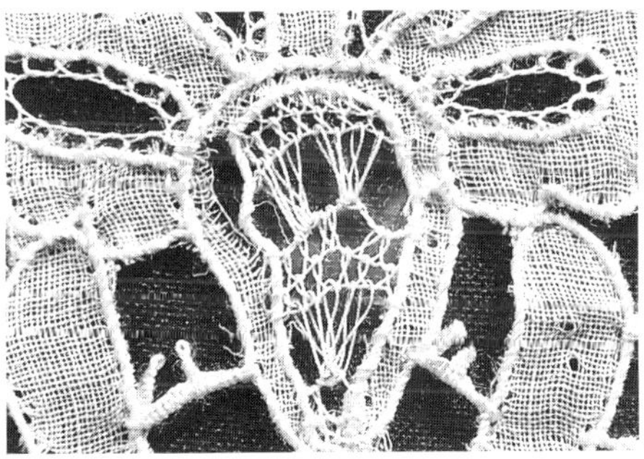

Stitch 4.

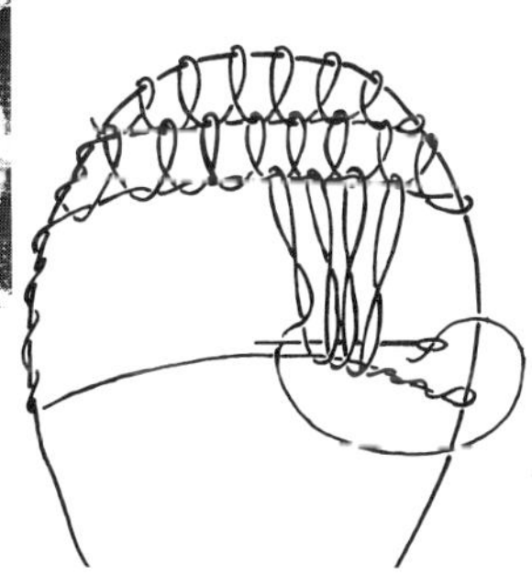

Left: binding the bunches of needleweaving stitches together at the base.

Stitch 5.

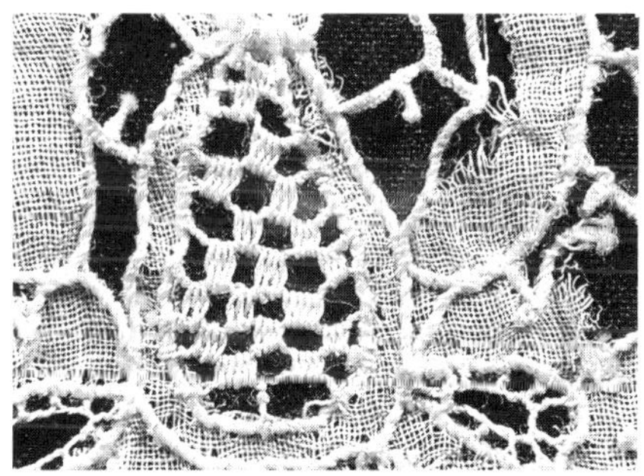

Stitch 5.

Apart from row 1, this stitch is worked entirely in needleweaving.
Row 1: LR. Work 2 dbs over the top outline cord, adding an extra twist to the sides to convert them into tbs.
Row 2: RL. Whip back over the two loops of row 1.
Row 3: Take the thread down the L outline cord, then carry it across to the R-hand side and secure it.
Row 4: Whip the thread back, working 4 needleweaving stitches over each loop of row 1.
Row 5: Repeat row 3.
Row 6: Whip the thread back, working 3 groups of needleweaving stitches so that their upper ends lie between those of row 4.
Continue to work in this manner, until the casket is filled.
Note: You can probably improve on the spacing shown in the photograph, of the original lace, by arranging the needlewoven groups more symmetrically.

Additional decoration

(a) The holed circle. This looks like a cherry with an exposed stone. Take the outline cord around the outer border, dip it inwards to make the central 'stone', then out again. Overlap the two ends of the cord slightly and bind then together with tight couching stitches.

(b) The barred leaf. The drawing shows the leaf outline in continuity with the petals of a flower. The arrows indicate the direction of working. When the outline is complete, make triple strands across the future vein, passing the ends around the cord. Buttonhole closely over the strands, taking care that the needle never passes through the organza.
Refer to the swan's neck, p.138.

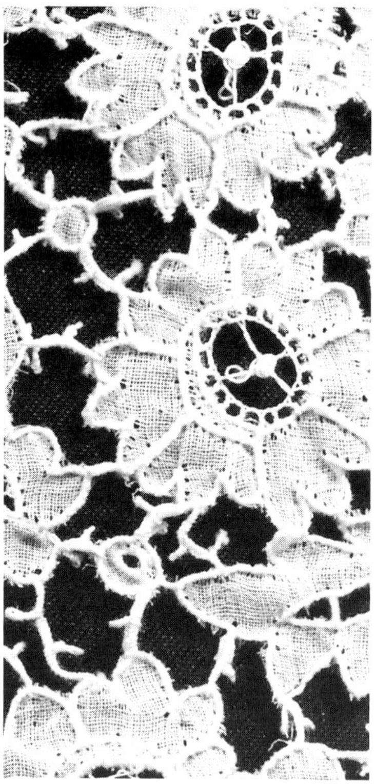

The photograph shows: below left, a holed circle; below right, a barred leaf; above, in the flower centres, a ring of whipped tbs centred by a darned wheel.

Below: working tbs, and whipping over the loops.

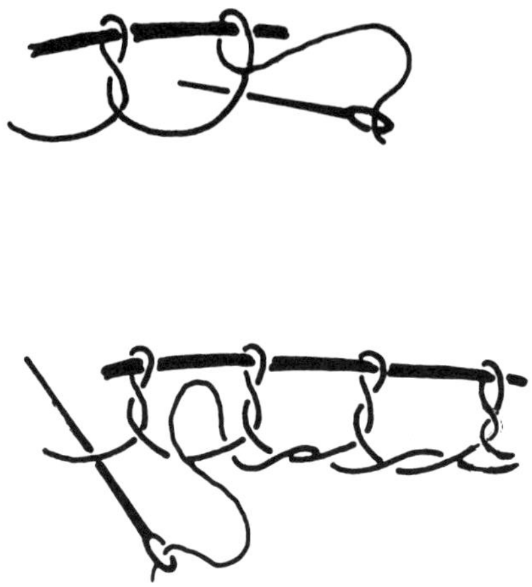

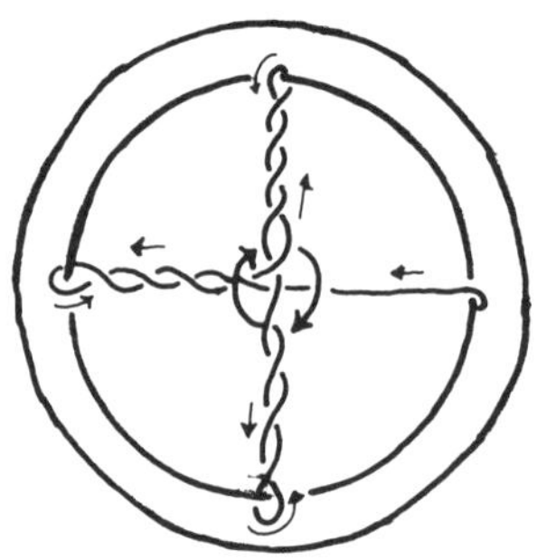

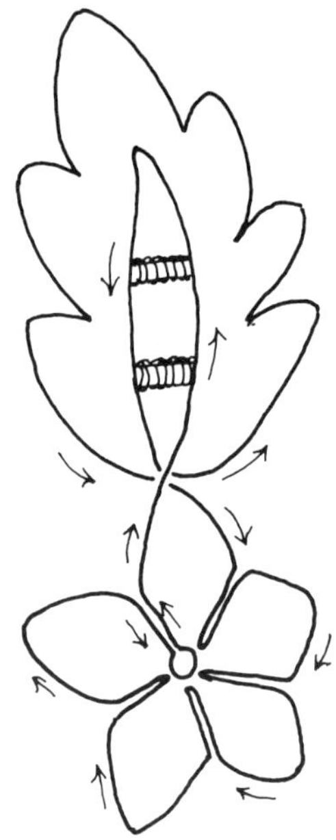

Left: working a darned wheel.
Above: continuity of outline between a barred leaf and a flower.

(c) Twisted buttonhole-stitch lining. Construct a ring of single tbs all round the inner border of the area, taking the thread over the outline cord with each stitch. Go round again, whipping the loops. Tighten the thread slightly for a crisp finish.

(d) Darned wheels. Line a circular shape with whipped tbs, as in (c). Without cutting the thread, take it across to the far side and secure it around a tbs loop. Return to the centre, twisting the needle twice around the thread. From the centre pass outwards at right angles, take the thread over a loop, twist back to the centre and continue across to the far side, then back again. You now have 3 whipped strands, and one single.
Weave the needle in and out of these four, in a circular manner, until the wheel is sufficiently large. Finish off by whipping the thread along the single strand to the outline cord.

Bar Ground

As can be seen from the photographs, not only is the entire ground of bars enlivened with picots, but picot whiskers bristle along the outer border, replacing the twirls. Work both bars and picots as described on p.138.

When all the embroidery is completed, cut away the unwanted fabric. To do this, turn the lace face down, raise a portion of the superfluous organza with the tip of a needle, snip a hole in it, and carefully cut back to the couched cord, taking the utmost care not to damage the bars (see p.126).

CUT CAMBRIC APPLIQUÉ WITH A CHAIN-STITCH OUTLINE

There is no documentary evidence that the couched cord outline typical of Irish cambric appliqué was ever, in Limerick, replaced by tambouring. Yet with the tambour industry so well established in that area, the adoption of this alternative would be a temptingly easy transition. Alternative geographical sources are Switzerland or Belgium for example (1) below, and Slovakia for (2). Because of the doubt attaching to the Irish attribution, they will be described only briefly:

1. In this quaint fall cap with ear pieces, the fabric is much denser, even completely opaque, suggesting that the pattern must in some way have been stamped upon it, since it could not have been seen through it. The design shows how particulate the fragments of cloth can become. Each tiny flower, and the entire beaded border, is tamboured around, and the excess material finally cut away from every single one of the miniature motifs. There are no decorative stitches.

2. The two-tiered fichu, c1840, has a semi-translucent fabric, but fixed beneath the net, instead of above. The flower centres are filled with either vertical figure-of-eight or horizontal cobweb stitches. Along the border the decoration is of holes distended by a stiletto, then whipped around to hold back the mesh sides.

The fall cap with ear pieces.

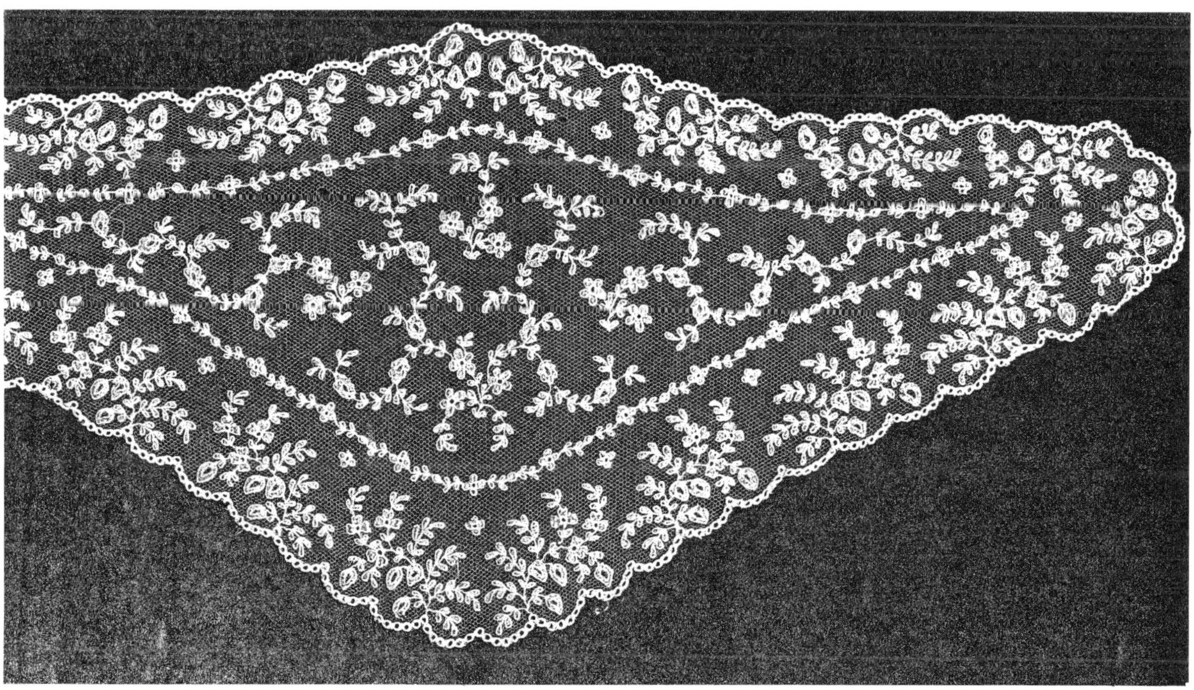

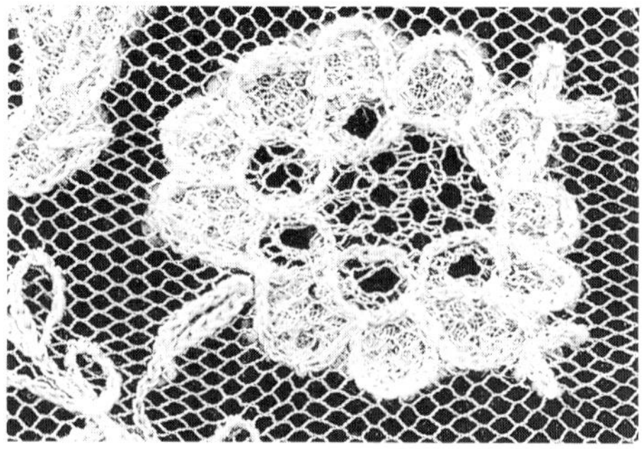

A border from the two-tiered fichu, and a single floret showing the net placed over the cloth, and joined to it by chain-stitch.

* * * *

Above: net-on-net, highlighting the grapes. A small tamboured ring encircles a pale area where the upper net has been cut away.

Above: a motif extracted from the net-on-net veil.

Right: the formal border and naturalistic flowers of the design.

CUT NET (NET-ON-NET) APPLIQUÉ

Whatever doubt there may be about the Limerick origin of cambric appliqué laces with chain-stitched outlines, contemporary records establish that, at least for a limited time, net appliqué with chain-stitched outlines was not only worked in, but was exclusive to, Limerick. Thus in 1886 [106]: 'In some instances net is applied to net. Of this however there have not been many specimens, though it has a beautiful effect'.

Needlecraft in 1902 added:

'Limerick appliqué is now very rarely seen. At one time it was reckoned amongst Limerick laces, and a very delicate appliqué of net upon net was also made. The wonder is that this latter style has been allowed to drop because it has a distinct beauty of its own, but it is only now and then that one meets with a piece of it' [107].

The background net was invariably square-meshed, while a layer of round net replaced the woven cloth. Since the square or diamond net is no longer available, net-on-net appliqué cannot now be made in its traditional form. However, one layer of round net over another can still produce an intriguing effect.

In this light and delicate-looking lace, twirls and pops are usually absent.

A VEIL

In the veil illustrated, the labour of drawing or stamping such an extensive pattern on to the upper net would be enormous, and a great deal of additional stitching would be needed to anchor the layers together securely enough to prevent any movement between them during the fairly vigorous plying of the tambour hook. On the other hand, if the pattern is below the net, tambour work as described on pp.78-79 becomes impossible.

The threads of which the two nets are composed are exceptionally clear, crisp and fine. That of the round net is slightly heavier than the square, and the embroidery thread slightly heavier again so that, although very limited in extent, the decorative stitches are clearly visible.

There are only two of them:

Horizontal cobweb (refer to stitch 2, p.33);

Diagonal figure-of-eight.

These are the same as in the fichu, p.150.

In the caskets, the upper layer of round net is snipped away, and the stitches worked on the square net.

With regard to the design, the neatly formalised border of swags and scrolls, and the central areas with their delicate suspension of dots like falling snow, might suggest a Brussels influence, but the naive form of the roses, tulips and daffodils, and the stencilled effect of the vine leaves, are distinctively Irish.

The cut edges of the round net are amazingly neat and this, together with the unwavering precision of the lines of tambour stitch, and the tightly crocheted outer border, demonstrate the superb professionalism of the work.

The sketches on pp.151-3 are isolated from the design. (x) and (o) mark the positions of the casket stitches.

A STREAMER

This elaborate net-on-net concoction is about 9 feet long. The finely tambioured design outline runs in ingenious continuity between leaves, petals, veins and stems. All the tiny dots, even those sprinkling the petal and leaf surfaces, are individually tamboured.

The upper net is restricted to relatively small areas, and extensive tambour work decorates the exposed lower layer.

A snake-like beading flanks the central cartouches and is reminiscent of the sinuous arcs of pops found in cambric appliqué. Beyond this lies a border of ivy leaves, and then a slashed margin where an open space is crossed by bars. The idea is similar to the neck decoration of the swan (p.138), but here the bars are tamboured over, not buttonholed. Both layers of net are cut away beneath them.

There are no casket stitches apart from the 'suspended point d'esprit' in some of the flower centres. This is reminiscent of Brussels bobbin laces, but the stitches are in fact minute darned wheels (see p.148).

The design also recalls Brussels tambour laces, but this does not rule out an Irish origin. Belgian nuns had settled there, Bruges bobbin laces were made at a number of centres, and Brussels laces were in any case a constant target for imitation.

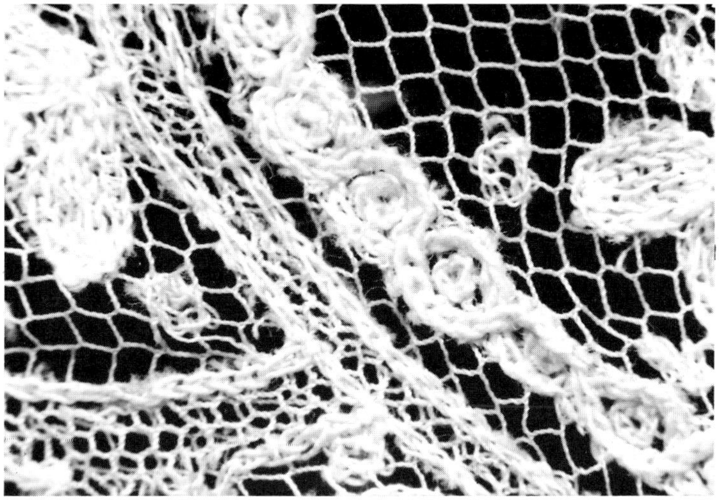

Net-on-net, intersecting circles trapped by chain-stitching.

CUT NET GUIPURE

This technique is not mentioned in the archives of Limerick laces, though it could be made in exactly the same way as cut cambric guipure, with the pattern situated below the net. The example illustrated is a boudoir cap from c1903. It was embroidered on the Schiffli machine, and the unwanted net then cut away by hand.

There are no bars as such and a certain spacing of the motifs is created by spirals of thread worked on the net. In the machine lace trade these are referred to as 'needlerun'.

As with all guipures, the cutting away is most easily done from the reverse side. As the spare fabric is cut back to the motifs, the blunt tip of the scissors must glide over the bars but never cut through them. Refer to p.150.

There are twirl-like loopings of the outline cord along the edges of the net.

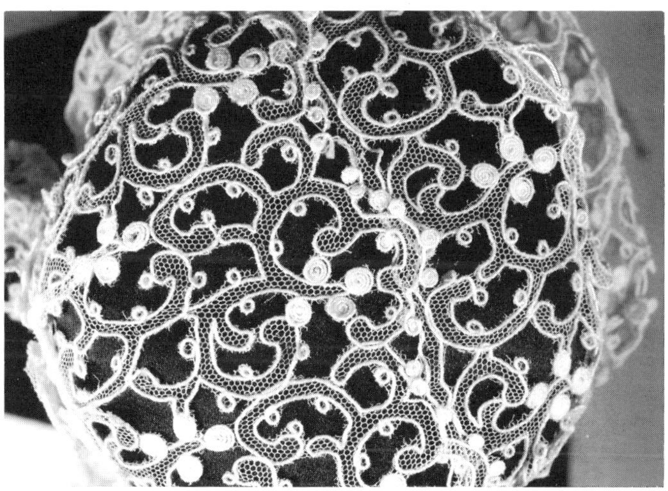

A LAPPET AND A HANDKERCHIEF

In these possibly unique pieces, the fabric is a leno-weave gauze. The design is outlined by a couched cord, and the numerous bars are buttonholed, with long branching picots.

Some of the petal, bud and leaf borders are highlighted by a 'relief' or 'cordonnet' - padding closely buttonholed over in the manner of three-dimensional needle laces.

Pops cluster like grapes or, singly, mark the centres of tiny cruciform flowers.

The superficial similarity of this lace to point de gaze might again indicate Brussels as the geographical source, but the general technique, and the casket fillings (dbs, tbs and needleweaving) with their strong similarity to cut cambric guipure, suggest Ireland. The typical sinuous beading is present, as well as the long open slashes traversed by bars.

The instructions given for making the casket stitches are based on the number of stitches per row and the size of the loops in the original lace, but both can be varied as necessary.

The finest threads are used in the caskets. The threads of the bars are twice as thick, and those of the cordonnet and pops twice as thick again. The cordonnet and pops are added only after the rest of the embroidery is complete.

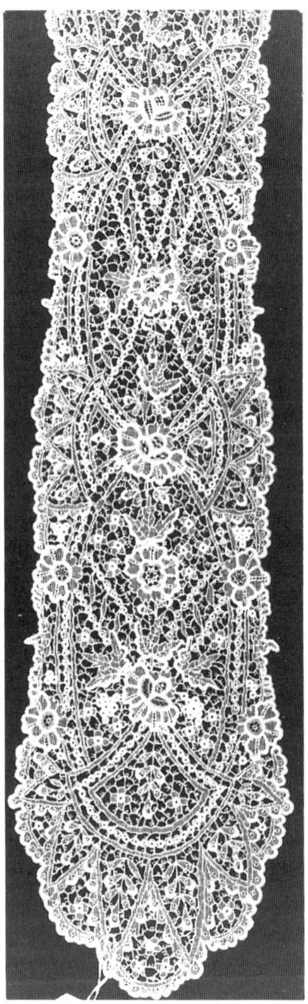

Left, the lappet. Below, the handkerchief (photo: R.B.).

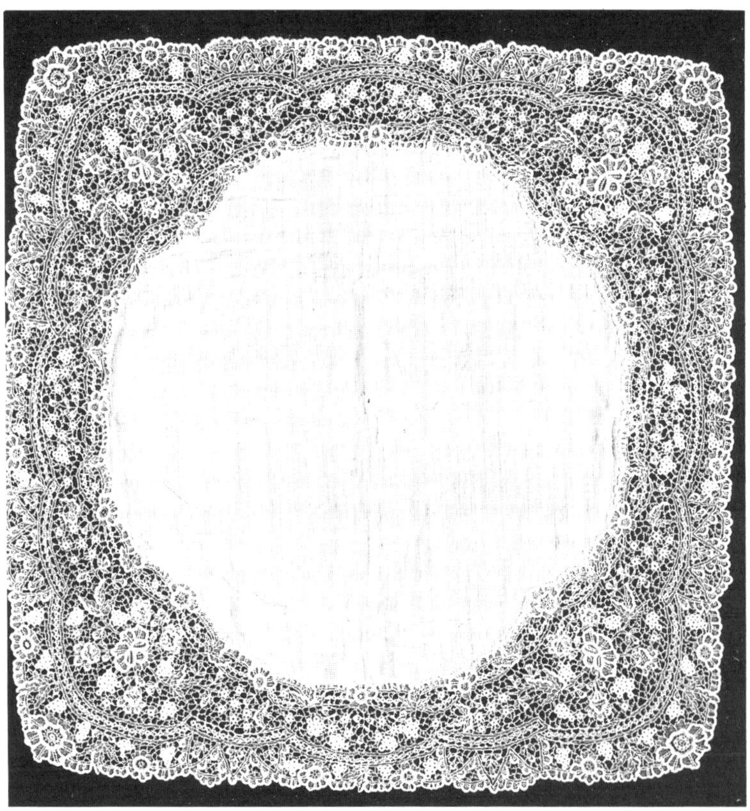

Stitch 1.

Row 1: LR. Work 10 dbs over the outline cord at the top of the casket.
Row 2: Carry the thread back from R to L, whipping it once over each loop of row 1.
Row 3: LR. Work 1 dbs over each whipped loop of row 1.
Row 4: RL. Whip along the loops of row 3.
Row 5: Work 2 dbs, miss the next 4 loops, work 3 dbs, miss 4 loops, 2 dbs.
Row 6: RL. Whip once over each loop of row 5.
Row 7: Take the thread down the L-hand cord, carry it across to the R-hand side, and pass it around the outlins.
Row 8: RL. Whip the thread around this strand for a short distance. Work 9 nedleweaving stitches, carrying their tops over the first long loop of row 5. Whip on until you are beneath the second long loop, and repeat.
Row 9: LR. As row 5, but arrange the long loops below the blocks of needleweaving.
Row 10: Whip the thread back to the L-hand side.
Row 11: LR. Work dbs right across, taking one stitch into each short loop of row 9, and two into each of the long loops.
Row 12: Whip the thread back.
Row 13: Work dbs right across.
Row 14: Whip the thread back.

In the centre, stitch 2; in the wings, stitch 1.

Stitch 2.

Note: Every alternate row - 2, 4 etc - is a whipped return, worked RL, with one whip per loop;
Row 1: Work 5 dbs.
Row 3: Work 6 dbs, one into each loop of row 1.
Row 5: Leave a long curved loop below the first 2 stitches, work dbs into the 3 central loops of row 3, leave a loop below the final 2 stitches.
Row 7: Work 4 dbs into each long loop of row 5, but not into the 2 short loops in the middle.
Row 9: Work 1 dbs into each short loop of row 7, and 3 into the middle long loop.
Row 11: Work dbs right across.
Rows 13, 15, 17, 19: Repeat rows 5, 7, 9, 11.

Stitch 3.

Row 1: LR. Work 6 dbs over the top outline cord.
Row 2: RL. Whip the thread back, once over each loop, to the L-hand side.
Rows 3 and 4: Repeat rows 1 and 2, adding additional stitches to make a total of 9.
Row 5: Work dbs into the first 2 loops of row 3. Miss 5 loops, work 2 dbs.
Row 6: Whip the thread back.
Row 7: Carry the thread down the L-hand side. Lay a strand across the casket and attach it to the R-hand cord.
Row 8: Whip around the first part of the strand, then work 10 very close needleweaving stitches, taking their tops over the long middle loop of row 5. Whip to the end.
Row 9: Work 3 dbs over the first long loop of row 7, pass the thread below the block of needleweaving, work 3 dbs over the second long loop.
Row 10: Whipped return.
Row 11: Work dbs right across, one into each dbs of row 9, with 4 into the long loop in the middle.
Row 12: Whipped return.
Row 13: Dbs right across.
Row 14: Whipped return.
Repeat rows 5 - 14 until the casket is filled.

In the centre, stitch 3; in the wings, stitch 4.

Stitch 4.

Odd-numbered rows are worked LR.
Row 1: Work 17 dbs over the top outline cord.
Row 2: Whip the thread back, passing once around each loop of row 1.
Row 3: Dbs right across, working one stitch into each whipped loop of row 1.
Row 4: Whipped return.
Row 5: Work 5 tbs with deep curved loops, one into every fourth stitch of row 3.
Row 6: Work 5 dbs with deep curved loops, one into every loop of row 5.
Row 7: Work 3 dbs into each loop of row 6.
Row 8: Whipped return.
Row 9: Dbs right across, one into each loop of row 7.
Row 10: Whipped return.
Row 11: Work a row of extended tbs (with additional twists in their sides) over alternate loops of row 9, anchoring their lower ends around the bottom outline cord.

Stitch 5.

Rows 1 - 4: Line the casket with a row of tbs (54 stitches). Go round again, whipping over the loops. Repeat these two rows.
Row 5: Work one tbs into every 6th stitch of row 4, leaving deep loops between them (total 9).
Row 6: Work 3 extended tbs into the centre of each of the 9 stitches of row 5, so that their inner ends form the circumference of an oval.
Row 7: Whip around these inner ends and tighten the thread. This sharpens the oval, and at the same time pulls on the tbs of row 5.
Row 8: Needleweave over the sides of the row 5 tbs to form 9 triangles.
Note: In the drawing, rows 1 - 4 have been simplified, for clarity. The sketches show further details.

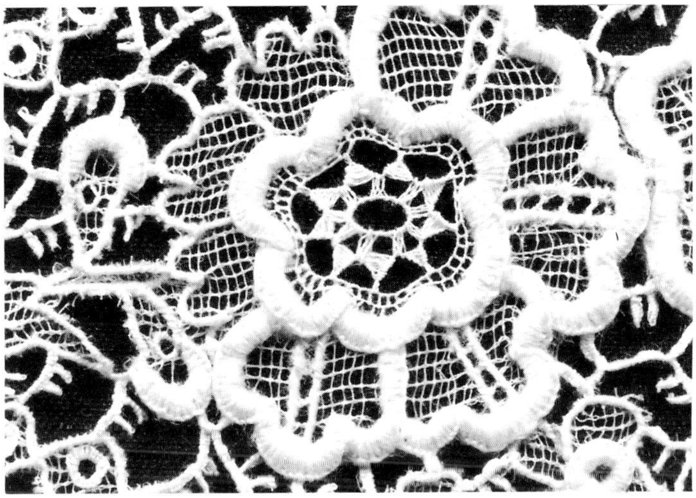

Note the overlapping segments of the cordonnet.

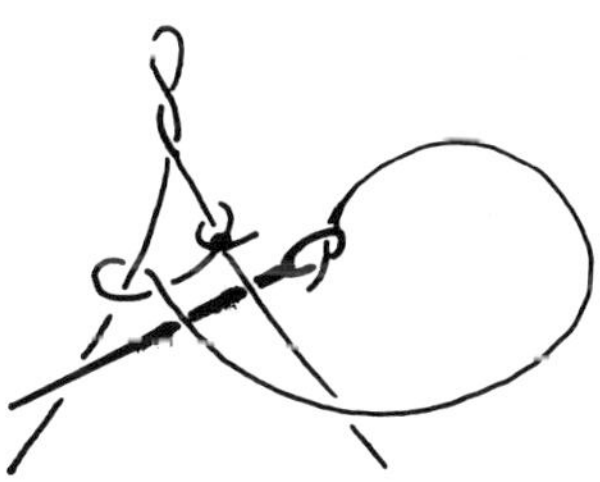

Working the central area.
Left: needleweaving the triangles.

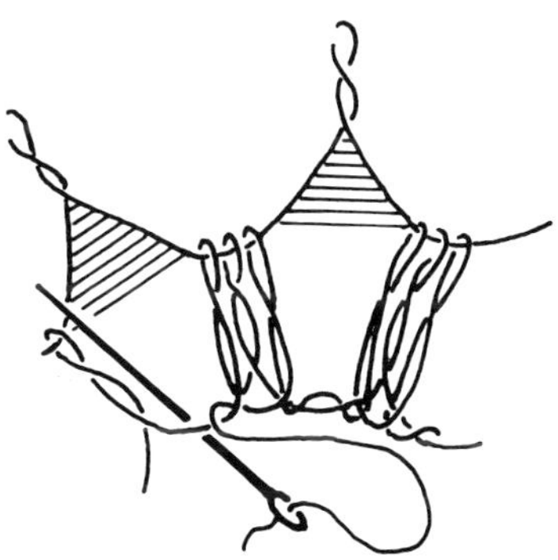

Left: whipping around the middle circle.

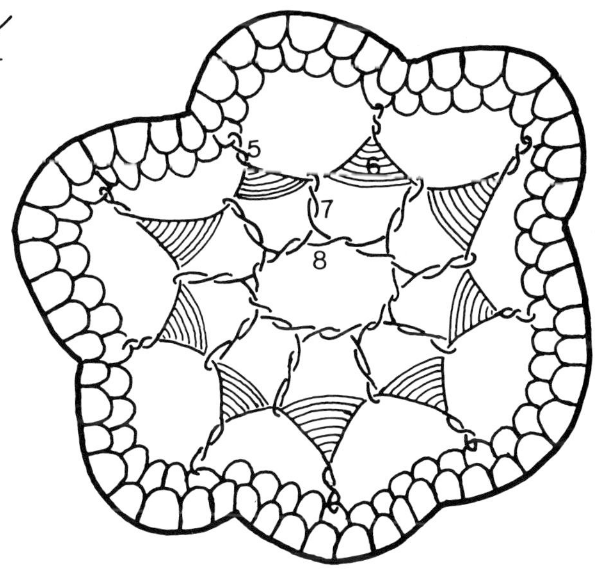

Stitch 6.

Round 1: Work one row of tbs all round the inner border of the casket (24 stitches), hooping the stitches over the outline cord.
Round 2: Continue in the same direction, whipping once over each loop.
Round 3: Work one tbs into every third stitch of row 2, leaving deep loops between them (total 8).
Round 4: Work 8 tbs, one over each loop of row 3.
Round 5: Whip around the 8 inner loops of row 4 to produce a firm circle.
Round 6: Work a series of darned wheels in the positions shown. The spiralling of the thread in and out of the spokes is reinforced by a topping of satin-stitch. Whip along the loops of row 4 from one wheel to the next.

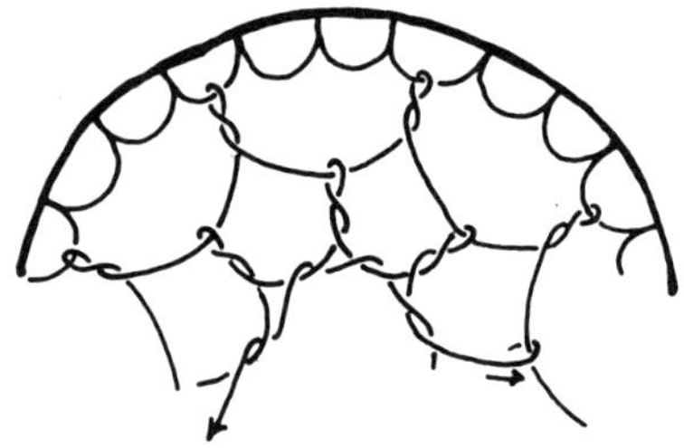

Stages in the working of stitch 6.

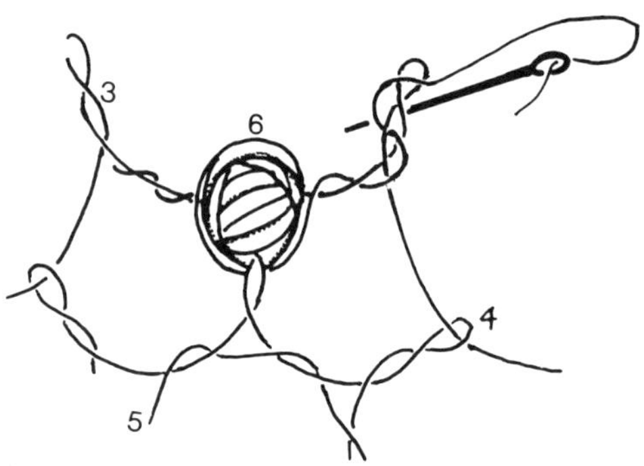

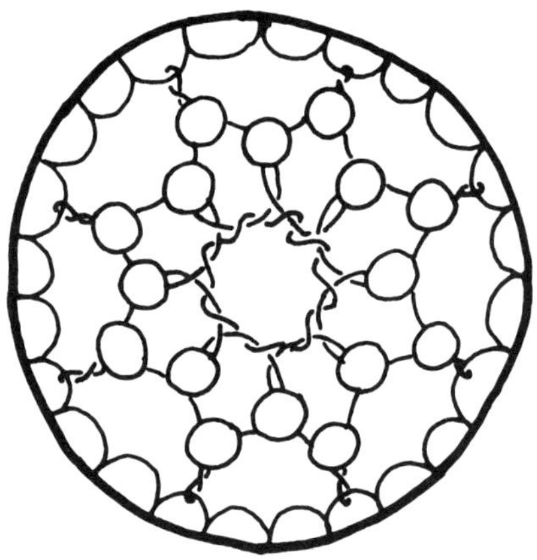

Pops (see p.138).

These are worked directly on to the gauze, which will remain beneath them when the excess fabric is cut away. To increase the height of the pops, run the embroidery thread several times in a circle before buttonholing over it (p.126).

Cordonnet

Lay padding strands in the required position, and whip them lightly to the gauze to fix them in place. Using the thickest thread, buttonhole over them, carefully tapering the ends, and passing the needle outwards from the centre. The exquisite expertise of the orignal is not altogether easy to achieve. Note for example the precision with which each scallop of the cordonnet overlaps the previous one.

Finishing off

1. Removal of backing. Cut through the tacking stitches between the support and the pattern. If there is no separate pattern, cut on the underside of the backing to avoid accidentally snipping the net.

Cut away both fabric and net along the outer border of the lace, beneath the twirls.

2. Removal of the excess cambric. *Appliqué*: Turn the lace face side up and cut around the outside of every motif, close to the cord but without weakening the couching stitches (see p.126). Little by little the distinction of the lace into solid design areas of woven organza, and openwork ground of machine net will be revealed.

For net-on-net appliqué, substitute 'top layer of net' for 'cambric'.

For *appliqué* where the fabric is below the net, cut away the excess from the reverse side.

For *guipure* laces, the hazard is not the possibility of damaging the net, but of cutting through the bars (see p.150).

3. Pull any loose ends of thread through to the reverse side. Using a very fine needle and thread, run them invisibly along the back of the outline cord.

The two ends of the outline cord itself can also be pulled through to the back. If they are overlapped and stitched there, instead of on the face side, the bulge will be less noticeable.

4. The outer border of the lace can be strengthened by additional overcasting of the couched cord.

Pops grouped into bunches of grapes.

Care Even though the net and muslin have been preshrunk, it would still be foolish to jeopardise the pristine condition of the lace by overenthusiastic washing. If it is necessary to clean or freshen it, use lukewarm water and a mild detergent, bounce gently in the water but do not rub. Rinse, then wrap the lace as flat as possible in a towel to dry, and iron carefully from the reverse side, pressing down on the twirls to keep them straight.

Cut cambric work has an inbuilt tendency to self-destruct, and there are few antique pieces which do not show damage of some kind. As *Needlecraft* commented at the beginning of the 20th century: 'Unfortunately there is a great deal of inferior quality in the market, which is likely to injure the reputation of the industry . . Either poor quality muslin or poor needlework will result in the outline cord failing to hold the cloth in place - the first time the lace is washed, the muslin pulls away'[108].

Caulfield and Saward recommended that the lace be cleaned, not washed, as 'it will not stand hard wear'.

Some can be adapted for either appliqué or guipure work (pp.162-164)

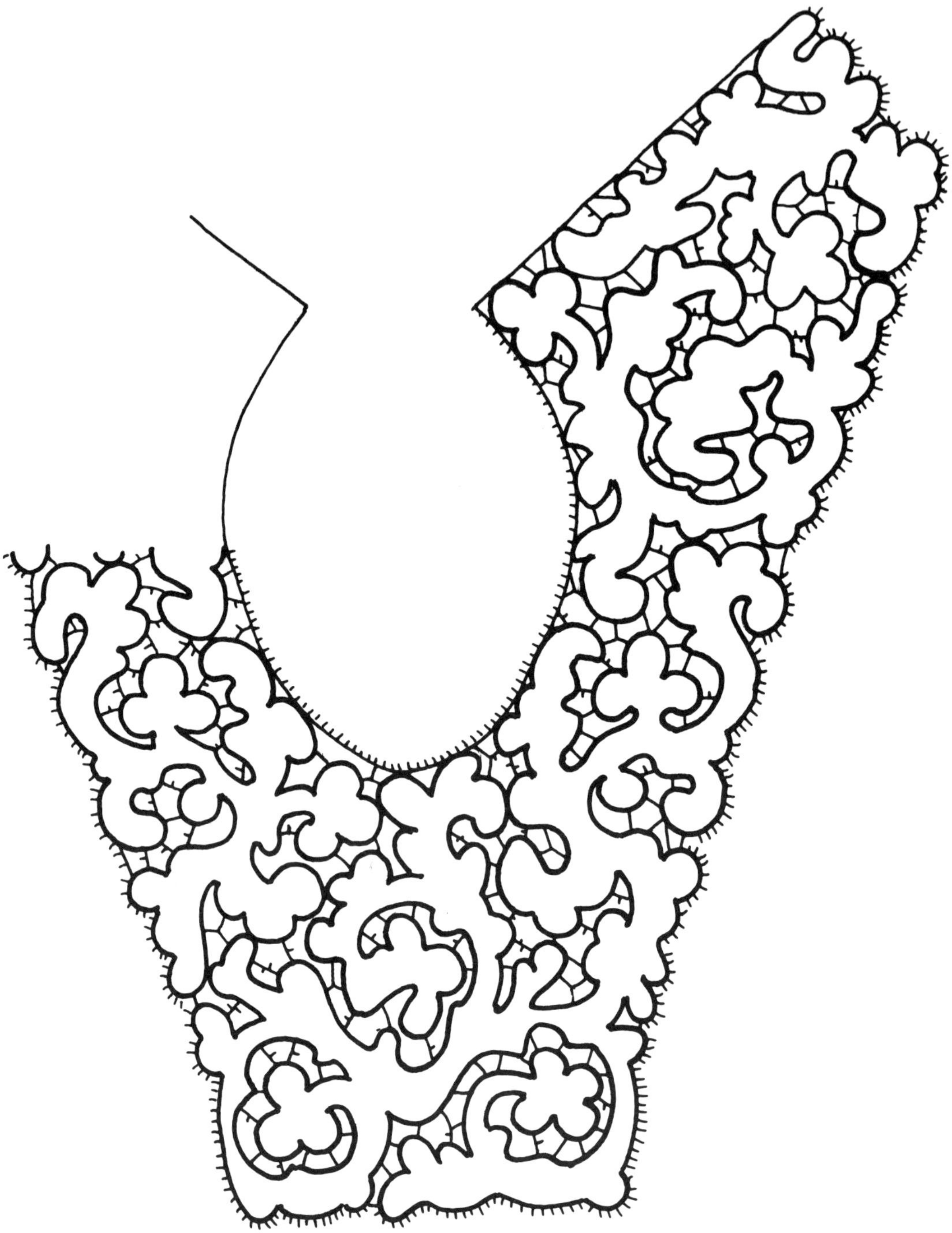

5. EMBROIDERED MACHINE NETS WORLDWIDE

The situation

Machine nets, of many different types, all coming plain off the machines, were embroidered by hand almost from the time they were first marketed near the end of the 18th century to the period when fashion became indifferent to them, around the middle of the 20th.

Throughout this 150 years, commercial competition from both hand-made or machine-patterned, laces was cut-throat. The development and partial automation of one machine after another threatened to gobble up the market with gluttonous speed, and while the embroidered nets had a price advantage over the totally hand-made bobbin and needle laces, they could not compete financially with the more rapidly produced, and sometimes prettier, products of the Pusher and Leavers machines.

One striking result of the vast outpouring of lace of all types was a rapid democratization of consumer demand and, in an attempt to satisfy it, a rapid expansion of the areas manufacturing simpler laces for a local or near-local market. Hand-in-hand with this went a widening interest in amateur needlework.

In consequence, embroidery on machine nets appeared at one time or another in almost every country in the world, some using one technique, some another, but none, like Limerick, producing all three of the major forms.

Spain, France, Belgium, Switzerland, Scotland and England aimed in varying degrees at high fashion and an export trade, while the Netherlands, Denmark, Norway, Germany, Austria, Hungary, Czechoslovakia, North America and Australia were more occupied with the home market and, where appropriate, with provincial costume in which substantiality of effect, linked to the work aspect of the wearer, sought isolation of style and visual discrimination from surrounding agricultural areas.

Middle class fashion on the other hand fostered a convergence of ideas, and to this end encouraged imitation and piracy. Thus pattern books from Germany, Denmark, the Isle of Wight and Nottingham contain almost identical designs of spiralling floral bouquets. Limerick, aiming higher yet, and bolstered in its happier times by opulent commissions and a flourishing export trade, could indulge in light, intricate and very decorative laces, immensely rich in casket stitches, and taking sometimes months to complete.

The spread of Limerick laces

The embroidered machine nets from Limerick were spread around the world not only intentionally, by the export trade, but also involuntarily.

The missionary zeal of Irish nuns drove them to all quarters of the globe conveying, along with Christianity, practical skills such as run and tambour laces which could provide a life-line for the poor and needy forgetting, their heads in the clouds, the effects their 'charity' might have on the poor and needy lacemakers at home, whose deprivation was only exacerbated by the tide of cheap laces pouring back to undercut their own. Run embroideries for wedding- and baby-gowns began to be made for example in India.

Emigration also played a part in the dissemination of Limerick laces. Between 1846 and 1851, during and immediately after the famine years, almost a million Irish fled to North America, doubtless taking both lace and lacemakers with them. Of those left behind, one-and-a-half million died of starvation and the resultant fever. The outward flow continued into the 20th century. Statistics of the Irish Department of Agriculture show that some 47,500 crossed the Atlantic in the twenty-four months between 1912 and 1914 [111].

In attempting to allocate a geographical origin to individual pieces of embroidered machine net, all these factors inevitably create problems, most especially when the past history of the lace is unknown, or regarded as inaccurate. Variations in the type of net, stitch, design, thread thickness, and the nature of the end product (bonnet, stole, apron etc) certainly provide pointers, but such pointers frequently direct us to speculation and an assessment of probabilities, rather than to a state of absolute certainty.

SPAIN and copies of its blonde laces

The glossy silk bobbin laces known as *blonde*, and associated with France, Spain and to a lesser extent Genoa, were extremely popular with the élite of society almost continuously from the mid-18th to the mid-19th centuries. They were worked in an intense black, shiny as coal, or in lustrous shades of pale gold. Their lawful importation into Britain was halted during the Napoleonic wars of 1800-1815, though thanks to the surreptitious activity of the smuggler longboats some clandestine trade continued. At the same time, attempts were made to find substitutes by copying the Spanish designs in run embroidery on machine net. Silk, during this period, was also in short supply in England, and the bonnet veils, small wedding veils and aprons produced in Nottingham were made of cotton net embroidered with cotton thread. Though superficially similar to the genuine article, they were unlikely to be mistaken for it.

Fashionable blonde, full dress, April 1830 (Courtesy: Chertsey Museum) (photo: R.B.).

Spanish blonde bobbin lace (photo: R.B.).

In Kells, County Down, a two-twist cotton bobbinet was run with silk embroidery thread. The founder of the Kells industry was Emma Colston, daughter of a Nottingham bobbinet manufacturer. While her brother, Henry Hubbert, continued the Nottingham business, Emma in 1825 established an associated industry in Kells for embroidering the net from his factory by run or tambour techniques. Production continued into the early 1850s. Apart from trimmings, a speciality appears to have been lace caps [112].

'Blonde' silk run lace, on point net, c1820 (Courtesy: D.A.R. Museum).

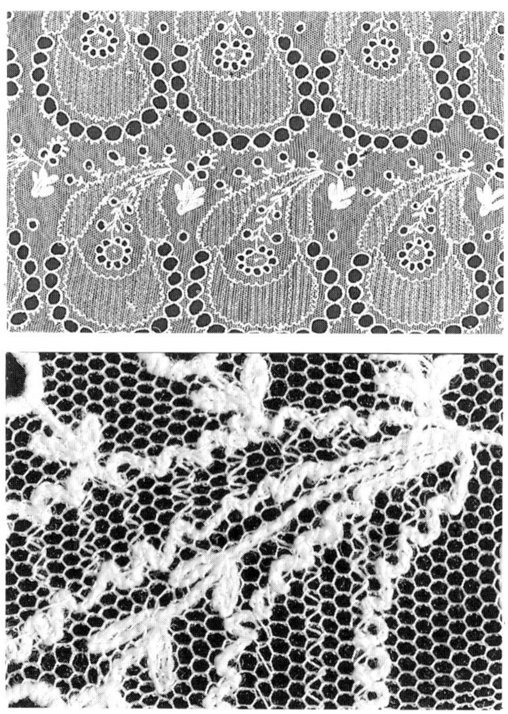

Silk patent or point nets had been manufactured in Spain from quite early in the 19th century, and by 1831 there were 500 'pin machines' turning out weft-knitted fabrics [113], but silk bobbinets appeared only with Heathcoat's improved reeling techniques in 1824, when he set up a factory in Paris. In 1825, Dognin of Lyons also began to produce silk bobbinet [114]. These permitted far more realistic facsimiles of blonde to be created, and no time was lost by Barcelona and Valencia in Spain, and Bayeux in Normandy, in producing and marketing 'run blondes', which were known as bordados.

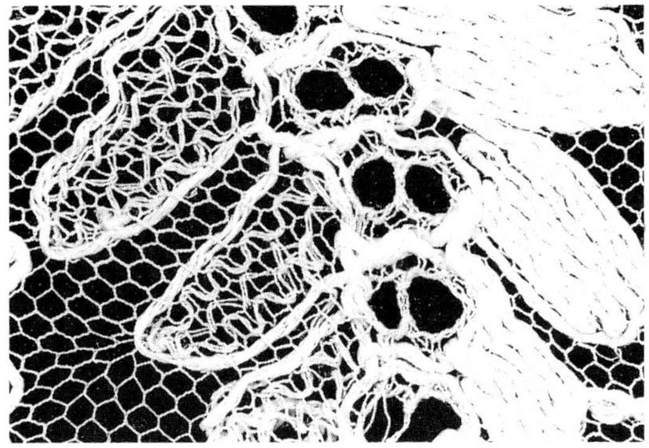

Run blonde in two thicknesses of thread: heavy run-stitch, lighter tent-stitch. Above: detail.

The heavily floral designs, the paucity of decorative stitches, and the powerful contrast between the thick loosely spun silk of the solid areas and the very fine bobbinet openwork, produced a rich effect and made it easy for copy to be passed off as original. By 1856 Spain had 80 bobbinet machines of its own [117].

It is interesting that Cole, in his article on 'Lacemaking in Ireland', 1890 (p.657), says that Walker taught the Irish 'how to embroider net in imitation of the patterned tulle *blonde*, and Brussels pillow [bobbin] laces, which were in considerable vogue at that time' i.e. c1830. Assuming that tulle means machine net, the patterning at that time could only have referred to hand embroidery such as the Nottingham run work which, according to tradition, sought to imitate the bobbin blondes. A later problem of terminology arose when lustrous silk stoles, scarves and mantillas, made on the Pusher machine, and imitating with exquisite competence the true Spanish bobbin laces, had the outlines of their motifs run in by hand and were in consequence - with a graceful elision of the truth - advertised and marketed as 'real hand-run laces' [118].

Below: Pusher machine-patterned scarves with hand-run outline.

The 'Silkworm' of the *Englishwoman's Domestic Magazine*, on her visit to Paris described a 'Bonnet of rice straw, the border ornamented with deep embroidered blonde arranged into a bow at the top. Upon this bow is placed a bunch of flowers', and another 'entirely made of embroidered blonde ornamented in front with a deep border of beaded blonde forming a drooping bandeau' [115].

The following year, in June 1869 she wrote - 'soft tulle or lace lappets fall gracefully round the face, veiling the advances of age or the *démarche* of illness in a manner highly agreeable to those who have parted with the *beauté du diable*' [116].

Below: a hand-run border, commemorative (Courtesy Joslyn Baker) (photo: R.B.).

Opposite page: a beaded gown from the Royal Embroidery Works, Paris and Brooklyn, N.Y., Oct. 1912.

Apart from its embroidered blondes, Spain also produced run laces of more original and representational designs, such as the vignettes surrounding this beautifully drawn two-headed eagle. The theme is a harking back to the days of Ferdinand of Aragon and Isabella of Castile, the scrolled motto *Tanto Monta Monta Tanto* indicating their equality of power. Apart from expelling the Moors, persecuting the Jews, setting up the Spanish Inquisition and supporting Christopher Columbus, they were grandparents to Charles V, elected Holy Roman Emperor in 1518. The variety of decorative stitches in this lace provides a richly jewelled effect.

FRANCE

Felkin wrote in 1867, 'Embroidery on white [i.e. cotton] lace net gives a considerable amount of employment in France to needle and crochet [tambour] workers *au passé* or *au plume*. Thus in Paris robes and elegant shawls and scarfs are produced'. His terms correspond to the *broderie au passé* and *broderie au plumetis* of Charles and Pagès. According to a Report in 1860, the French embroidered net industry then amounted to £64,000 a year, the embroiderers being paid approximately 9.5d [4p] per day [119].

The Department of Vosges was quoted separately. Famous for its textiles, and lying just south of Lunéville, its embroidery was valued at £160,000 in 1867, 2.5 times that of the rest of France.

Lunéville near Nancy, in eastern France, close to the border with Germany, is frequently named as the cradle of all run embroideries on 'net' [= filet]. Its industry can be traced back to the 1300s, but until the late 18th century the net was hand-knotted and made to the size required rather than cut to measure. The traditional run technique of the *broderies de Lunéville* was transferred, around 1800, to the early silk knitted nets, and then to the cotton bobbinets, though the embroidery of filet still continued.

About 1850, tambour embroidery, then in high demand, was added to the run, using a rich creamy silk thread on a ground of either round or square silk net of similar tint.

The illustration shows a lady's cravat or neck-tie in 'application'. Described as 'one of the very prettiest of Parisian novelties' in The Ladies' Department of *The Illustrated Exhibitor* for 1852, p.108, it required only some rather stiff white net, three-quarters of a yard of broad glacé silk, a few bugle beads and sewing silk for attaching the net and fabric together using fine chain-stitch, the design having been pounced on the silk. Grey ribbon with black net, beads, and bugles, and worked in black silk was recommended; or water tints such as rose, cerise and dark blue, chain-stitched in black.

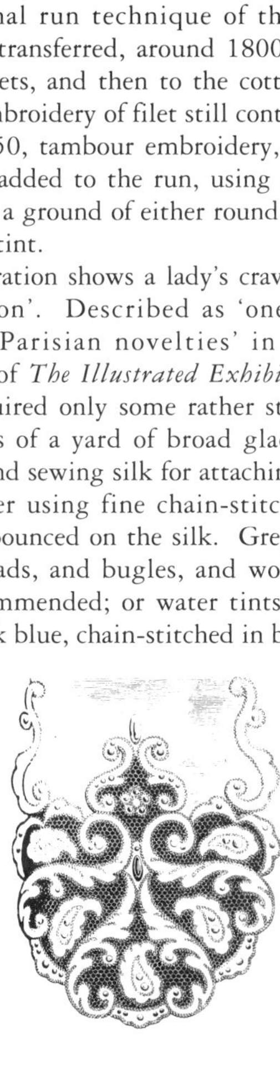

By 1867, bobbinet was being embroidered at Lunéville 'in every class of article, chiefly by the needle' [120].

About that time, a new and exclusive luxury trade was introduced - the attachment of extraneous objects such as jet and pearls, by chain-stitch, to the net, by tambouring. By the end of the century designs were not only outlined but filled in with additional artifacts such as cut and polished gem stones, gold beads, bugles and metal sequins creating scintillating evening gowns that flashed and glittered in the artificial light. Wood, cork and leather are also said to have been used [121]. The method of attachment is described on p.116. The lines to be followed were pounced on the net.

The general practise was for the pearls to be bought by the kilo, each kilo containing 12 strings. Sequins or paillets were supplied in 'pompoms', each consisting of 10 threads holding 1,000. These were more difficult to attach since, to present a smoothly glinting surface they had to be stitched on so that they faced in the same direction.

Such fabrics were expensive. The bead-embroiderer underwent a ten-year apprenticeship before acquiring the 'eyes in the ends of her fingers' essential for perfect workmanship. The slow care required by these exacting high fashion agglomerations might involve hundreds, even thousands, of hours, bearing no comparison with the lightening speed of straighforward tambouring [122].

Nevertheless, the hand-industry survived the fierce competition of the Cornely machine, which could also string beads, though with far less variety and luxuriance of effect. By the 1930s there was increasing competition from the Far East with its lower labour costs, but for those who must have the best, the extravagantly decorated Lunéville nets remained supreme.

The town of **Tulle**, in central France, was also famed for its hand-knotted linen filet: 'the good wives spend their entire lives in doing nothing else but produce these patternless bands ready for others to embroider in fine darning stitch'. The common machine-made embroidered cotton tulle which drugs the market is copied from it and is very pretty for window curtains and furniture, although its use on clothes is decidedly inartistic' [123].

The 'cotton tulle' referred to with such contempt (the Italians much preferred their own linen, hand-spun in the country areas) was not bobbinet - the fashionability of which was never in question - but the plain net from the Curtain machine, produced in the second half of the 19th century for domestic embroiderers to imitate filet designs without the labour of constructing the net. The transference of the town name to its products is just one of the many unfortunate terminological malapropisms which muddy the waters of lace language.

Bayeux and its connection with the embroidered silk *blondes* has already been mentioned (p.10).

Calais and Lyons both became in the 19th century very important centres for machine-lace production. A pin-machine for the manufacture of silk weft-knitted nets had been smuggled into Calais from England in 1786, and the embroidery of its products was soon centred in Lyons, already famous for its silk trade. By 1800 there were 2,000 pin machines in the city and the net was being exported to Nottingham to be run there.

These machines continued to be operative in France, with its easy access to silk, long after they had ceased in England, and in 1840 there were still 3,500 of them producing single press point net. Even as late as 1867 the product, then sometimes called tricot openwork silk net, continued to be made in large quantities [124].

On the other hand, France, less advantageously placed for cotton than for silk, was obliged for some years to import bobbinets from Nottingham for embroidery by hand, in the Calais area.

In 1815 parts of a bobbinet machine - at that time only a metre in width, and known as the Old Loughborough - were smuggled into France and set up by English workmen at Douai, near Valenciennes[125]. By the end of the year the machine was actively productive, and in 1816 a dress of French bobbinet, embroidered by Madame Ward, was presented to the Duchesse d'Angoulême, daughter of Marie Antoinette [126]. In 1821 there were 124 embroiderers of plain net around Calais, and the number of machines had increased to 38 [127].

France continued to lag behind England in its cotton-spinning. In 1819 the finest thread was a no.40. This was not an insurmountable problem as higher contraband counts flowed in easily enough, the smugglers charging only a 4% excess, and the French spinners then substituting their own trade marks before distribution to their customers [128].

In 1834 the English finally allowed cotton threads of counts above 170 into France, though at a duty of 35%, an exorbitant charge which stimulated the French to work more diligently at their own production. By 1836 they could spin 170 counts themselves, by 1840 a 210, and by 1850 a no.250[129]. A year later, Houldsworth of Manchester was exhibiting - for demonstration only - a 2,150 yarn, 'a single pound of which would reach the extraordinary length of 1,026 miles' [130].

In 1819 and again in 1827, bobbinets embroidered in silk and cotton, and intended for gowns and costume accessories, were exhibited at the Louvre, one of their attractions being that with almost no loss of beauty they cost only one-tenth as much as fully hand-made linen laces [131].

Bullet-hole nets were a first experiment in patterning the bobbinets. Slightly enlarged holes were created at regular intervals, on the machine, the holes being subsequently outlined by run-stitch. The technique was invented in 1823 by Samuel Ferguson at his factory in France.

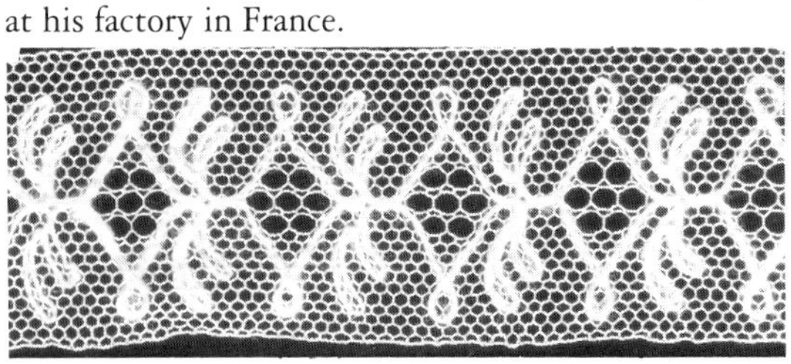

A bullet-hole net with run outline, imitating Lille.

Ferguson's 'wire ground net', similar to point de Paris, was also invented about this time, and was occasionally used for run embroidery [132].

Between 1837 and 1841 the Jacquard automatic patterning device was successfully applied to both the Pusher and the Leavers machines, and a secondary industry grew up - the addition of hand embellishments such as run outlines or three-dimensional highlights to machine-regulated designs.

By 1840 the number of embroiderers had increased to between 3,000 and 5,000 [133], and France could now produce its own two-twist bobbinets in silk and cotton, as well as some four-twist (square or diamond) nets, but it never manufactured the untraversed three-twist net [134].

Old habits died hard, however, and even in 1860, of the 150,000 kilos of cotton thread consumed in Calais, 60,000 were spun in Lille, 30,000 imported, and the remaining 60,000 smuggled [135].

The close connection, legal or illegal, between the Nottingham and French machinists during the early decades of the Calais lace industry is indicated by the frequent appearance of English names in the documents of that time.

From the 1860s, Calais concentrated on the production of patterned machine laces, and bobbinet manufacture was transferred to Lyons where mainly silk was used. At the same time, with the invasion of the industry by the Cornely machine, production was gradually switched from tamboured yardages for gowns and accessories to tamboured window curtains, and other soft furnishings, formerly a monopoly of the Swiss [136].

Tarare, just north of Lyons, was also an important centre for tamboured curtains.

Lace coifs at Quimper, Brittany. Birket Foster, 1878.

Coifs and sabots at Dinan, Brittany. Birket Foster, 1878.

Brittany coiffes were almost entirely composed of dramatic cut-out work, or of filet embroidered with satin-stitch, or of heavily patterned crochet, though both run embroideries on tulle, and cut cambric appliqués also appeared. These last straddled the borderline between Richelieu embroidery with its heavy bars, and cut fabric guipure sufficiently fine to qualify as lace. The motifs were grounded with delicate spider stitch, and the woven cambric was sometimes replaced by net.

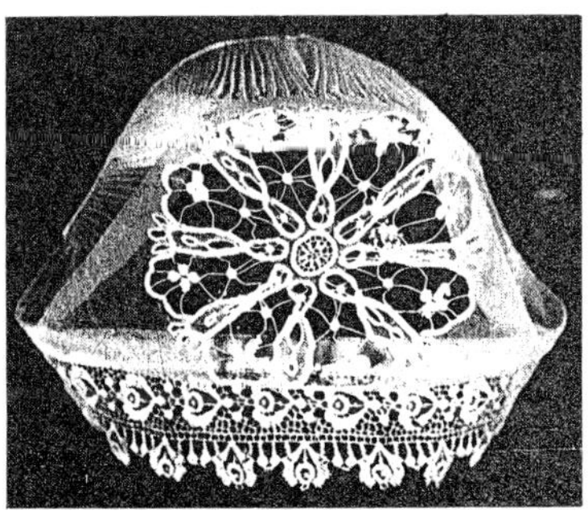

Typical cut fabric openwork from South Brittany.

During the 1930s the lace decorating some of the caps was embroidered by the Schiffli machine in two thicknesses of thread, a heavy outline and a light infilling. This was known as *dentelle de Bretonne*.

The name had already been applied to a Leavers machine lace manufactured by Hénon in 1878, though having no obvious connection with the geographical area.

In the *Journal des Ouvrages de Dames*, 1913, instructions were given for a collar in 'Breton embroidery on tulle'. The design was transferred or printed on to a fine tulle (known as Bretonne net), and the tulle then stitched to moleskin (a shaved cotton fustian, dyed dark) for contrast and support. The outline was run with a no.30 thread, then the fancy stitches added. Finally a thick cord was whipped over the entire contour, making a cordonnet.

Broderie Bretonne sur tulle.

Mme Hardouin in her *Album de Broderies sur Tulle* describes a heavy net worked almost exclusively in run-stitch (broderie au passé) using a thick soft thread locked into small geometric shapes such as stars, circles, diamonds, arcs and the simplest possible representation of flowers and fruits with no tendency towards those truly naturalistic forms which contributed so much to the charm of Irish work. The embroidery thread was sometimes doubled or even trebled, and the outline thread omitted, producing bold plainish pieces appropriate for church use. A similar technique on fine net, with the embroidery thread single instead of multiplied, produced a daintier lace suitable for lingerie, and referred to as *la dentelle dite Bretonne*.

Similar stitches illustrated by Cousine Claire, though slightly more adventurous, were still formal, lacking entirely Limerick's imaginative liveliness.

After Mme Hardouin.

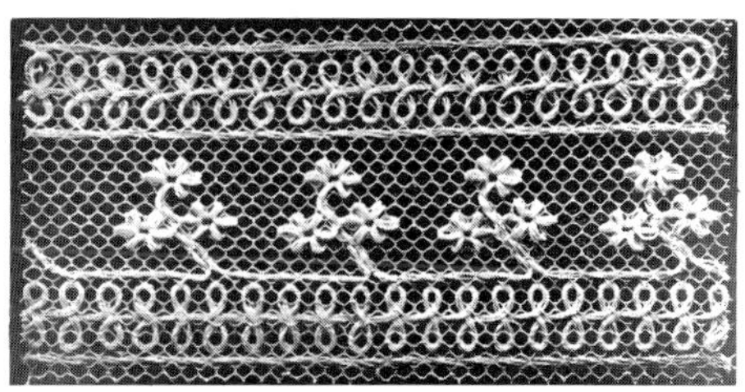

Margot lace was, like Bretonne lace, embroidered by the Schiffli machine. However the net was of silk, and the cotton embroidery thread of a single thickness throughout. Occasionally rayon was used, worked so thickly in parts that it produced raised shiny patches. Much of this work was done at **Tours**, where naively hand-run boudoir caps and 'Dutch' collars were also made.

Margot in rayon.

Other areas of France were also involved in the hand-embroidery of machine nets. In **Alençon**, embroidery on tulle was practised from 1820, enticing workers away from the slower and more exacting Alençon needle lace, and the no-longer-in-demand cutwork *(vélin)*. The workshop employed 200 children, and when in 1827 it was visited by Madame la Dauphine (= Princess Royal) they 'humbly offered her a superb bed cover embroidered on net, with the royal monogram in a crown of flowers' [137].

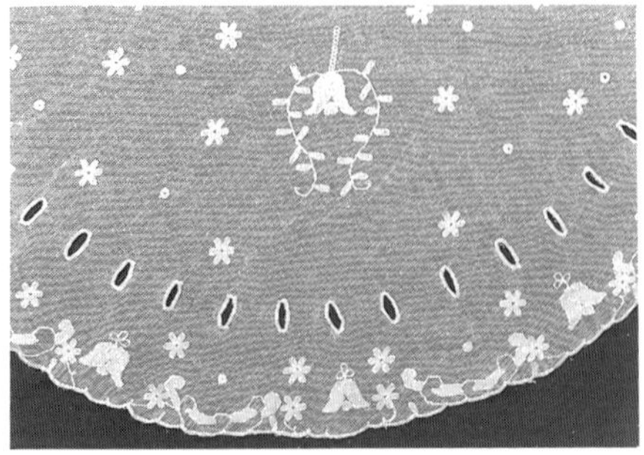

Above: Plain angular run embroidery, a boudoir cap slotted for ribbon, probably Tours.

A Bordeaux trade card showing a tambourer at work (Courtesy: Metropolitan, Gift of Bella C. Landauer).

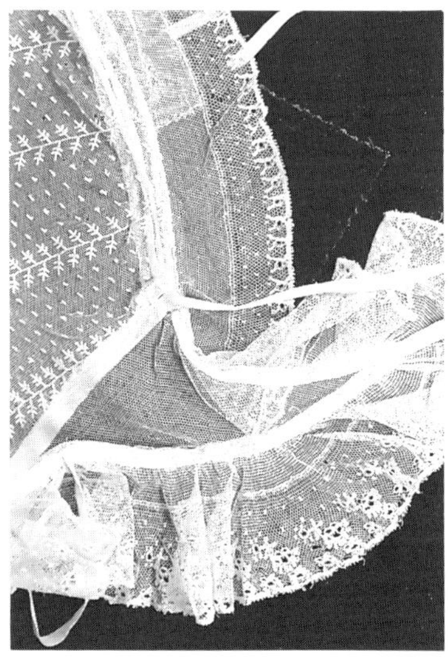

Right: a Normandy regional bonnet with characteristic fronds and dashes in simple run work.

Left: an Alençon design in solid tambour work. Polishing, as in the original needle lace, has indented every leaf and petal, increasing the resemblance.

The stitches and designs illustrated in the three French instructional books - Dillmont, Hardouin and Cousine Claire - were more concerned with encouraging domestic do-it-yourself enthusiasts than with presenting a historical record of professional techniques.

BELGIUM

The nine provinces which now make up Belgium were united as an independent kingdom in 1831. The capital city, Brussels, had already given its name to both bobbin and needle laces of such excellency that, for much of the 18th century, they dominated the royal and noble fashion market. It was a Brussels lace of hand-made motifs applied to a ground of hexagonal bobbin-made meshes (droschel) which reputedly inspired Lady O'Brien to experiment with the application of woven cambric onto machine net to produce cut cambric appliqué (see ch.4).

Machine nets are known to have been imported into Antwerp before 1815, through an Englishman, Mr William Wood, though their nature is uncertain. An example of black silk knitted net, decorated with chain-stitch motifs, was exhibited at the Provincial Museum of Antwerp in 1987. Its origin is unknown.

In 1817, following the lead of France, eight bobbinet machines were set up at Termonde, in the triangle enclosed by Brussels, Antwerp and Ghent. By 1828, twenty were working in Ghent itself [138]. As in France, the quality of the net was inferior to that of England, and the cotton thread was imported.

In 1831 a new technique, invented by Sewell, resulted in the production of untraversed 3-twist nets made in such fine English-spun cotton that they were scarcely visible. This 'Brussels net' surpassed in its filmy translucency the 4-twist traversed nets developed at much the same time. In 1831, a Mr F Washer had eight three-twist net machines built in Brussels, and during the succeeding decades their products were extensively used for the application of fine quality hand-made Brussels sprigs, displacing droschel.

Examples of Washer's net, shown at the Great Exhibition of 1851, were made of the finest cotton then available, nos.400 to 530. By 1867 Washer had 16 Brussels net machines, 8 Mechlin machines, and a further 4 Mechlin machines at St Josse ten-Voode[139].

Tambour

Using the three-twist net, a vast deluge of tambour work, marketed internationally under the trade name 'Brussels', cascaded from professional fingers in the form of stoles, flounces, veils, shawls, scarves, jackets and dresses 'for the attire of ladies of the highest circles'.

As ever, the geographical name - Brussels - did not necessarily indicate the main centre of production. The tambour industry, intended for high fashion and the export market, was in fact centred at Lierre, just north of Mechlin. Embroiderers are listed in the census records as early as 1817 [140], and it is possibly from this area that net-tambouring was taken to Coggeshall. But the whole tale of Mr Drago has an apocryphal air about it. He may have been French or Flemish, he may have emigrated in 1812 or 1820 or at any time in between, even in 1812 tambouring on muslin had already been established in Essex for at least two decades, and any bobbinet used in Flanders prior to 1817 would have been imported from England.

The designs appearing in 'Brussels tambour' lace consisted of formal scrolls, wreaths and flower heads caught to a ground so delicate that only a deluge of spots confirmed its existence. Fancy stitches were rare and consisted more of varied shapes created by the threads than of any organised chain-stitch formations. Artistic contrasts of heavier outer and lighter inner embroidery sometimes provided all the variety that the design required.

The basic method of working was similar to that already described for Limerick tambour, with minor variations. The net was supported on a wooden frame and the pattern copied on to it by means of a brush dipped in a grey oily substance that was equally visible on black or white net. A white or green cloth was placed beneath the work to make the outline, thread and stitches show up more easily. On the left side, one or more spools of embroidery thread were mounted on nails so that they could turn freely. Sitting in front of the frame, the worker attached the thread to the net at one end of the design, and while the left hand controlled the thread and held it against the net, the right pulled it across the meshes with a hook, following the grey outline. For large pieces, as soon as the net was full, the completed part was rolled up and the pattern traced on again. The embroidery thread was usually the same colour as the net, but tinted silks, gold, silver or steel threads of varying thickness were also used [141].

As in Lunéville, the hand industry suffered from the higher speed and lower cost of Cornely machine productions. But the work lacked durability and the articles - sometimes sold as hand-made - 'could not survive three washings' [142]. Some laces bridged the gap by having the design outlined by machine, and the decorative fillings entered by hand.

On the Cornely, the net was backed by paper to reduce its elasticity, and mounted in a frame which was moved beneath the needle as it flashed up and down making the stitches. Occasionally on the reverse side of a large tamboured piece the tiniest fragments of grey cloth are found caught in the stitches, suggesting that a woven fabric had been used instead of paper for support. This must have been pre-treated in such a way as to make it easily degradable, by a blast of hot air for example, or some other of the many techniques used to create the Schiffli-machine guipures.

Details from a Cornely-embroidered tunic c1922 when waists dropped into insignificance. The grotesque animals, and the foliage like hands with spread fingers, make a fascinating design (Courtesy: Avril Heron).

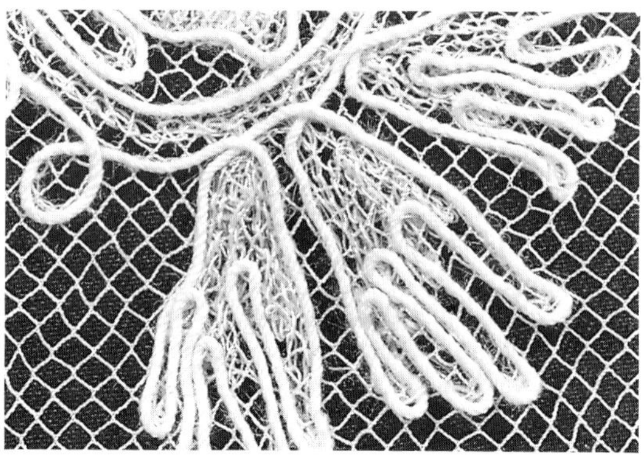

The upper (face) side shows back-stitching within a prominent cord attached to the net by invisible chain-stitching (on the reverse side). The stitches are made as described on p.117, the cord fed from above in the same way as beads.

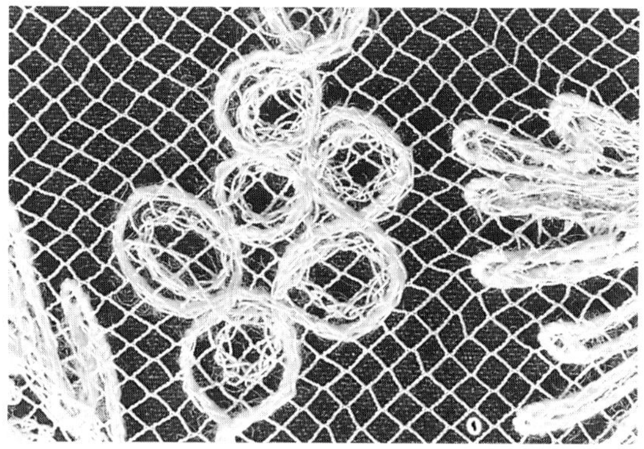

The reverse side shows chain-stitching beneath the cord.

The tamboured attachment of glass beads and sequins was introduced from Lunéville about 1890, replacing the formerly fashionable jet. As described on p.116, the beads appear on the upper future face side, and the chain-stitches on the under future reverse side.

At the same time, a down-market hand-tamboured industry, cottage style as opposed to truly professional, flourished along the border with the Netherlands, and was aimed at the wives of rich farmers. Its early history is obscure, but it is known to have been active at the beginning of the 20th century when, in 1902, there were 1,100 workers in Lierre and 45 in Antwerp. A further centre at Maldegem, near the city of Bruges, was established about 1900, and by 1902 had 30 workers. A stout two-twist bobbinet was used instead of the delicate, impractical and far too expensive three-twist. [143].

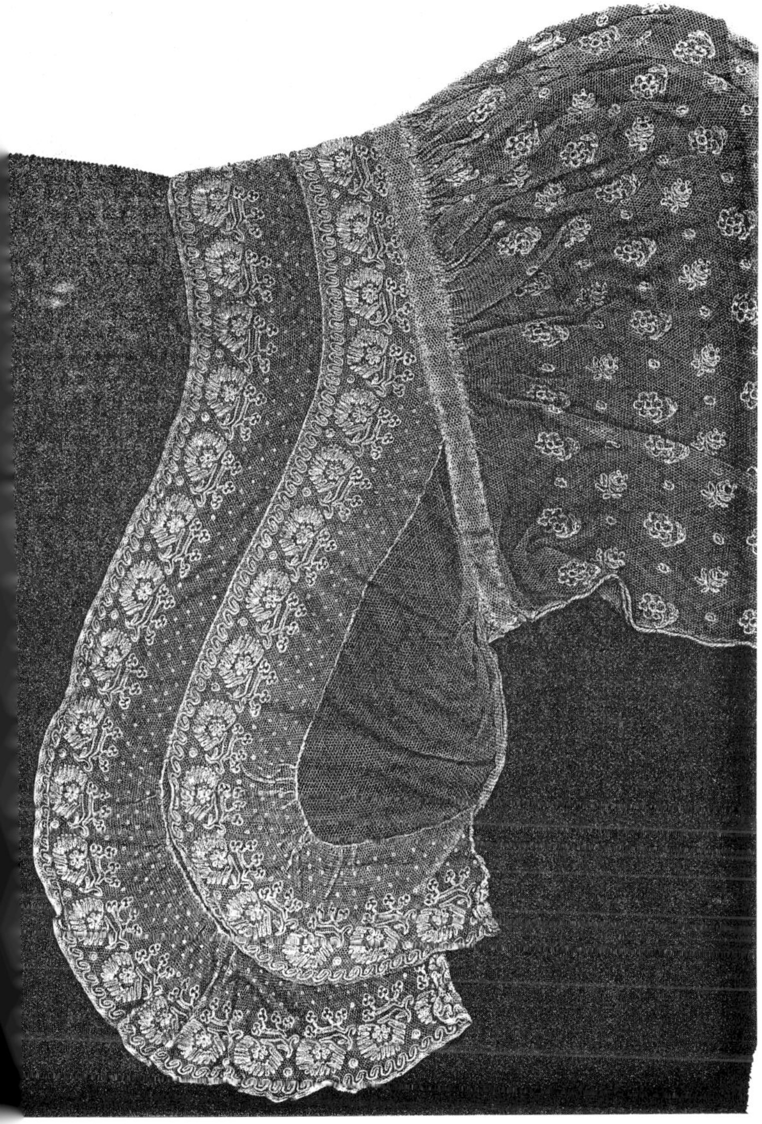

Left: a neatly tamboured border imitates a simple bobbin lace design.

Above: back of bonnet type (b).

Back of bonnet type (c) (after Verhaegen).

Run embroidery was located over a wideish area of West Brabant near the Netherlands border, an area which included the old lace-making centres of Beveren, Antwerp and Ghent. The industry was provincial in nature and intended for the Dutch markets of North and South Holland and North Brabant. The history of the industry is not well documented, though the general organisation appears to have been much the same as for cottage industries in any other lace-making area, with a merchant distributing materials, and collecting the completed work. One such merchant in East Flanders employed 300 net embroiderers as well as a large number of bobbin lacemakers. Run work was considered to be slower and more tiring to the eyes than tambour [144].

Since the customers were in the main farmers' wives, demand might be expected to vary with agricultural prosperity. This seems to have reached a peak between 1900 and 1920 though, according to Verhaegen, there were only some 400 old ladies still actively at work around Lierre, in 1902 (p.145).

Four distinct forms of run embroidery can be distinguished, the setting-up of each being basically the same as for Limerick run. The pieces to be embroidered were on the whole small, and the work was hand-held, supported on grey or brown paper, no frame being used [146].

(a) Imitations of the much coveted bobbin laces of Beveren, which resembled a rich Lille, provided substitutes along the bonnet borders. Since the original was worked with a half-stitch ground, identical to two-twist bobbinet, the copies could be gratifyingly deceptive. Such sturdy decorations might be used on the working caps with which the women protected their hair from dust and their heads from cold as they laboured in the chilly dawn.

(b) Beautiful 'Sunday' bonnets said to derive from the Lierre district, and referred to as flat-work, had long ear-pieces hanging forward over the cheek-bones and a frill shadowing the forehead. Loops along the neckline allowed the net to be tightened by cords. The decoration was worked in fine thread on a square of net, destined to form the balloon-like crown. Central flowers were outlined with spots and flanked by formalised leaves. The design flaunted its own clearly visible heirarchy of cost and prestige, a single complex flower carrying a higher status value than three more simple [147].

Similar embroidery was done, on a small scale, in the Netherlands itself, for example at Eindhoven, but the industry ended with the second world war.

(c) Cap backs covered with roundels of flowers - 25 to 145 per bonnet, and varying in size from 2 to 6.5 cm - were produced near Antwerp [148]. According to Verhaegen this work was also done at Lierre, though it was more substantial altogether than the flat-work described above. The motifs were outlined in thickish thread and decoratively filled with thinner, but again the stitches were restricted almost entirely to run. There was no counting of meshes, and an experienced 'runner' could work the flowers from memory, though if necessary a drawn pattern could be tacked beneath the net. The direction of working in relation to the cross-sides remained important. Such a freedom of approach is possible only with relatively simple thread movements, and could never be applied to Limerick's precision-worked trellises where symmetry is essential. Indeed the emphasis placed on accurate counting of holes is painfully illustrated in the instructions for a domestic embroidery, set out in *The Young Ladies' Journal* of 1871: 'Each diamond has 44 holes of the net on the straight side . . the stripes between the diamonds are 13 holes of the net in width . . ' [149].

(d) Another form, known by the broad term 'plumetis', was a needle embroidery, dependent on satin-stitch formations rather than on plain run-work. As with Lierre's tambour laces (see p.173), the decorative stitches of plumetis, though clearly defined, were more in the nature of a rearrangement of normal run stitches in circles, stars etc than creative jugglings such as those described for Limerick. They were known by rustic names such as fat paws, double rosebuds, little knots, triangles, coffee beans, small heart, star-shaped blooms and windmill [150]. They differ markedly from the French 'stitches' of Madame Hardouin.

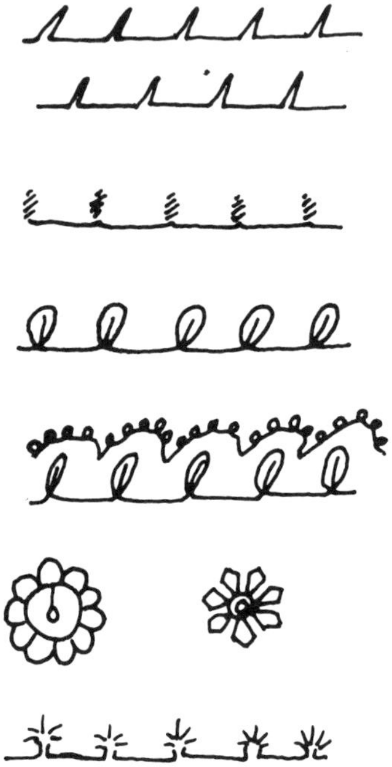

'Stitches' or thread-motifs identified by Trees van Ierssel-Ruygers in a bonnet of type (a).
Below: plumetis-type embroidery on a baby cap. Origin uncertain.

Appliqué

Cut cambric appliqué, outlined with chain-stitch, also goes by the name Brussels, though it is difficult to distinguish from similar work produced in Switzerland (see p.183).

Plumetis was sometimes combined with small areas of muslin appliqué, and Lierre was an important centre for this work from about 1880 [151]. Some net-on-net appliqué similar in design and technique to that associated with Limerick was produced there in the second half of the 19th century [152].

The motifs of tambour laces were sometimes enlivened by raised (appliqué) petals, or by pieces of cloth stitched to the reverse side of the worked flowers, i.e. with the net above and the fabric below[153].

A further form of appliqué, using shaped tapes (inaccurately called braids) developed around 1900. Though its similarity to Limerick appliqué is distinctly marginal, nevertheless within its design areas 'light and dainty darning stitches' were worked 'to relieve the bareness and monotony of the plain net'.

A two-twist (round) bobbinet was used with the advice that since 'the [square] nets for Limerick and Carrickmacross laces were of not quite the same form of mesh, it would not be possible to employ quite all the same stitches' [154].

Belgian Princess appliqué lace, c1906.

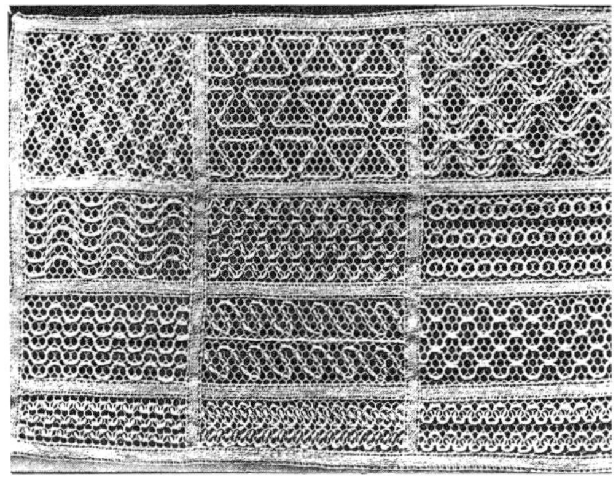

A selection of run stitches used in Princess appliqué lace (after *Needlecraft*, no.59).

The technique was called Princess, or Princess Louise, lace at the beginning of the 20th century when its market was not only among domestic embroiderers but extended to middle-class fashion and international trade. The variations of running stitch used for the fillings were fairly mundane.

NETHERLANDS

Considering the popularity and wide range of embroidered bonnets and other accessories found amongst farming communities in the Netherlands, it may seem surprising that nearly all of it was imported from Belgium.

Between the 1850s and 1930s a number of lace schools were set up, their initial aim being to produce marketable products, but they failed to develop the necessary commercial efficiency, and achieved no international recognition.

In the last quarter of the nineteenth century, training in design and a Diploma in Fine Needlework were introduced, and the curriculum extended to include embroidery on tulle. Inspiration was drawn from a variety of sources, and little was produced which could be identified as indisputably Dutch.

In an Exhibition at Rotterdam in September 1992, two run-stitch samplers, dating from around 1900, had respectively 33 and 36 stitch variations, worked in fine thread, and indistinguishable from the Irish. A further sampler contained 48 stitches, while a strip of run lace showed motifs which could equally well be from Limerick.

A sampler of run stitches, worked by Geesje Smid, c 1890-1900.

Some very effective coarse run was made in the French style, such as that described in Cousine Claire, using shapes as simple as a folded ribbon where overlapping run stitches, worked diagonally, crossed each other at the folds to give the darker effect of darning.

An example of cut cambric appliqué c1900, also in the Exhibition, was made of crisply woven fabric attached to a two-twist bobbinet by chain-stitching and garnished with ropes of satin-stitch pearls, the border emphasized by thick buttonholing. Smaller openings within the motifs were unfilled, the plain net thus exposed lightening what might otherwise have been a rather heavy effect.

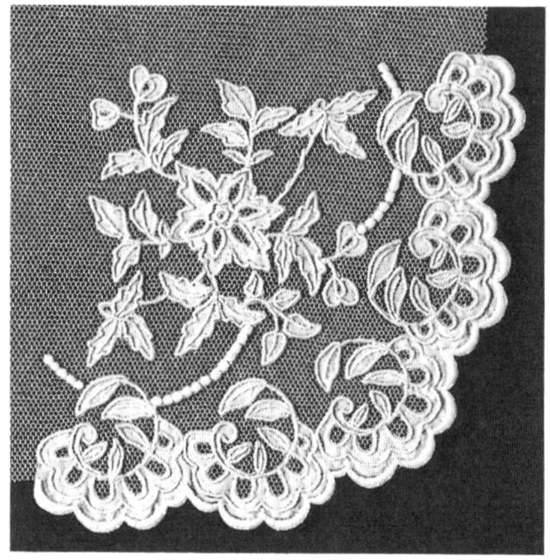

Cut cambric appliqué by Geesje Smid, c1890-1900.

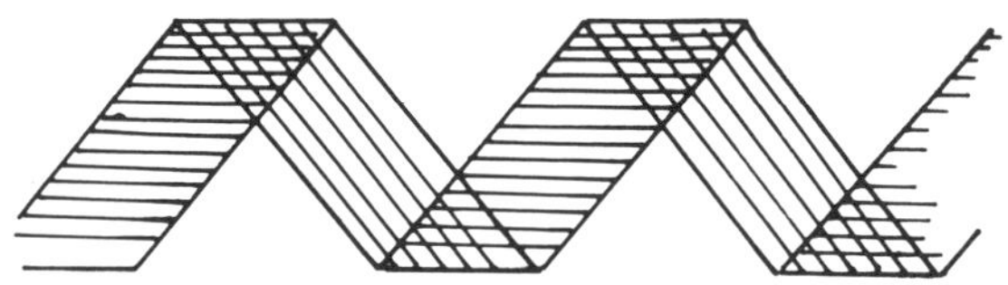

(All the photographs on this page by Courtesy of the Rijksmuseum, Amsterdam).

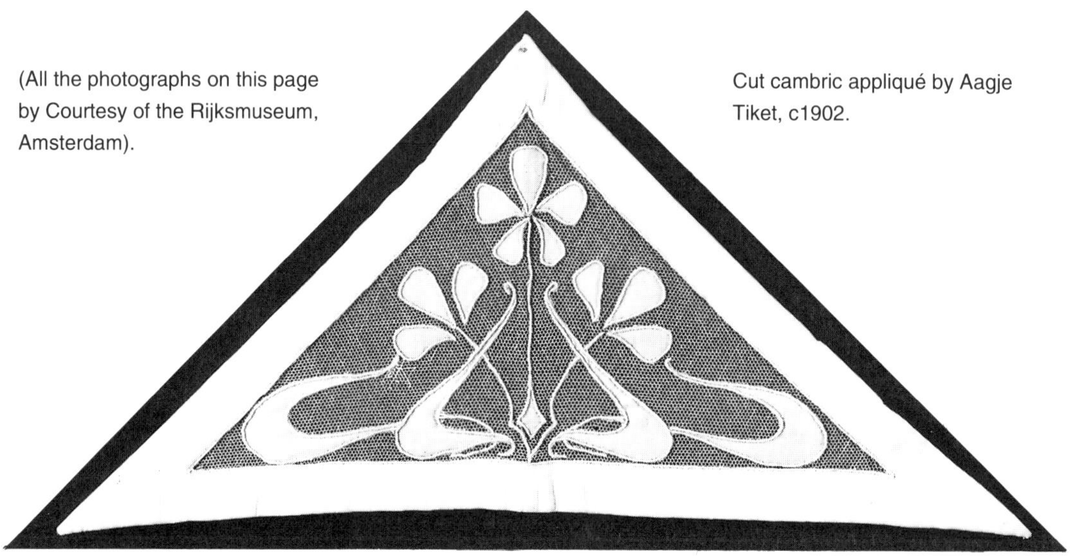

Cut cambric appliqué by Aagje Tiket, c1902.

A guipure example showed the closely couched cord, which bound the edge of the cambric, coiled into a multitude of twirls not only along the outer borders but also around all the outlines of the inner motifs.

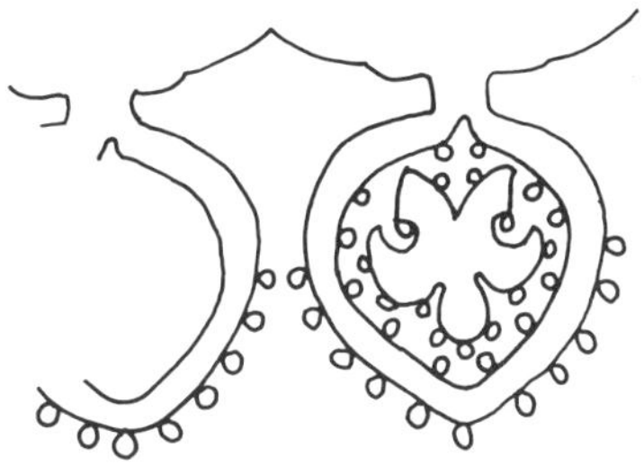

A second appliqué had a couched cord untrammeled by any excrescences, its plain lines and absence of casket stitches clarifying the purity of its art nouveau form.

A third, with the outline tamboured, had caskets of run fillings worked on the net, and in other areas even the net cut away, the holes then being filled with needlelace stitches.

Taken together, these examples incorporate elements of Swiss and Hungarian, as well as Irish, laces.

Though pure tambour work was not in evidence, chain-stitching was obviously practised, and tambour work with run fillings was included in the curriculum of an Amsterdam school in 1888-9 and again in 1916-17 [155].

In the early 20th century, small home industries in embroidered net, for provincial bonnets, existed at Eindhoven in North Brabant, and at several places in Zeeland [156].

The crown of a baby cap, fabric appliqué attached to net with stout buttonhole-stitching. Origin uncertain.

A simple run embroidery by Geesje Smid, c1890-1900 (Courtesy Rijksmuseum).

DENMARK

The main production was of bobbin laces which, in a predominantly agricultural community, formed an important prestige symbol on provincial bonnets. Named after the town of Tønder, these laces originated from the west Frisian area, being distributed from there throughout Denmark, though they were expensive for country women to buy. There were 12,000 lacemakers in 1812 [157].

Like the Beveren and Lille laces of Belgium, these bobbin laces had a Bucks point or half-stitch ground, identical to the two-twist bobbinet, which was introduced from England about 1820. Run embroideries became centred in the eastern parts of Schleswig, such as Als and Sønderborg [158]. Their designs copied those of Tønder, and cost only a fraction of the price. Particular favourites were stringently stylised representations of strawberries, carnations, oak leaves and marguerites. The more southerly part of Schleswig was annexed to Prussia in 1866.

This cottage industry functioned in much the same way as any other, coordinated by travelling merchants who solicited and then placed the orders, paid the women and marketed the goods. Some of their books of samples survive showing dentate borders, and a fairly sparing and inconspicuous use of decorative run stitches, though variations of cobweb-, cross-, tent-, and satin-stitch rosettes or stars do appear. The industry ended soon after 1880.

It appears to have been replaced, to some extent, by domestically orientated embroideries where the instructions and illustrations are amazingly similar to the geometric formations of France. Worked in relatively heavy floss linen thread, these coarser embroideries were used for headwear and kerchiefs, as well as for the decoration of household linens.

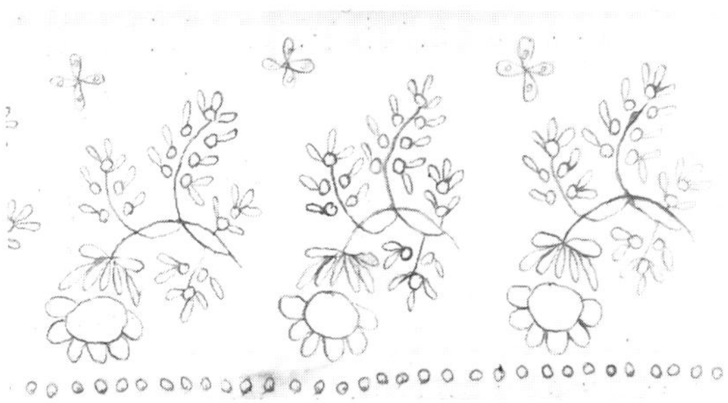

Top: a Tønder bobbin lace below, a copy of the design by run embroidery, above.
Middle: an inked pattern for run work.
Below: support, pattern and net set up for working. Note the cloth folded over protectively along the top and the roll of net on the right.
(These three photographs courtesy of the Museum of Sønderborg Slot).

Samples from a merchant's pattern book (Courtesy: Annette Kappel).

Tambour embroidery on net was not practised in Denmark, but a few examples of cut cambric appliqué have been found, with raised satin-stitch crescents decorating the petal and leaf borders as described in the practical instruction books of the late 19th century [159]. It is hard to tell whether they are connected in any real way with a Danish tradition.

An example of tambour embroidery on black silk weft-knitted net survives in the collection of the Kunstindustrimuseet in Copenhagen, but its geographical origin is unknown.

Above: a sampler of Danish stitch motifs (Courtesy: Museum of Sønderborg Slot).

A voil in run embroidery (Museum of Sønderborg Slot).

Below: Echoes of this French style of run work appear in some Norwegian examples (after Mme Hardouin, p.40).

NORWAY and SWEDEN

Examples of French-style run embroideries are known from the south-east of Norway near the border with Sweden, and from Rogaland in the extreme south-west [160]. An example from the Nordiska Museum in Stockholm shows the typical encicling spears of blossom so characteristic of Limerick laces, while collections of samples thought to be from Bergen have run-stitch formations almost identical with those from Denmark. There does not appear to have been any organised industry.

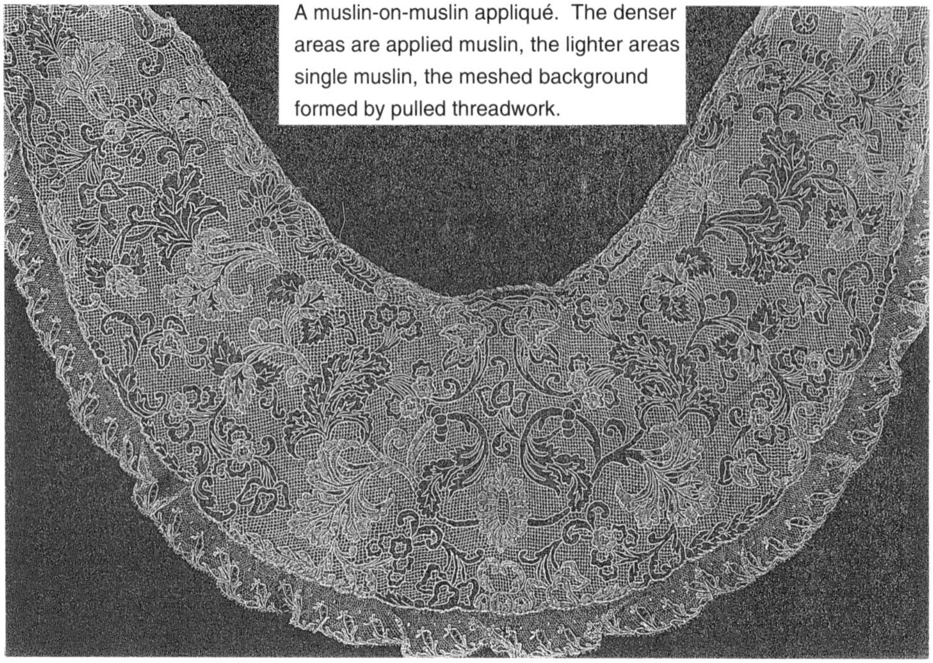

A muslin-on-muslin appliqué. The denser areas are applied muslin, the lighter areas single muslin, the meshed background formed by pulled threadwork.

SWITZERLAND

Fine quality hand embroideries on muslin were produced in the Appenzell and St Gallen areas of north-eastern Switzerland from the mid-18th century. They included muslin-on-muslin appliqué in which the motifs lay in an open ground formed by the repeated separation of minute strands of warp and weft in a regular manner, to give the appearance of meshes. Some tamboured muslin was also worked.

Though this was for the most part a cottage industry, the products were luxury articles, for commercial rather than domestic use. By 1794 there were 40,000 embroiderers. Satin-stitch was introduced about 1800.

Thus the historical precedents of tulle embroidery were similar to those of Limerick, Hamilton and Coggeshall.

Although there is no record of modified stocking frames producing silk knitted nets in Switzerland, they were active in neighbouring Austria, and one embroidery on warp-knitted net is in the collection of the Textilmuseum, St Gallen.

The first bobbinet machines to arrive in Switzerland were brought by an Englishman called Read. Their heavy bodies had been made in Calais, and their delicate insides by the firm of Hilton, in Nottingham. From this time on, bobbinet began to replace muslin as a base for embroidery. By 1856 the number of machines had increased to 80 [161].

Below: tambour embroidery on warp-knitted net (Courtesy Textilmuseum, St Gallen) (photo: Marianne Gachter-Weber).

A curtain sample, mixed hand- and machine-embroidery (Courtesy Textilbibliothek, St Gallen).

During the 1820s and 30s, Swiss tamboured shawls, pelerines, collars and neckerchiefs were extensively exported to North America. At the same time, the famous Swiss curtains began to be manufactured, incorporating not only tambour work on net but cut cambric appliqué with the layers attached by chain-stitching, and occasional run decoration [162].

Twirls sometimes clung along the border, also attached by chain-stitch. They could not, obviously, be worked in the Limerick style as extensions of the outline cord, since there was none. At times they were made of contrasting thread. In the richest curtains, crowns of picots encrusted the motifs, like the ebullient outbursts of 17th-century Venetian gros points.

The *Art Journal Illustrated Catalogue of the Industry of All Nations, 1851*, though of the opinion that 'In lace and embroidery, and analogous work, the Exhibition makes no such display as to demand any detailed examination', yet admitted that 'Several good specimens of curtains were exhibited by J J Sutter of Buhler, in Switzerland' (p.XX***).

'Swiss' lace, 1852.

In *The Illustrated Exhibitor* of 1852, instructions were given for making a 'deep border for a veil, in Swiss lace'. The article continues, 'Swiss lace is (as many of our readers are aware) the name which has been given ever since the Great Exhibition, to muslin [cambric] appliqué on net'. The materials required were: 'Brussels net, and French cambric or the finest widow's lawn'. In spite of the name Swiss, the attachment of fabric to net in these instructions was not by chain-stitch but by a couched cord, for which an embroidery cotton no.10 was recommended, held down by a no.120 couching thread. Both fabric and net were cut away in the areas to be decorated, and the holes filled with 'point lace' stitches (that is the decorative stitches found in needle laces) worked in cottons of varying thickness - nos.70, 100 or 160.

To make a strong edge, two strands of thread were run through the net, following the shape of the border, then neatly buttonholed over. The superfluous fabric and net were cut away, and a 'very good pearl-edging' added [163].

Neither the buttonholed border nor the point-lace stitches appear in Limerick laces.

Apart from bobbinet, the thin wiry Mechlin net, made entirely of warps so that there were no cross sides, was also used for tambour embroidery. In this net, the sides of the meshes are twisted three times to make a 4-sided hole, or five times to make a 6-sided hole. This is done so tightly, and the net so stiffly 'dressed' that the mesh shape is retained throughout the embroidery process.

A fine tambour embroidery in silk on Mechlin net (Courtesy Textilmuseum, St Gallen) (photo: Marianne Gachter-Weber).

Design in general had received some harsh comments in the 1851 *Art Journal* Report:
'There is nothing new in the Exhibition in ornamental design; not a scheme, not a detail that has not been treated over and over again in ages that are gone' (p.V***).
'The prevailing gorgeous [in the sense of engorged, swollen, bloated] taste of the present day, which gives the eye no resting place . . . ' (p.V***);
'one curious picture presented to the mind by this assemblage of the world's industry is the wide-spread influence of France, in spite of the most debased taste in design ever tolerated' (p.VI***).
'as yet it may safely be assumed that high class design has been little identified with the [present day] manufacture of lace in any country' (XX***).

Some personal bias may have been operative here, but in any case, there is seldom any direct connection between popularity and good taste, whether of design, presentation or whatever. Certainly, although no school of design was set up in St Gallen until 1866, the popularity of Swiss curtains was immense. By 1860 the embroidery of 'white lace net' amounted to £48,000 a year [164], the actual embroiderers being, as ever, poorly rewarded.

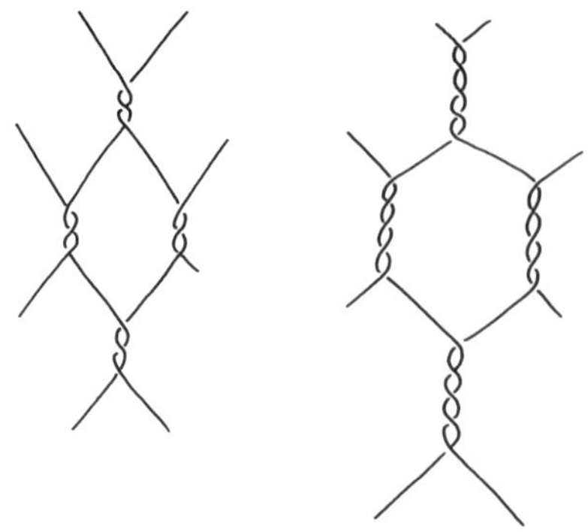

Mechlin net had no bobbin threads, and was made on a 'warp machine', from warp threads only. Two twists gave a diamond-shaped mesh. As the number of twists increased the holes became more hexagonal (round).

But portents of doom were already on the horizon. From the Lace Curtain and Madras machines, active in both Nottingham and Ayrshire from the late 1840s, came richly ornamented curtains rolling off at a speed of two yards an hour, their novelty and comparative cheapness threatening the formerly secure, undoubtedly superior, but more labour intensive products of Switzerland.

At the same time, French curtains of darned filet held their place, while France combined with Germany and Italy to exclude Swiss goods by imposing high import duties.

In addition, the ingenuity of the Swiss in the development and use of embroidery machines was soon to damage their own hand-industry.

In 1828, Josue Heilmann invented the handmachine, thus making possible excellent copies of run embroidery, though using thickish thread. The handmachine could also imitate fabric appliqué, using whip-stitch to join fabric and net together. Though the general impression was passable, this was, overall, an amateurish and fugitive production.

Cut fabric work made on the handmachine.

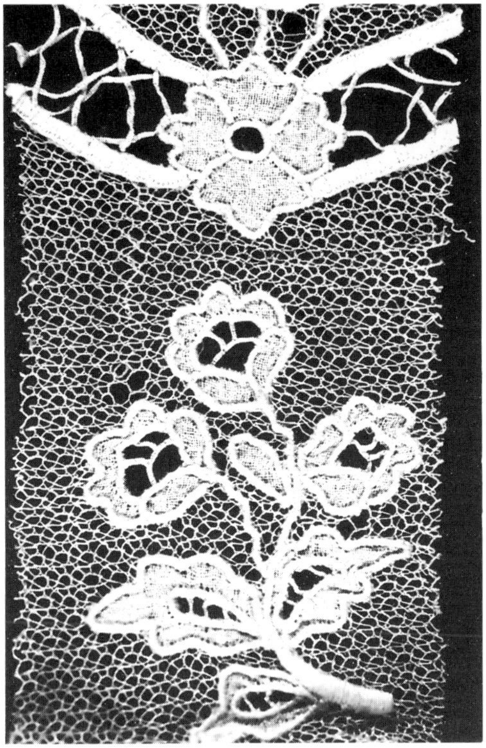

Cloth appliqué with a whipped outline (no cord) on craquelé (crackly) net (both photographs Courtesy Textilmuseum, St Gallen) (photos: Marianne Gachter-Weber).

A greater threat to Swiss hand-work was the invention of the chain-stitch machines. Experiments with this technique had begun in the early 19th century.

The disadvantages of such machines, in contrast to the virtues of Wheeler and Wilson's 'Silent-Working Sewing Machines' is extolled by an advertisement in *The Englishwoman's Domestic Magazine* of May 1, 1869 (No.109, vol.6). The ease of unravelling of both the single- and double-thread chain-stitches was awesomely illustrated, while the security, flatness and versatility (it could be used for gathering) of the lockstitch was loudly proclaimed.

But these were sewing, not embroidering, machines. It was the invention, by Bonnaz of France in 1865, of a machine in which a hooked needle was poised above the fabric as in hand-tambour work, and the thread fed from below using a looper to pull it forward to the required position for the hook to pick up, which both posed a threat to hand-work and provided at the same time serious possibilities for novel developments in the Swiss embroidery industry. The chain-stitches appeared on the upper surface of the net where their effect could immediately be assessed.

Problems of the hooked needle snagging the threads as it passed rapidly in and out of the meshes, led to the substitution, for the hook, of an eye-threaded needle above, the hook then being placed below. This machine, marketed in Paris by Cornely, and known by that name, could reproduce chain-stitch, fluderstitch (loose chain-stitching) and moss-stitch, as well as apply cord or braid to the surface (top braiding) using a universal feed metal reel. The chain-stitches now appeared on the underside and the cord or other decorations such as beads, bugles and paillets on the upper.

Thicker or thinner threads could be used by changing the needle and hook size, so imitating the characteristic heavier outlines and finer fillings of hand tambour. The Cornely's productions of fabric and net appliqués outlined with chain- stitch were slightly coarse, as befitted soft furnishings. In the less expensive forms, the hand-cutting of the excess fabric was not always well done.

Cloth appliqué with a chain-stitch outline, on bobbinet.

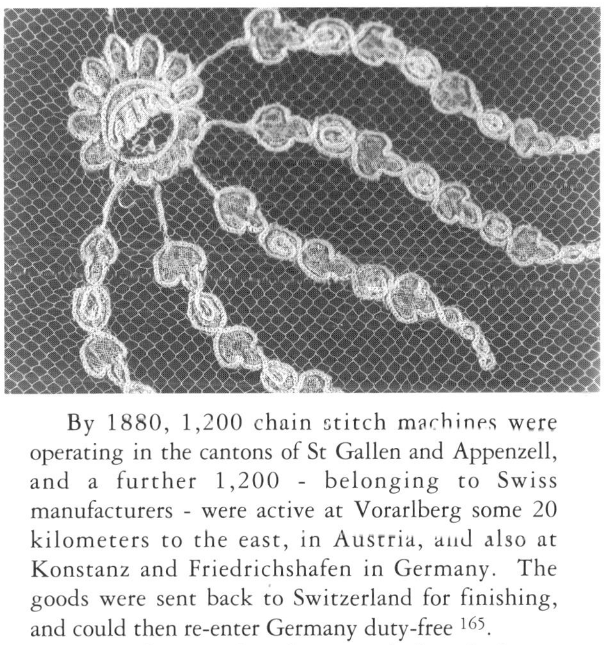

By 1880, 1,200 chain stitch machines were operating in the cantons of St Gallen and Appenzell, and a further 1,200 - belonging to Swiss manufacturers - were active at Vorarlberg some 20 kilometers to the east, in Austria, and also at Konstanz and Friedrichshafen in Germany. The goods were sent back to Switzerland for finishing, and could then re-enter Germany duty-free [165].

Many of the hand workers switched to the better paid machine embroidery, but the most excellent curtains still required extensive hand finishing and it was not unusual for the large round tambour frames to share house space with the small Cornely. While the outworkers functioned mainly in isolation, the most intensive and highest quality pieces required several women working together to achieve uniformity of effect.

Fabric appliqué, attachment and decorative stitches by the Schiffli machine. Possibly Plauen (Courtesy: Waltraud Schönebeck).

Quantities of hand-embroiderers continued to be employed in trimming back loose threads, cutting away surplus fabrics, attaching layers of petals to flat flowers, and inserting Schiffli motifs into, rather than applying them on to, the net - work which was less creative but also a good deal less demanding.

Efficient and versatile as the Cornely was, a new Swiss machine shortly appeared which was to pose a threat to hand-, machine- and net-embroidery alike. Invented in the 1860s, the Schiffli was to become fast, steam-driven, huge and eventually computerised. Though its basic blatt stitch, produced by the interconnection of needle and shuttle threads, resembled nothing made by hand, its facility for imitating designs convincingly and for attaching fabric securely to net, made it a formidable competitor.

It is hardly surprising that with such skilful and relatively inexpensive competition, hand-embroidered Swiss curtains - though remaining popular almost to the end of the century - did not progress. Leopold Iklé in his report on the 1883 Exhibition in Zurich complained that chain-stitch curtains did not advance with the times, the general picture was the same as at all the earlier exhibitions 'with few exceptions, we find innovations neither in the materials used nor in the techniques. The patterns remain traditional'. Experiments with Gothic representations of angels, symbolic figures and mythical beasts - popular in filets - did not work well with chain-stitching, and failed to attract buyers. Iklé concluded, 'The prospect for the chain-stitch industry is indeed not at all rosy' [166].

The Schiffli embroideries on two-twist bobbinet flourished first at Plauen in Saxony, just west of the Erzegeberg Mountains and close to the border between Germany and Slovakia. This was a historic area for hand embroideries, those of Dresden being particularly famous, but their skills had for some time lain dormant.

It was a further development in the Schiffli production, the invention during the 1880s of a chemical process to degrade the base fabric once the embroidery was completed, which particularly caught the imagination of the St Gallen manufacturers. Since there was no net, the removal by aetzing, or corrosion, of the fabric on which the embroidery had been done, left the entire lace consisting of nothing more than embroidery stitches. Only bars held the motifs together, earning the products the name of guipures (Swiss embroidered guipure laces). In the trade, the Plauen stitchings on net are classed as 'embroideries' and the St Gallen guipures as 'laces', though the techniques were not exclusive to either area.

After 1900, with changes in architectural fashions, reduction in window size, and gradual simplification of life style, especially after the first world war, interest in Swiss curtains came to an end.

Swiss and Limerick laces seem to have developed independently, their influence upon each other being minimal. Although 'pantograph machines' (handmachines) accompanied by Swiss workers were imported into Ulster at the beginning of the 20th century [167], they were most probably for white embroidery, a traditional industry of that area, rather than for any decoration of machine net. One link with Ireland is to be found in the non-Limerick three-dimensional Irish crochet, high fashion in Paris around the turn of the century. It was one of the first designs to be imitated by the Schiffli chemical lace technique.

In 1912, *The Lace and Embroidery Review* reported: 'The only new lace, or rather old lace, brought into use again, is the Carrickmacross. The real is so very high-priced that it can never be exceedingly popular, but St Gall and Plauen are trying to bring out some very beautiful workmanship in this article'. [168]. 'Carrickmacross' is here used as a generic term for cut cambric appliqué and guipure, and the reference is to machine-made imitations.

ITALY

In the 19th century, Italy had a number of point net machines, producing knitted nets [169]. What was done with these nets is uncertain, but there are records of run and appliqué work on round-meshed bobbinets, and of the importation of thousands of kilos of cotton net at the beginning of the 20th century.

The *Nuova Enciclopedia dei Lavori Femminili*, published probably in the 1920s, contains instructions for run embroidery in heavy thread which differ in no significant way from similar handbooks from France, Denmark or Germany. The simple attractive infillings are not, with a few exceptions, similar to those of Limerick.

In one of the designs, enlarged holes are encircled with thick overcasting, making them more conspicuously eye-like. Solid areas packed very closely with run-stitch filling every mesh, give dramatic emphasis, while petals lightly shaded with diagonal darning provide a pleasing diversion. The run border is sometimes strengthened with firm buttonhole-stitching, sometimes by a neat diagonal running.

As far as it is possible to glean any distinguishing characteristics from the designs and stitches shown in the encyclopaedia, these are: the frequent use of the same thread for both outlines and infillings, and the occasional omission of the outline where the stitching is particularly dense. One very effective design, of a row of angels for an altar cloth, consists almost entirely of just an outline, with a minimum of run stitches used to stripe their robes.

Two techniques of cut fabric appliqué on bobbinet are also illustrated. In one, cloth and net are attached by whipping over a close row of small tacking stitches instead of a cord; in the other the fabric is cut little by little all around the motifs, a few millimetres outside the actual border. This margin is then turned under and overcast - a type of work found also in Hungary.

The two techniques of attaching cloth to net: above, cut edges turned under; below, overcasting with thick thread.

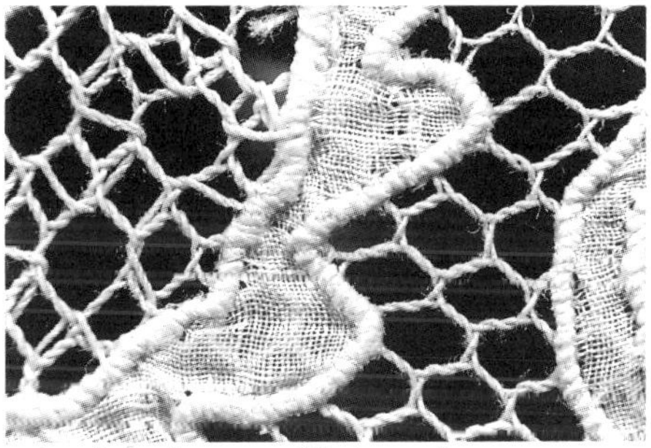

Insets of net into cut fabric, and even a geometric art deco style net-on-net appliqué, are again stitched together by whipping over an outline of run stitches. For the edge of the lace, two additional strands of thread are laid on top of the short tacking stitches and all three buttonholed over to make a very substantial rim.

As with all instructions intended for domestic embroiderers it is difficult to know how far the techniques and stitches described represent traditional work in that area.

Somewhat different in character is a cotta c1849 recorded from Sicily. It is said to have belonged to Cardinal Giuseppe Guarino and may have been made near the seminary in Agrigento prior to his ordination in 1849. The simple design and cottage-style execution are contra-indications of professional work [170].

A sampler from Turin contains over 50 run-stitch variations, many of them similar to those illustrated in the *Enciclopedia*.

A Sicilian cotta, run work (Courtesy Doretta Davanzo-Poli).

AUSTRIA

Modified stocking frames known as pin-machines, for making weft-knitted nets, were introduced into Austria from France in the early 19th century, and continued to operate for the next fifty years [171]. It might be assumed that they were used as a base for hand embroidery, as elsewhere.

Bobbinet machines followed and there were already 60 of the small straight-bolt variety in 1814. By 1856 the number had grown to 500. Later, three of the improved and much larger 'circular' bobbinet machines were imported from Calais [172].

In spite of references to the hand embroidering of net in Vienna in 1873, only ten years later Iklé recorded 1,200 Cornely machines operating at Vorarlberg (see p.185).

The size and functioning of Austria's tulle industry throughout the 19th century is obscure, but in the early 20th century, a direct and aggressive competition with Irish goods, for the American market, was in in full swing (see p.13).

The art nouveau revival which saw the production of many quite lovely bobbin and needle laces in Vienna, seems also to have touched the tulle embroideries. Innovative designs showed striking originality combined with a sensitive lightness of touch. Sharp clear outlines of beautifully balanced shapes were enhanced by light infillings. Though the period of gaudy colours was to some extent past, experiments in white thread on black net, or in gold and silver thread on black or white net were very effective.

GERMANY

Germany is known to have had between 50 and 70 bobbinet machines by 1856 [173] but, as in Italy, Denmark and some other continental European countries the embroidery of this net appears to have been for home consumption and, in the absence of any international trade, there is little factual information relating to it.

The earliest record is a pattern book *Sammlung neuer Muster zum Stocken in Plattstich und Tambourin* (New Designs for Run-stitch and Tambouring), published in Hamburg in 1827.

Two instructional books from Darmstadt and Leipzig were issued in 1909 and 1922 respectively, and there are others more recent. Most of the work illustrated is run embroidery, with counted meshes. There are some similarities here both to Dillmont and to the Italian *Enciclopedia*, but in spite of the general convergence of ideas for domestic embroideries it may be possible, when a whole lot from one area is viewed together, to discover some national or regional characteristics. For isolated pieces, identification of the geographical source remains difficult.

The boldness of the run stitches is created not by a single thick thread but by several thin threads used together. The borders of the lace are often buttonholed on top of short run-stitches, while in the filling stitches the basic straight-run defines simple shapes rather than being modified into quite different formations. Designs tend to be geometric.

Application of muslin in small areas on the underside of the net, with needle lace stitches filling cut-out holes is also found [174], and is very similar to some Hungarian embroidery still made today. Pea tulle with its wider hexagonal mesh was used for the heavier work. The embroidery was sometimes in floss silk.

A net with a 4-sided mesh similar to the bobbin-made torchon ground appears in some modern pieces and may have been made on the Barmen machine.

Above: a very simple run pattern in thick thread makes an effective border.

Above: a sampler of run-stitch motifs used to make geometric patterns (Courtesy: Gudrun Borck).

Below: a sampler of run-stitches more in casket style (Courtesy: Lieselotte Gotz).

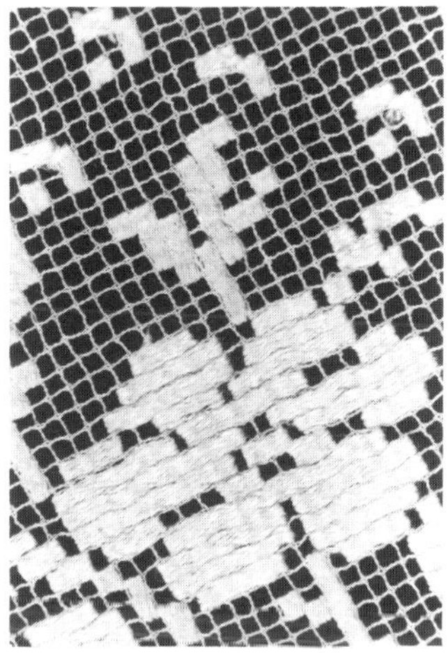

Above: the square-meshed torchon ground gives the design an angularity reminiscent of filet (photo: R.B.).

Below: a pattern from the Hamburg book of 1827 (Courtesy: Metropolitan).

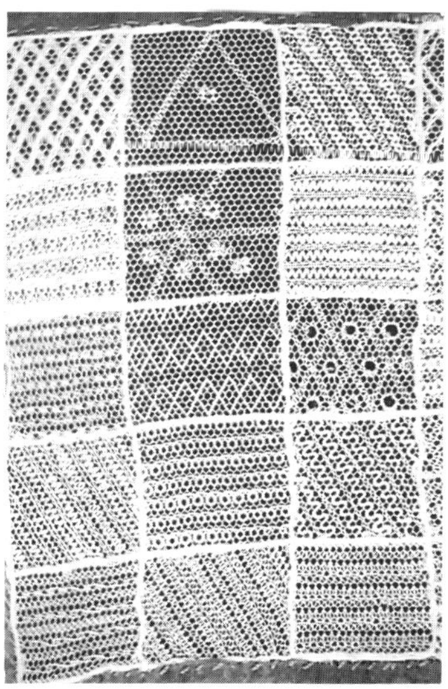

There is little record of any tambour work except for that connected with bead embroidery. Workers from Lunéville established an industry at Seligenstad in 1882, following a request from the local leather trade which sought a novel means of decorating purses. The embroidery was soon extended from leather to machine net. This was a cottage industry, the workers being supplied, and paid, by travelling merchants.

As in Lunéville, the beads were bought in one kilo hanks, transferred from their strings to the feed thread, and offered to the hook one at a time, as required.

The pattern was drawn on tracing paper, and then the entire outline pricked using a mechanical pricker motivated initially by foot power (a treadle) and then by electricity, the active needle being supported by struts on either side, or by a celluloid shield beneath, which had a small central hole through which the needle could operate. The holes were placed less than 1mm apart. The pattern was then pounced onto the net using blue chalk - 'Bestes Pauspulven' supplied by Josef Schraml of Munich. A block of thick felt backed with wood was used to rub the powder through the holes.

When the tracing paper was removed, petrol fumes were blown across the powder deposited on the net to fix it. The petrol was held in a small can fitted with a special mouthpiece so that the operative could direct the vapour without himself being contaminated. Once fixed, the dye was not water-soluble, but the close attachment of the beads or sequins covered the markings.

Three Plauen laces: left, a net guipure; middle, a net-on-net appliqué; right, two samples of coloured 'run' embroidery by W. Bierbaum and Co. (Courtesy: Textilbibliothek, St Gallen).

More open patterns, spreading vermicilli-like across the net, could be worked freehand [175].

HUNGARY

Hungary has a long tradition of regional embroideries - on leather, felt, linen and filet, but although there are veils and aprons of run embroidery on machine net at the National Museum in Budapest, no information about the industry is available.

The only exception is a very decorative form of muslin appliqué on net. This began during the 19th century, and the designs are often copied from the rather abstract forms of felt-on-felt appliqué. Information about its making is taken from *Muskatli*, a magazine edited by Zulawsky Eleméné and published by the School of Applied Art in Budapest, in October 1931.

As with the cut cambric work of Limerick, both the cross-sides of the net and the threads of the woven fabric were carefully aligned with the pattern (and with the frame if one was used), to prevent bias pull and fabric distortion. The net was a fine round-mesh of smooth silky texture. The muslin was insufficiently transparent for a pattern to show through from below, and it was customary for the embroiderer to draw it on the cloth, using concentric circles and diameters as guide lines.

Above: part of a table cover in muslin appliqué on net.

Below: a design for muslin appliqué based on a felt-on-felt appliqué pattern (after *Muskatli*).

A neat row of small stitches was run approximately 4mm inside the outline of the design to hold the muslin in place. These stitches would be removed later. When the attachment was complete, the surface material would be cut away, a bit at a time, leaving a 3mm margin, and using scissors with one blunt tip to avoid snagging the net below.

As the cutting progressed, the raw edge was tucked under and sewn in place, keeping the stitches small and at right angles to the outline, and pointing the needle inwards, as recommended for Irish cambric appliqué.

On the table cover illustrated, the outermost border of the cloth is overstitched with fine, close, and almost invisible buttonholing. The stag design is another Hungarian favourite.

CZECHOSLOVAKIA

An independent republic only since 1916, Czechoslovakia included the uneasily unified regions of Bohemia and Moravia to the west and Slovakia to the east. They lie at the heart of central Europe, encircled by Germany, Austria, Hungary, Poland and the Ukraine, and have strong cultural and racial ties with each.

Czechoslovakia's embroideries on machine net, intended for the local agricultural communities rather than for international consumption, show an earthy exuberance expressed in bold designs ablaze with primal red, orange and yellow cottons or silks. Their air of joyful abandon makes the elegant lightness and delicacy of the white veils, collars and flounces from other areas, intended for the richer classes of society, appear almost anaemic, and even the multicoloured beads, sequins, silver and gold thread of the early 20th century, by comparison pale and restrained.

In Bohemia, *tamboured* net shawls were worn on the head, or round the shoulders and crossed over the breast. In Slovakia, bobbin laces might be imitated for dress trimmings, and women's caps made of strong machine net were embroidered in thick cotton, sometimes white, sometimes in bold colours. This tambouring was done in the traditional manner, using a frame but, as with some of the Hungarian embroideries on cloth, there was no fixed pattern, the women using their memories and their imagination to create the design.

Areas of net within the flowers, intended to be cut away, were first stabilised with close buttonhole stitching.

The decoration of the coifs and shawls of Slovakia often included some applied cambric or batiste. The abstract designs, marked on the semi-transparent linen, were not dissimilar to those of Hungary. The linen was most commonly stretched on top of the net, and the two layers joined as described on pp.187 and 190. The thread used for the hemming-stitch was very fine and the stitches placed close together to give the required neatness and solidity. The outer border of the article was secured by satin-stitch, worked in white or yellow silk. Net and fabric were sometimes cut away in the centres of the motifs, and the holes filled in with needle lace stitches of the 'point lace' type.

Additional richness was added to the appliqués by surface embroidery of the net, using yellow silk and silver metal thread. At times the batiste was placed beneath the net instead of above.

A primitive design of birds and flowers from a pastoral area.

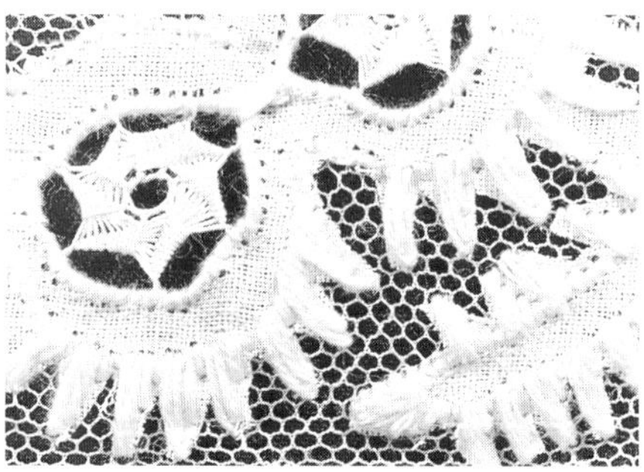

Left: the corner of a shawl or kerchief, fabric appliqué, the two layers attached by overcasting, without a cord. Prague (Courtesy: Gudrun Borck).
Middle and above, face and reverse: two details from a similar piece, showing unusual filling stitches which occur also in Hungary (Courtesy: Wolfgang Haüsle).
Below left: surface embroidery with pulled threadwork above; run embroidery, below.

In the area of Slovakia just north of the Hungarian border, run or tambour embroideries showed lively representations of naive birds and flowers enriched with a variety of decorative fillings.

On another level - lace art as opposed to craft - Czechoslovakia has long been noted for the originality, beauty and professionalism of its designs and workmanship. An intriguing piece, shown at the Exposition of Arts Decoratifs in Paris in 1925, has the entire pattern consisting of no more than a thin cord attached by overcasting to the net beneath, a type of work known as *soutache*, which lacks entirely any solid and decorative areas, or even stitches. This technique enables a thicker cord to be used (than would be possible if the outline was run) so that it stands out sharp and clear above the surface of the net[176].

Above: a fine soutache embroidery by Mme E. Mildeova-Palichova, 1925 (after Dormoy, Pl.39).
Left: a run collar with raised satin stitch around the flower centres, Sudetenland, c1930 (Courtesy: Lieselotte Gotz).

POLAND

In spite of its long border with Czechoslovakia, there are few close similarities between the regional costumes of the two countries. Indeed since provincial dress tends to be isolationist - in marked contrast to high fashion - it might be more surprising to find that they were alike.

A book on Polish folk art, published in Warsaw in 1958, has some interesting illustrations of embroidered machine nets. In particular there is a bonnet of silk knitted net, embroidered in plumetis fashion, from the Museum of Wloclawek.

Another cap back, worked on bobbinet, shows carnations with spiky petals, and veined leaves framed by long spiky fronds.

Several other styles are distinguishable, such as a round-meshed bobbinet bordered with the familiar flower clusters of Limerick and elsewhere, but closely worked in an inelegant run-stitch which slopes haphazardly in different directions. The execution of the design is primitive and ill-proportioned, the stems being too lumpily thick, the attachment of the leaves gauche, and the flower sprays generally stodgy. Such white laces, modelled perhaps unsuccessfully on more up-market fashions, appear to have been worn copiously frilled as aprons, overskirts, capes, closely-gathered ruff-like collars and triple-frilled coifs, in such a way that their defects would be concealed and only the general somewhat overwhelming voluptuousness on view.

More interesting are the enormous and dramatic geometric designs worked with masterly precision in thick close tambouring, on caps and cap-strings, or arranged vertically on elbow-length capes.

A third form, inserted as a V-shaped band into a cloth of heavy hand-spun linen, shows a neat but conventional design of symmetrically arranged and spotted vine leaves with bunches of grapes, worked with thickish thread in chain-stitch. Open areas of the net are filled with stars [177].

A neo-classical design in black silk tambour work (Courtesy: Hamilton District Museum).

SCOTLAND

Scotland was a pioneer in the production of tamboured muslin in the British Isles. Skilled weavers were able, in the prevailing dampness of the western lowlands, to produce misty wraiths of muslin which rivalled those of India for spectral weightlessness. Skill in embroidering the cloth (initially by pulled threadwork) was introduced by Luigi Ruffini from Piemonte in north Italy. From the 1790s, tambouring flourished in the Edinburgh area.

In the 1820s a tamboured muslin industry was established on the estate of the Duke of Hamilton, south of Glasgow, where earlier experiments in bobbin lace had been instigated in the 18th century by the then Duchess. The fashion, in the 1820s, for white muslin gowns and accessories, lightly embroidered, created a huge demand.

Tambouring was soon extended from muslin to bobbinet. By 1835 there were 2,500 workers in the area, and many training schools and factories sprang up. By 1846, 7,000 veils, scarves and shawls a week were being produced, including quite a number of black ones, some for the American colonies.

In 1844 a veil embroidered with devices including the arms of the royal family, of the Duke, and of the town, was prepared as a gift to Queen Victoria [178].

A change of fashion, away from muslin, in the late 1830s caused 36% unemployment among the weavers, but tambouring on net continued to flourish for another ten years until, as in Limerick, it began to decline. Although in 1859 there were more than 20 lace manufacturers, agents, washers, dressers and pressers in the town, fashion continued its fickle pathway. At the same time, the patterning Leavers and Pusher machines were making their competition felt, and the Cornely and the Handmachine were about to reinforce the destructive (to hand-work) influence. The number of workers fell to 900, and the demand for tambouring diminished still further until it was no longer a truly commercial enterprise, but a pin-money occupation for a few ladies continuing to embroider for pleasure, and a meagre reward.

Hamilton tambouring appears to have been entirely white-on-white or black-on-black, but for the 1851 Exhibition more adventurous Glasgow firms experimented with colour [179]. Thus there is mentioned a 'new and very pretty style of hand-tambouring on net by Messrs M'Quarrie, Fisher and Co of Glasgow, including an apron tamboured in violet silk on net of the same hue, lined with green and violet glacé silk', and another apron 'the outline tamboured in gold thread and filled in with green silk [tambouring], the material is black net and the lining lilac silk, finished with green cord and tassels'. A parasol, embroidered in white and gold on white net over a light blue silk mount, was accompanied by matching sleeves, bertha and cap. A dress of crimson silk run embroidery on black net was shown by the firm of M'Arthurs [180].

During the most prosperous period, the large wooden tambour frames were rectangular, holding a full width of net, and enabling several women at a time to work on the same piece. Each frame, nearly 4 feet in length, would hold two rollers, one for the worked net, the other for the net yet to be worked. The rollers rested in cast-iron brackets at the top and bottom of the frame, while ratchets enabled the net to be either tightened to make it taut for working, or turned onwards when the area had been completed[181].

Such frames were located in small factories, but there was also an extensive cottage industry using circular frames about 2ft in diameter and supported on stout wooden legs. It is said there was no house in Hamilton without its tambour frame whether for gain or amusement [182].

Detail of a white tamboured wedding veil (Courtesy: Hamilton District Museum).

The rim of the inner frame was wrapped with velvet so that the stretched net could cling and not slip as the outer hoop was fitted over it. The outline of the design was lightly printed on the net, using carved blocks of sycamore wood.

As usual, after the outline was worked, the more solid areas were filled with ordinary chain-stitch, and the flower centres with fancy chain-stitch groupings.

Another form of machine net decoration, known initially as Royal Victoria lace, appeared in the Illustrated Exhibitor of 1851 (pp.274-6). It was a form of soutache, consisting of Adelaide or Coronation braid, attached by overcasting the thin connections between the elongated seed-like ovals. Arrangements of this braid, by themselves, formed the design.

NOTTINGHAM

Nottingham has long been famous as the centre of the machine lace industry. The Stocking Frame was invented at nearby Calverton in 1600. Its conversion into a pin-machine and others capable of producing weft-knitted nets was in the city area. Its further conversion to the production of warp-knitted nets, though begun in the Spitalfields area of London, was continued and finally perfected in Nottingham. In Loughborough, about 15 miles to the south, John Heathcoat developed in 1808 his two-twist bobbinet. In Nottingham also the first Pusher, Leavers and Lace Curtain machines were set up and, later, their patterning automated by the application of the French Jacquard apparatus.

Each development in turn offered competition to its predecessor, and the peaceful co-existence of a static hand-embroidery industry with a constantly and frenetically changing machine industry, in the same area, soon became untenable. In the 1850s the hand-industry suffered an early demise.

Decoration was almost entirely by run work. Indeed, in England, Nottingham had a near monopoly of this technique, and it was from Nottingham that Walker initially derived his ideas for the run embroideries of Limerick.

Prior to the appearance of bobbinet on the market in 1809, thousands of women and children were already engaged in simple or more complex embroideries of the silk weft- or warp-knitted nets. Especially popular was the single press point net, invented in England, then smuggled into France, where it was converted into such a superb product that vast quantities were imported back into Nottingham to satisfy society's demands.

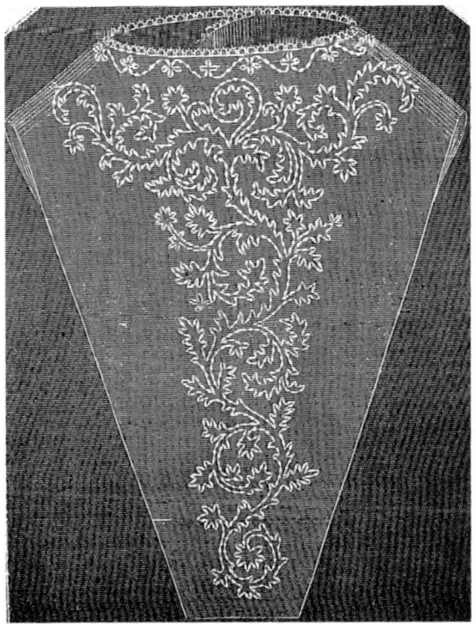

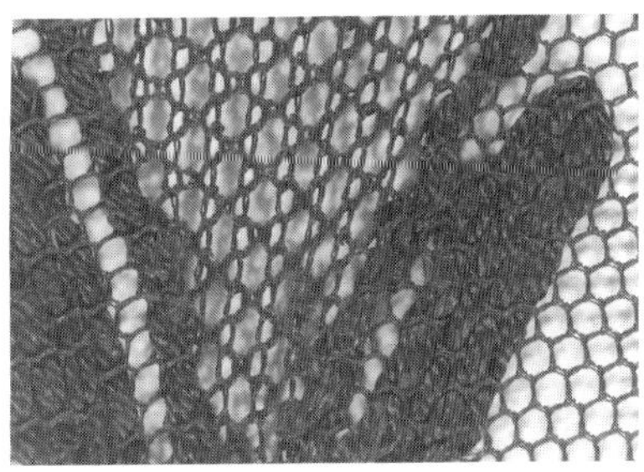

Above: black run on black silk point net. The already thick sewing thread is passed twice through each mesh, creating totally solid blocks ('fining'). The interior ('openwork') is decorated with tight cobweb-stitches (Courtesy: Nottingham).

Left: a chemisette in Royal Victoria lace, 1851.

It was in an attempt to stem this ruinous flow, and reverse a humiliating situation, that Crowder, following the introduction of silk bobbinets in 1824, attempted to make a net which would not only compete with, but surpass, the French single press. He constructed in 1830 a bobbinet of the finest white Italian silk, stiffened yet pliable, but so expensive that it needed a launching to present it to the market place.

Queen Adelaide, wife of William IV, was persuaded to grant her patronage by wearing the dress at the Juvenile Ball given on May 24, 1831, in honour of the Princess Victoria of Kent, heiress presumptive to the Crown. The dress was prepared by Messrs Train and Wesson. 'The pattern was made in stripes of nine inches, and consisted of an elegant star, having a large openwork in the middle, beautifully worked with the needle, encircled with a series of roses; the whole appearance of the dress was of the most brilliant description imaginable and absolutely dazzled the eye. The appearance was that of flowers of brilliantly shining silk worked upon a gossamer light fabric, having more the appearance of net made of mother of pearl than of any other substance'. For the occasion, the gown was further ornamented with silver and worn over a white satin dress' [183].

Below: a typical black blonde design. The stitches are very simple: close run in heavy thread, and more open tent in finer thread. In addition, the holes create in themselves an attractive design.

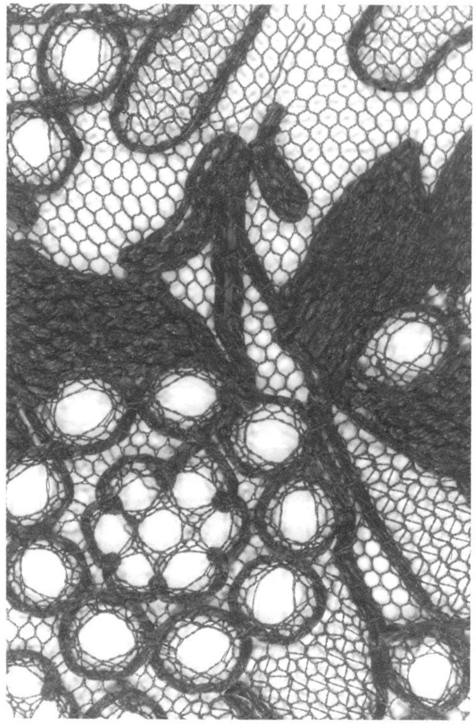

Detail: the sides of the enlarged holes are caught by thin overcasting, while a compact looping around the central rosette binds the net neatly into six radii (both Courtesy: Nottingham).

Meanwhile the run designs of the cheaper cotton bobbinets continued to show a strong if superficial resemblance to the Spanish blondes, still socially desirable in spite of the prohibition on their importation resulting from the conflict with Napoleon I during the first 15 years of the century. The angular sprays and bouquets of flowers, following at first too rigidly the lines of the mesh sides, were stiffly stylised. The solidity of the main design areas (known locally as fining) was provided by embroidery threads so thick they entirely filled the mesh holes. The tightly-packed run-stitches passed always over two mesh side and under one, and always in the same direction, with no attempt to follow the contours of the leaves or petals, giving the work an inert quality quite at variance with the lively freedom of Limerick.

The decorative areas (known as openwork) occured mainly in the flower centres. The fancy stitches were severely restricted to cobweb and tent-stitch, worked in a finer thread, giving an impression of the honeycomb and katstitch fillings of East Midlands bobbin laces. The stitches were sometimes arranged diagonally instead of in horizontal lines.

The earlier products were mainly bonnet veils, those 'curtains' which, attached around the crown of poke bonnets, could be modestly draped forward to conceal the face, or coquettishly thrown back - indeed offering a multiplicity of devices according to the occasion. The general arrangement of the motifs was a broad lower border, narrow side borders, a centre lightly sprigged so that the wearer could still see through it, and an arrangement for a draw-cord along the top.

An embroidered patent net veil c1801 (Courtesy: Chertsey Museum).

When these went out of fashion in the 1840s, many of the 150,000 runners (run embroiderers) active in 1831 were put out of work. Attempts to compete with the newly popular broderie anglaise by making lots of large holes did not work too well, as the texture was totally inappropriate.

Running continued in the area for a time, less creatively, by adding the 'liner' to the early patterned machine laces, until Leavers developed a technique for inserting the liner on the machine, in 1841. Although the Pusher never managed to do this, some of its outlines were attached as a raised cord by the Cornely, which was faster and less labour-intensive. The coup de grace for the Nottingham runners was the transference of the Pusher industry to Lyons, in the 1870s.

Hand-embroidery did not, as in Limerick, switch to the square net when this was invented in the 1830s. For one thing, the designs continued to be based on Spanish blonde bobbin laces and, later, those of Chantilly, both with a half-stitch ground identical to the two-twist (round) bobbinet. In addition, the constant search for improvements to the

A bonnet veil c1830 (both photos: R.B.).

Leavers and Pusher machines discouraged any attempt to develop a truly individual style, and in a rather depressed economic climate there was little incentive to change the workers' habits - the adaptation of stitches from the round to the square net, though possible, was not always easy, and the extra time involved would probably be uneconomic unless there were sound prospects of future prosperity.

An account of how the work was done is given in M. Digby Wyatt in *The Industrial Arts of the Nineteenth Century at the Great Exhibition, 1851*, in his description of Plate XXX, a lithographic engraving of a black lace flounce shown by the Nottingham firm of Greasely and Hopcroft. The pattern was printed on the net using carved wood blocks. These were prepared from the original design by 'quasi-artists' who combined artistic ability with a keen awareness of prevailing fashion styles. The transference on to the net of the entire outline to be worked, along with the overall shape of the piece - collar, cape, veil, bertha etc - was followed by a pattern-setter deciding the distribution of the various stitches to be used in its execution. What dye was used that would show up clearly on the black net is not explained.

The lace-runners might work in their own 'humble homes', or space in a large house might be rented out. Then seven or eight women could work in one room, sharing a meal for the sake of economy, and sitting at the large rectangular frames for twelve to fourteen hours a day [184].

As usual, the outline was run in first, the left hand apparently resting beneath the net to hold it steady so that it could not wobble and dizzy the eyesight.

The distinction between fining and openwork was ingeniously explained by Wyatt - the former bringing the meshes of the net 'to a smaller size' (by filling the holes with thread), the latter bringing the meshes 'to a larger size', presumably by the use of a bodkin to expand the appropriate holes and then bind them in position by tightening the threads of the decorative stitches.

When the embroidery was complete, the work was cleaned, stiffened with gum arabic, the excess net cut away around the border, and an edging (presumably machine picot) sewn on to prevent fraying.

This description relates to current practise in 1851, when the industry was already near its end. What arrangements may have existed in earlier years, before the invention of machine picots for example, are not recorded, though some laces are known with hand-made bobbin picot borders.

The piece described by Wyatt post-dates the fashion for Spanish blondes, and is an excellent imitation of the hand-made Chantilly, even to the half-stitch effect apparently simulated by lightweight cross-stitch infillings.

Digby Wyatt said of this example: 'It combines all the qualities of openness and clarity of design

The lace described by Digby Wyatt (Courtesy: Nottingham).

required by lace . . Lace is obviously best displayed when the openness of the ground permits the surface over which it is worn to be distinctly seen through the meshes, so as to assist in relieving or "throwing up" the flowers or ornaments worked in the net. Too often this principle of contrast is overlooked. Scroll work, shells and every description of enrichment are heaped together so as to completely cover the transparent field, rending the work unpleasantly full and *tourmenté*.

This statement, even in its powerful condemnation of the era's 'gorgeous' taste, spotlights the lack of accord between the delicate refinement of embroidered nets, and common popular preference, and helps to explain the difficult times experienced by Limerick in the mid-century.

The large frames could be used even for small items, when several were printed close together on one expanse of net to avoid wastage.

In spite of various brief references to a tamboured net industry, little is known about it. Caulfield and Saward, writing in 1882, say that tambouring was 'an article of manufacture sixty years ago, and gave employment to the poorer classes in Middlesex, Nottingham and Ireland' [185].

ISLE OF WIGHT

Although this was a fairly minor industry, it is placed next to Nottingham because its embroidery was also mainly, perhaps exclusively, run. Caulfield and Saward in 1882 described it as a run lace 'of no particular value'.

Its development followed the invention of a Traverse Warp machine by John Brown of New Radford, near Nottingham. Patented in March 1810, and also in April 1811, it worked by reversing the principle of the Old Loughborough. The threads were mounted above the machine, and the net collected below, so that the crosses and twists were carried downwards instead of upwards as in Heathcoat's machine. The diagonal passage (traversing) was made by the warp, not the bobbin, threads. Brown had started his experiments because he wanted to improve on the Loughborough net which was 'too thick and heavy'.

Surviving a period of great hardship when he had to sell his furniture to finance his research, Brown eventually succeeded in his aim. His net ground was thinner, and the selvages better. Heathcoat himself declared it to be 'a great improvement, enabling very narrow breadths to be made, though slowly, of excellent texture, indeed perhaps superior to any other' [186].

An all-over handkerchief border, with 'Mechlin' roses (Courtesy: V and A).

Following the death of Brown in 1819, the business was taken over by his two associates, George Freeman and H.W.Nunn. Around 1826, half their machines were transported to Whippingham on the Isle of Wight [187], where Freeman, a silk stocking weaver (i.e. working with Stocking Frames) had already established a branch of his business under the management of his brother, W. Freeman.

According to a report in 1853, the machines in the Isle of Wight remained primitive, operated by hand and foot instead of by any outside source of power. It seems likely, therefore that they produced only the round net. By the mid-1860s they had almost disappeared, victims of the constantly advancing Leavers [188].

Felkin records 99 bobbinet machines in the Isle of Wight in 1826, and though this seems an enormous number, they were at this time quite narrow. Their presence however leaves open the question of just how much of the island's embroidery was on Heathcoat's bobbinet and how much on Brown's traverse warp.

Surviving laces in the Museum of Carisbrooke Castle, near to Newport, include one or two with diamond net, but there is no way of telling if such laces were made locally, or imported.

Characteristics designs included small roses of the type often found in late 18th/early 19th-century Mechlin bobbin laces. Oak leaves, cornflowers and sunflowers were also popular. There was a double-thread outline, the solid design areas were filled with a light run stitch, and the caskets with a small range of run-stitch variations. A distinctive feature was the tendency towards all-over designs, as opposed to the sparsely embroidered centres of Nottingham and Limerick veils, stoles and other pieces. Palliser mentions examples where the interior was sprinkled with tiny quartets of dots [189].

Two designs of a sunflower and similar composites from the pattern book at the Museum of Carisbrooke Castle.

A charming book of patterns at the Museum may or may not relate, in whole or in part, to the embroidered net industry. Some of the examples appear so exquisitely drawn and so professional that they scarcely fit with the image of a purely local cottage-style occupation.

Indeed but for Queen Victoria's passion for the island, it seems likely that its embroidered nets might have atrophied at an early stage. The considerable profits made by Freeman and Nunn were the fruits not of embroidery but of patterned machine laces, most particularly a silk blonde which they successfully passed off as hand-made 'during a whole season' [190]. Although Nunn distributed through warehouses in London, the nature of the goods is not disclosed.

The Princess Royal at the drawing room, on her first presentation wore a dress of Newport lace, her train trimmed with the same. Queen Victoria had several lace tippets (fichus) made of Isle of Wight lace for the Royal children, and always chose the Mechlin style of rose pattern [191].

By 1901 there were only two or three old women embroiderers left. *Needlecraft* no.59 commented in 1906, 'Isle of Wight lace and the pretty old-fashioned needle run laces, which are so seldom seen in these days . . '.

COGGESHALL

In Coggeshall, as in Scotland, tambour work on net was preceded by a commercial enterprise of tambour work on fine muslin, also introduced by a foreigner, the legendary Drago, who entered the country between 1810 and 1823, though his name does not appear in parish records of that time. He is said to have lodged, on his arrival, with a Mr Thomas Johnson, listed in 1823 as a silk manufacturer or 'engine weaver'.

Above: a baby gown of very fine muslin with tambour embroidery, thought to be Coggeshall.
Left: detail.
Below: A panel of tambour embroidery by a still-active Coggeshall worker, late 19th century (Courtesy: V and A).

Drago and his daughters hired a three-storied mill or factory, and set up a room where local girls from the age of seven were taught the art of tambouring. The intention was that these children should remain attached to the factory, earning after three years 6d a week, after five years 3s6d, and up to 7s6d when grown up - quite a normal wage for such labour at that time [192].

At what date the tamboured net industry, as opposed to that of tamboured muslin, commenced is uncertain. Possibly it was in the year 1823 when Heathcoat's patent expired and the price of bobbinet dropped quite rapidly from £5 per square yard to 18d, a x66 reduction.

Compared with the Nottingham run industry, the Essex industry - though the main centre for tambour work in England - was minute, and small compared even with Hamilton. Its rise and fall followed the same patterns of fashion experienced elsewhere. In 1820 a scarf was made for the Dean of Bocking's daughter, Bocking being a centre for the lace trade, carried on until about 1830 by the firm of Messrs Powell and Cooper. In 1832, when embroidered nets were the height of fashion, a Coggeshall gown was made for Queen Adelaide, presumably following the Nottingham one made for her the previous year [193].

A perusal of a sample of 150 tambourers from the 247 listed in the census records of the Greater Coggeshall area in 1841, shows 94 of them declared as 15 years, or under. One was aged 5, and six were 8 years old. The preponderance of the number 15 is explained by the regulations: 'Write the age of everyone under 15, then ages at 5 year intervals, taking the lower figure, for example for the ages 15-20, write 15'. Also listed for that year were two lace manufacturers, William Crosby aged 28, and John Byng Banks - who also had an establishment in London - aged 55. James Popham, aged 30, was described as a tambour worker.

In 1851 there were 382 tambourers; and by 1855, 450. The number of lace manufacturers listed in *Pigot's Directory* for that year had increased to seven, including James and William Spurge, who married the daughters of Thomas Johnson, with whom Drago had originally lodged. At this point the lace business of the two families was amalgamated.

During this decade there were experiments in rainbow colours of both thread and net, and perhaps Coggeshall was the source of the 'trial of black lace embroidered with colours' so strongly disapproved of by *The Illustrated Exhibitor* of 1851, and destined only 'to obtain some success in the colonies' [194].

The number of workers then fell rapidly, to a mere 84 in 1861. Tambourers in 1866 were described as makers of 'Honiton' lace 'by which probably only the more elaborate form of tambour work is intended, both laces being of a similar nature although made in a different way' [195].

In 1871, 449 tambourers are listed, but the decline then continued. At the 10-year intervals of the census the numbers were 222, 156, and 149 in 1901. The decline was accompanied by deterioration in character and quality. Though the older women were still capable of exquisite work, fine execution was often sacrificed to speed and laces 'hardly to be distinguished from chain-stitch sewing-machine work' were produced in large quantities.

In *Kelly's Directory* for Essex, in 1874, eight lace manufacturers for tambour lace are listed, all at Coggeshall.

Improvements in design followed. The Misses Spurge, in charge of the business in 1906, tried to reform the designs with copies of beautiful old Italian designs, in tambour work, while their brother, an art teacher, contributed a number of elegant ideas.

In spite of the small size of the work force, the laces were said to have found a ready sale in London and on the Continent: 'Exquisitely wrought dresses and shawls have been manufactured by the Coggeshall toilers, and have been worn at her Majety's Drawing-rooms and at other state assemblies'.

As at Lierre and Lunéville, richer effects were introduced in the second half of the 19th century, the nets being decorated not only with white cotton but with coloured silks, beads, and gold and silver tinsel. Workers were given a paper pattern which they fixed beneath the net while they copied it on to the surface, using pipe clay. Bead and spangle decoration continued through the Edwardian era and on to the introduction of flapper dresses in the 1920s. Skilled workers at that time could earn £1 a week [196].

The situation in 1906 is summarised in *The History of Essex*: 'A considerable trade is now done at Coggeshall in sequin and spangled net for dress fronts, flounces or sleeves, as well as in hat crowns and brim pieces worked in a combination of sequins and chenille . . . It is a demand of fashion not appreciated by the work people who complain that their eyes are injured by the glitter' [197].

Apart from the extraordinary appellation 'Honiton', until 1906 the tamboured nets were referred to as Essex lace, or British lace. At the 1851 Exhibition they appeared as Nottingham lace, and in the London stores in the early 20th century as

Limerick or Brussels [198]. They were even described as 'a form of point lace'. All of which brings something of a crisis of identity not only to the actual laces but to records concerning them.

The Misses Spurge were the chief employers, marketing via a London distributor [199]. Numerous tambourers also worked independently, marketing directly through salesladies in London [200]. By 1910 there were 50 workers, and the laces were sufficiently well thought of for Queen Alexandra to place a number of orders [201].

Above: a cuff of Coggeshall lace, bought at Liberty's in 1921.
Right: a rubbing of a design for a modesty, (ex-Spurge collection, courtesy: Phyllis Wood).

In 1921, Liberty's were selling 'Coggeshall tambour laces', in the form of collar and cuff sets. On June 1, 1923, laces from the Coggeshall Tambour Lace Makers were exhibited at the Khey Kamp Gallery at the Hague, at the Stedelijk Museum in Amsterdam, and at the Corn Exchange in Arnhem. The numerous visitors at the Hague included the Queen and the Queen Mother of Holland. A number of examples of the lace were sold, raising the possibility of Coggeshall laces appearing in Netherlands collections [202].

Liberty's lace exhibition of 1926 included specimens of Essex, Spanish (needlerun?) and South India laces [203].

The industry continued on a small scale, with fine pieces still being worked to special commission, until the outbreak of world war 2, in 1939, when the necessary materials were no longer available [204].

In 1951 a Miss Annie Spurge was still living in the 16th-century family house in Church Street. A tambour room with windows on three sides and space for 50 frames had been created there, but this was closed when business began to falter in the early 1860s, and the work was from that time transferred to the homes of the workers. In 1951 there was only one skilled embroiderer remaining. The industry was in effect dead.

The origin of Limerick tambour laces from the Coggeshall industry has been described in ch.1 (p.3).

For working the lace only two-twist bobbinet is said to have been used, but rubbed patterns preserved in Coggeshall, from the Spurge collection, specify both Brussels and Bretonne nets, the latter being particularly small-meshed. The practise of rubbing was economical of time and inventiveness, but was not really the best way to preserve the outline's pristine sharpness.

The circular tambour frame was early discarded, possibly at the time of the change from muslin to a net ground. A rectangular frame was set up, the reel secured, and the hook propelled in the usual manner. A firmly-twisted embroidery thread was preferred as giving a crisper finish and less likely to be split by the pull of the hook. A surviving sampler shows 35 decorative tambour stitches, but no run fillings were employed. The net was sometimes cut away within the motifs to provide wider openings.

For black lace, silk bobbinet and a black silk embroidery thread were used.

The designs of the muslin tambour work were exceptionally delicate, often no more than single lines of fine chain-stitching lightly traced into flowers and leaves bordered with more formal crenallations. These are excellent for practising simple tambour-stitching. The patterns are said to have been drawn on cards in ink and fixed in a hinged manner beneath the net so that they could be brought up to check against the line of stitches already worked.

Styles in the tamboured nets were sometimes fine, sometimes extremely heavy, rich and full-blown, such as the child's apron which, intriguingly, presents a harp and what might be construed as single overgrown shamrocks. This example is typically Coggeshall of the mid-century, though the design is reminiscent of Limerick - but designs can be copied any time, anywhere, unless fashion puts a restraining hand on them. Many of the patterns covered the greater part of the net, as in Isle of Wight laces, instead of being restricted to the borders.

In this rural industry, wild-flowers such as sheep's parsley, bluebells, primroses, ivy leaves and berries featured in many of the designs. The motifs were often carefully arranged to fit neatly into the shape of the article.

Distribution of materials, and marketing were most commonly in the hands of middlemen. It was they who made up the articles, carried out repairs, and sewed on the purl edge - in fact the usual arrangements for a cottage industry were in operation.

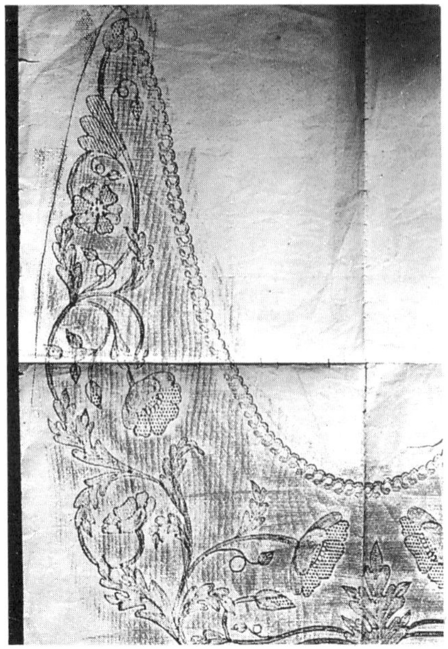

Above: a rubbing of an art deco design of poppies (ex-Spurge collection, Courtesy: Phyllis Wood). Similar designs are found in Ireland.
Below left: a christening apron in complex tambour work. In spite of the two harps in the border, and the plumed shamrock leaves, the rather congested style is more typical of Essex (Courtesy: Adrienne Thompson).
Below: two of the elaborate tamboured infillings.

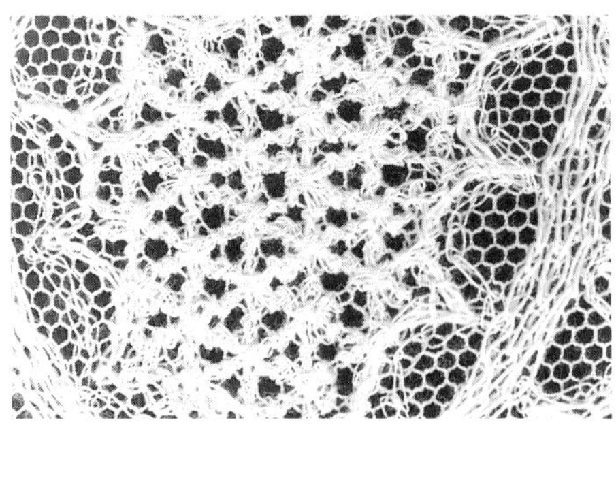

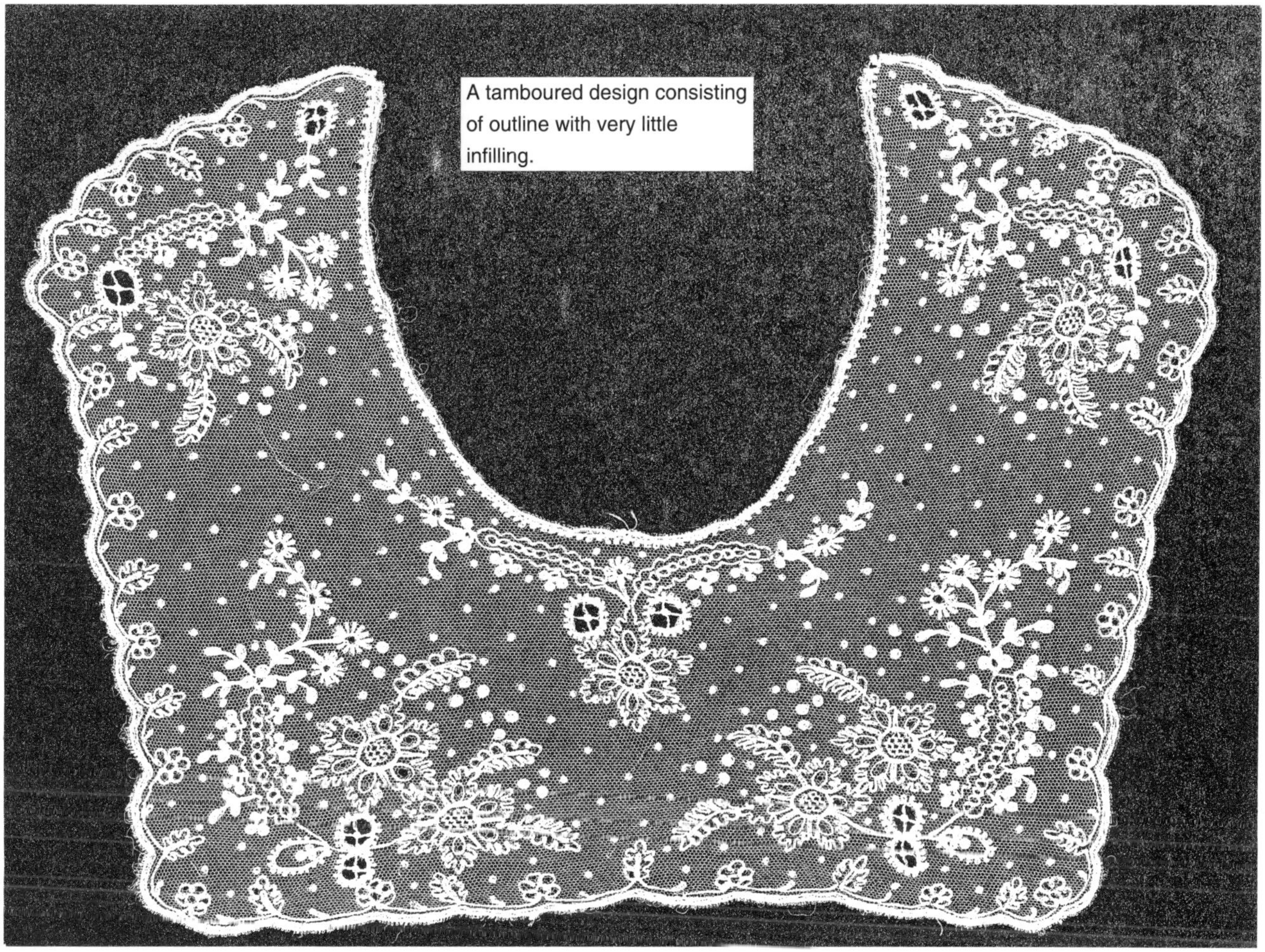

A tamboured design consisting of outline with very little infilling.

Below: a rubbing from a completed lace showing paired cornucopia sprouting poppy-like flowers entwined with spiralling ribbons (Courtesy: Phyllis Wood).

ISLINGTON

The Jury Report on the 1851 Exhibition (p.1014) mentions tambour work of Islington manufacture shown by the London firm of James Howell and Co, specialists in 'point lace'. Caulfield and Saward (1882) also refer to tambour work being made in 'Middlesex, Nottingham and Ireland', as though these centres were more important than the Essex manufacture.

Middlesex was a county bounded on the south by the River Thames. It formerly encompassed what was known as the Metropolitan area, now incorporated in Greater London. It included the parishes of Islington, Hackney and Stoke Newington, some three to five miles north of Spitalfields in the East City, an area famous not only for its ancient silk industry but also as an important centre for the production of silk knitted nets on modified Stocking Frames.

The term 'point lace', although apparently it could, in 1851, mean tambour work, also refers to laces of 'pre-made tapes' which were constructed by folding and gathering yardages of machine tapes or braids into pattern shapes, then joining them together by either needle-made bars or meshes. The motif centres were filled in by hand with needle lace stitches, often of particular types which were not necessarily the traditional ones of antiquity (see p.183). Assuming the term 'point lace' did in fact convey this more orthodox meaning in the reference to Howell and Co, the term 'Honiton' is then easier to understand, since shaped tapes indented like strings of sausages were known as Honiton tapes, and the laces made from them abbreviated to 'Honiton', never mind that a bobbin lace had long since claimed this title.

On the face of it, such documentation as there is provides no real evidence of any substantial tambour industry in the area, nothing to clarify the organisation, the number and ages of the workers, the conditions and hours of work, the date of origin, or details of internal or external trade.

However a study of various statistical records reveals the following:

Pigot's Directory for London, Middlesex and Essex in 1826-7 shows 7 lace cleaners and menders, 48 lace dealers, and 63 lace manufacturers and retail warehouses. Among these latter is the name of Charles Walker, who took the techniques of run and tambour embroidery on net to Ireland in 1829, and there established the Limerick lace industry. Though his main factory remained in Ireland, it was the practise for provincial manufacturers to have a distributing centre in London. Walker's office was in Cheapside, and so, interestingly, in the same year, was that of John Heathcoat, the inventor of bobbinet. Heathcoat's factory was by then at Tiverton in Devon, where he had moved from Loughborough in 1816, after the Luddite attacks.

Peter Bond, a 'British [= Coggeshall] lace' manufacturer is listed in the slightly more distant parish of West Ham. He is thought to have been one of Walker's apprentices [205].

In 1839, *Pigot's Directory* lists John Byng Banks, lace manufacturer and tambour worker, of Coggeshall and Cheapside, along with 64 further lace manufacturers and wholesale warehouses, 13 of them in Cheapside, and 1 in Whitechapel. Of a total of 46 lace dealers, 5 were in Islington, 2 in Stoke Newington and 1 in Finsbury.

Islington records of 1851 show embroiderers from Nottingham, tambour workers from St Giles, Middlesex, and bobbin lacemakers from the East Midlands active in the area, as well as a lace merchant, Francis Weedon and his son Thomas. From Finsbury there were two more lacemen: Harry Horton, originating from Derby, and a William Green born in City Road, in the parish of St Lukes; as well as a tambour worker, aged 16, born in Spitalfields.

The only reference to a lace industry in Hackney occurs in 1860, in the name of William Tritton, laceman, of Church Street (renamed Mare Street in 1868). A further laceman, William Norman, lived in Stoke Newington. Only three women, from Stratford and West Ham, are listed as lacemakers.

The 1870 trade directory for Islington lists 5 lace and linen warehouses.

Kelly's Directory of 1888 for London lists a total of 56 lace agents, all situated in the East City area, except for two in the West End. There are in addition 120 lace manufacturers and warehousemen, 29 retail Lacemen (including Peter Robinson, Marshall and Snelgrove, Haywards of Oxford Street, Dickens and Jones, and Debenham and Freebody), and 33 lace cleaners described as 'with menders and clear starchers'.

These entries to some extent illustrate the vicissitudes of the lace market, and help to pinpoint the areas of manufacture and storage, but they give no real idea of the type of lace being handled - whether hand, machine, hand-embroidery on net, or machine-embroidery on net.

In this context, the generic term 'lace' cries out for interpretation. The London warehouses could have received and distributed lace of any kind from anywhere in the country. In Middlesex, only by a process of elimination can we conclude that lace meant tambour embroidery on machine net - no other type is mentioned by name, with the sole exception of 'Honiton point lace mittens' and a 'point lace handkerchief' submitted at the 1864 North London Working Classes Industrial Exhibition, and the slightly dubious meaning of this term has already been mentioned.

Three of the enchanting run designs preserved at the D.A.R. Museum.
The middle piece shows an ingenious clustering of densely rounded berries, raised by repeated oversewing into glossy bulges which are highlighted by tiny circles of plain net, giving the appearance of reflected light.

NORTH AMERICA

The making of both run and tambour embroideries on machine net in North America, mainly along the eastern seaboard, flourished through the 19th- and into the 20th-century.

Its origin was certainly in the first place European, and the influence of the Irish, at least from the 1840s, was very strong, but even prior to this, from the late 18th century, Germans had settled in Pennsylvania. Their name 'Deutsch' was rendered 'Dutch' by the locals - the kind of semantic confusion which fits well into the general inconsistencies of lace terminology. A German pattern book dated 1827 is in the collection of the Metropolitan Museum in New York (see p.189).

From the last quarter of the 18th century, until about the mid-19th, tambour work on muslin was one of the most fashionable needlework techniques in use. An advertisement in the *South Carolina Gazette*, based in Charleston, on September 19, 1774, carried Ann Sage's notice that she would teach 'Reading, Tambour, Embroidery and all kinds of needlework'[206].

On May 20, 1775, the *Pennsylvania Ledger and Weekly Advertiser*, on May 20, 1775, announced that Elizabeth Wilson would give instruction in 'Dresden work, Pointing, Bobbing and Netting lace' [207].

By 1800 practically every advertisement offering instructions to young ladies mentioned tambour work. The tambour hook was sometimes referred to as a shepherd's crook [208]. As in Europe, the transition from muslin to knitted net, and then to bobbinet, would not be difficult.

European countries other than Germany and Ireland were undoubtedly represented, so that French, Belgian, Danish, Spanish, Swiss and Italian influences are all likely to appear. Bearing in mind also the huge quantities of lace imported, brought back from European travels, or sent as gifts, plus the vast size of the country, it would seem unlikely that any common diagnostic features could be found. Thus, while it would be perfectly possible to mistake an embroidered machine net made in North America for one made in Ireland, it would be difficult to identify any piece from North America as having with certainty been made there.

Two important collections which throw some light on this problem are those of the Museum of the Daughters of the American Revolution, Washington DC, and the Museum of the Litchfield Historical Society in Connecticut. The embroidered machine nets of the former were donated by members, and in some cases have family histories. They include both white cotton and black silk laces. There is one example of cambric appliqué, the fabric attached to the net by chain-stitching. A run embroidery features cornucopia, the ground studded with small stars - and there are many others. One very interesting example is a tambour embroidery on silk point net made by Ellen Ursula Bowie Belton, who married in 1822 (see p.166).

At the Litchfield Museum are relics of the Ladies' Academy, where needlework, painting and other skills were taught, the average age of the pupils being between 14 and 16, though the full range was from 7 to 28 years.

Here there are three point net veils, all of black silk decorated by tambour work. The predominance of black tambour work over run, may be explained by it being less strain on the eyes, the meshes not needing to be counted whereas in run they did - or at least observed closely. The first is filled with a beautiful all-over pattern and is from the trousseau of Elizabeth Tallmadge who attended the Academy c1810. The second has a large black neo-classical design, also indicating an early date. There is no information concerning the history of either the second or third examples.

Another interesting piece is a run composition of floss silk carnations tinted pink and green on a black net ground. This was worked by Caroline Canfield, date unknown.

Cloth appliqué on net is rare in America, but an experimental piece survives in the Litchfield collection. It consists not of cotton or linen but translucent pina cloth (pineapple fibre) mounted underneath a fine two-twist bobbinet. Below the cloth is an ink-drawn pattern, the whole being supported on brown paper. In the completed area, cloth and net are attached by chain-stitching. This must have been worked with a needle since with no space beneath the net tambouring would have been impossible. The design is quite original, incorporating irregular curving outlines which might be strange clouds or the complex contours of jigsaw puzzle pieces. It was the work of Maryann Bacon, of Roxbury, Connecticut. Born on February 9, 1787, she attended the Young Ladies' Academy in 1802, and died in 1869. It is uncertain what technique may have inspired this piece.

Of the bobbinets, almost all are of the round-mesh type. A two-twist bobbinet industry existed in Medway, Massachusetts, as early as 1818, again from smuggled machinery [209], but it is likely that the best quality nets were imported from England, since America's cotton spinning prior to the 1830s was not particularly good, and the more expensive four-twist square nets were probably never available locally.

In the tambour laces, the caskets are sometimes filled with tambour stitches, sometimes with run.

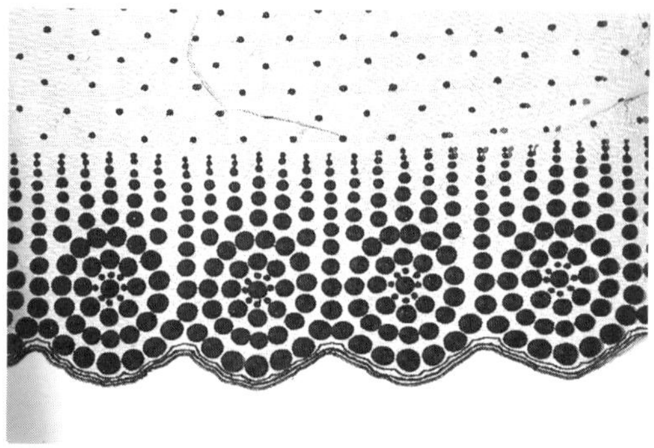

Detail of two tamboured veils and a pelerine collar run in white thread (all Courtesy: D.A.R.).

Opposite: an unfinished panel of run work made by Mrs Lucinda Vail Moser of Brooklyn, NY, c1825 (Courtesy: Metropolitan).

Many of the laces in the two collections are of excellent quality. An unfinished piece of run work at the Metropolitan Museum was worked by Mrs Lucinda Moser, in Brooklyn, c1825. The very subtle shading effect in the flower petals and leaves appears to have been achieved by the use of restricted areas of thicker thread worked over the thinner.

A finely tamboured veil, 46 x 42 inches was worked in white cotton on round net by Catherine Roosevelt Kinnon who attended the Moravian School at Bethlehem, Pennsylvania. She married in 1822, but the date of the veil itself is unknown.

A tamboured collar at the Indianapolis Museum of Art approaches perfection both in design and workmanship. Its origin is unknown (see pp.83-4).

All the laces recorded appear to have been made for personal or family use. There is no evidence of an organised industry in the embroidery of machine nets, but some of the poorer workers set up one-woman businesses, or tiny cooperatives, to make and market goods for the near-locality.

Encouragement of domestic needleworkers was, as in Europe, promoted by magazines - such as *The Young Ladies' Journal* which in 1871 illustrated a number of fancy stitches to be used in run work [210]. In February 1870 the same journal had pictured an antimacasser in Carrickmacross appliqué, describing it as 'suitable for machine work' (using chain-stitching). *The Delineator* of October 1905 gave instructions for a handkerchief border in Carrickmacross appliqué and a 'chemisette with stock collar' (= plastron with choker collar) in Carrickmacross guipure [211]. The designs were by Sara Hadley of West 22nd Street, New York.

The Modern Priscilla in September 1908 gave instructions for an attractive stock collar in 'Limerick run lace' (referred to as 'darned', a Belgian term for run stitch), by Mona Elizabeth Fargher; while in the same magazine, in August 1910, Nellie C Brown told how to make a yoke with a choker collar attached. Both pieces were worked, for the article, by Cora Anderson. In the first, the cut edge of the net is concealed by close buttonholing, in the second by an attachment of machine tape. A machine picot border is not used.

Women sufficiently leisured could vary instructions and designs with total freedom. Those aiming to sell had to take market demand into account.

In 1911 the Gaelic League Exhibition of Art Industries, from Sligo in Ireland, sent a small team of lacemakers to tour the large department stores throughout the U.S.A. and demonstrate, among other things, the techniques of embroidery on machine net. Proceeds from the laces sold were dedicated to 'the education of the masses', in Ireland. At that time, according to the article, both Limerick and Carrickmacross laces were being made extensively 'in the convents all over Ireland' [212].

The Modern Priscilla, Aug.1910.

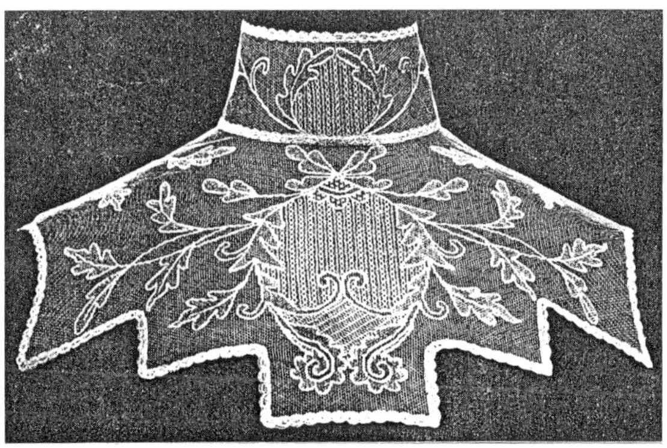

AUSTRALIA

Australia, like North America, benefitted from Irish immigrants, though their numbers were small in comparison (700 between January and September 1913).

The work produced was almost entirely functional and closely related to the needs of everyday life, but there were aspirations towards needle art for its own sake whenever sufficient leisure could be procured.

A retrospective Exhibition of Woman's Domestic Fancywork, *The D'Oyley Show*, held in Sydney in October 1979, was a tardy attempt to display and comment on the crochet, filet, tatting, Teneriffe work, Honiton and Limerick laces that Australians had made during the 19th and early 20th centuries. An article in the Exhibition catalogue [213] combined a vivid picture of the appalling conditions of sweated labour and its rewards, described as pitiful enough 'to make a stone image weep' [214]. The writer also made a spirited attack on a 'male-dominated culture' which denigrated needlework and trivialised women's labour.

Advantages of this quasi-practical pastime were that the patterns were mass-produced and therefore inexpensive, the techniques were within the capabilities of most women, the work could be made to express community attitudes, and it did not rely on literary skills (i.e. on literacy).

As elsewhere, pattern books and journals, including *Weldon's Fancywork Series*, sent in the late 19th century by mail order to isolated homesteads in the backblocks, were undoubtedly well used, and though hundreds of thousands must have been distributed, few of the books, or indeed of the laces themselves, have survived. Thus almost nothing definitive is known about Australian lace and embroidery. Again, as in North America, the records are confused by importations although, with a smaller fashion-conscious society, these form a less significant proportion of the whole.

It is known that, as early as the 1830s, alien European motifs were being replaced in the decorative and applied arts by Australian flora and fauna. Textiles however remained conventional and unaffected by this movement until the turn of the century when vestiges of an art nouveau influence began to appear.

In 1903, 'real Limerick lace' was being made in Melbourne, Victoria, the fashion being for young ladies to embroider the net for their own bridal veils[215].

Lucy Eyres of South Australia composed an attractive and original design based on the long pinnate leaves and leguminous flowers of the sturt desert pea. In the 1920s, Australian patterns began to feature strongly in women's journals, and by 1928 they had attained an impressive 40% of the total in *The Australian Woman's Mirror* [216].

These 'native' designs introduced not only flora and fauna, but representations of Queensland and other landscapes into the laces. Eucalyptus nuts and leaves, kookaburras, golden wattle, koala bears, waratahs (crimson flowered shrubs), and rosellas (brightly coloured parakeets), as well as modifications of aboriginal artifacts, all appeared. One designer, Mary Card (1861-1940), born in Victoria, moved to New York at the age of 56 and set up a company to produce her own patterns there, work which she continued when she moved to Wokingham, England, as the chief designer for Weldon's.

Though her almost total concern was crochet rather than embroidered machine nets, her story indicates the fine mix of influences which can make the geographical origin of individual laces so hazardous to determine.

As in North America, there was no properly commercial embroidered net industry, but some individuals advertised their work in appropriate magazines or, with luck, became regular outworkers for city stores. Dressmakers were, however, well organised. The larger employed over 100 hands, and in 1900 a magnificent bridal gown heavily trimmed with Limerick lace cost £200.

Run work appears most commonly. Its preponderance may perhaps be explained by the influx of defaulting Nottingham lacemakers, deportation to Australia being the recognised punishment for anyone who illegally exported lace machinery. In addition, Nottingham twist-hands who had gone to France in the hope of escaping the industrial espionage and cut-throat rivalry of the city, and of making a quick fortune, took fright at the Revolution of 1848 when Louis Philippe d'Orleans was deposed, fled back to England, were immediately arrested for their defection, and removed to Australia[217]. Livesey, one of the inventors of the Lace Curtain machine, also grew tired of his ill-treatment in Nottingham, and emigrated down under in 1867[218].

Tambour work does not appear in the records, all preoccupation with the hook being, one might suppose, satisfied by crochet. A single tambour example, 'collected in New South Wales' and shown in the D'Oyley exhibition, may well have been

A run embroidery of a native plant, possibly worked at the Convent of the Good Shepherd, Sandy Bay (Courtesy: Olwyn Scott).

imported. Appliqué also is rarely mentioned.

In 1913, Ethel Oates of Melbourne made a net stole run with ribbon bows and starry flowers with thin branching fronds of leaves, for her daughter's wedding, tying it with a lover's knot around her waist. A baby's bonnet which she made in 1922 has a quite different character. The run embroidery, on a heavy net, is augmented by raised satin stitch [219].

More is known of the lace workers of Tasmania, thanks to a generous donation of original pieces by Miss Ada Grey Wilson in 1938, and to the survival at the Tasmanian Museum and Art Gallery, Hobart, of the records of a projected School of Lace Design. A preliminary exhibition to raise funds was sponsored by a select committee of influential ladies, and held on September 15, 1910 [220]. The Exhibition contained several sections, including one for needlerun and one for Carrickmacross

In the minutes of their meeting in 1908, the ladies concluded that 'Lacemaking has now attained a fair degree of proficiency in Tasmania, and the time has come to endow it with a distinctive style. The value of all Arts and Crafts is greatly enhanced when given a national or local character'.

In spite of these admirable sentiments, and the excellent needlework it stimulated, the exhibition in fact lasted only one afternoon, the Tasmanian School of Lace Design was never set up, and with the outbreak of war in 1914 the whole cultural effort ground to a halt.

These endeavours were a natural sequel to the annual exhibitions sponsored by the Arts and Crafts Society of Hobart, founded in 1903. *The Weekly Courier* on September 19 of that year referred to a display of Limerick vestment lace, made by the nuns of The Good Shepherd at Mount St Canice in Lower Sandy Bay, as 'exquisitely worked'.

The 1904 exhibition included examples of Limerick and Carrickmacross from the same source, secular as well as religious, such as collars, handkerchiefs and flounces 'beautifully worked in silk or thread' [thread usually means linen].

By 1913, lace had become the *sine qua non* of fashion. Entire dresses, jackets, tea gowns, jabots, gloves, hats and bows were all made of lace. In addition lace curtains and lace lampshade covers were much in demand - a situation echoing that described in *McCalls* (USA) in 1905.

'Inspired by Tasmania's wild flowers, insects and animals, these Hobart lacemakers formed part of an energetic group of Australians who since Federation [the proclaimed union of the six states in September 1900] had been using native forms as a source for the creation of designs - sponsored by women of genteel society during the second half of the 19th century' [221].

In the 1910 exhibition, 109 of the 500 laces designed and worked for that occasion featured native forms. First prize for design in the needlerun section was won by Ada Grey Wilson, daughter of a former Premier of Tasmania, and a leading organiser. On

A postcard sent by Ada Grey Wilson in 1910.

one of the excellent photographic postcards showing her exhibit, she wrote to her sister in England, on November 7, 1910:

> 'A merry Christmas to you all. I only wish I could be with you to see the fun. I call this lace Adam and Eve eating the Kangaroo Apple, and the flower of it is all around. I won Lady Edeline Strickland's prize. No letter to answer this week from any of you. I am anxious for next week's mail to know you are well. Livy is much better for the change but a little upset by spring weather. I also feel better in my walking powers the last few days. Very much love from always your affectionate sister Ada'.

The fan, 38cm across, was designed by Miss Patty Mault. Tasmania's apple industry was flourishing by 1910, and by mid-September the orchards would have been a sea of white blossom.

The winner of the Carrickmacross section was Miss F. Downing for her 'Ti-tree' [222].

Although silk had been permitted at the exhibitions of 1903 and 1904, Mrs Nevill Jackson's strongly conservative views were now being felt, following the publication of her book in 1900. Coloured lace was considered 'always more or less barbaric', écru 'the colour of dirt', and any tinting of threads an 'outrage of good taste'. This dogma was not so much the outcome of an historical ignorance of the use of colour through the ages, but rather of a tenacious clinging to the Victorian tradition of white linen underwear and white cotton dress trimmings, combined with a revulsion for the blues, pinks, beurres, and buttercups with which the commonly popular Leavers bobbin finings, made of wood [artificial] silk thread, and the Cornely's bright fabrics glinting with coloured glass beads, sequins and 'sparkling', were flooding the market.

Nevertheless a mat run-embroidered with daffodils in yellow and green did appear in the D'Oyley show of 1979. Its date is not given, and it could have been quite recent.

Tasmanian wild flowers, birds, insects, animals, figures and scenes were said to lend themselves well to 'conventional treatment', implying that the old techniques had not been modified, only the designs had changed. Not only were these themes recommended for edgings, berthas and collars, but also such innovative ideas as landscapes and people, which appeared as handkerchief borders and church vestments. Such extreme representationalism would not have been approved of by the commentator on the 1851 Exhibition who referred acidly to 'the command which forbids our making the likeness of anything in heaven above, or in the earth beneath, or in the waters under the earth' [223].

Although the history of embroidered machine nets after their success in 1928 is largely unrecorded, it is known that by 1954 over 50% of the exports of net from Heathcoat's Tiverton plant were taken by the Australian market. This prompted his decision to open up a factory in Melbourne, an act described by Gore Allen as 'an unparalleled adventure' and 'a contemporary act of courage' [224].

NOTES

1. Longfield p.57.
2. *Needlecraft* no.31, p.3.
3. Palliser p.438.
4. Swain, *Embroideress* p.21.
5. Caulfield and Saward p.11.
6. Palliser p.438.
7. Riaño p.273.
8. Moore p.52.
9. Robertson p.23.
10. Maguire p.186.
11. Maguire p.475.
12. Moore p.52.
13. Page pp.485-6.
14. Felkin p.400.
15. Maguire p.475.
16. Ibid.
17. Felkin p.400.
18. Quoted by Maguire pp.474-6.
19. Maguire p.475.
20. Palliser p.441.
21. Maguire p.48.
22. Martin p.69.
23. Lindsey p.30.
24. Boyle p.39.
25. Maguire p.49.
26. Boyle p.84.
27. *The Illustrated Exhibitor*, 1852, p.47.
28. Ibid p.32.
29. Waring p.144.
30. Lindsey p.20.
31. Meredith p.47.
32. Meredith p.24.
33. *The Modern Priscilla*, 1910, p.9.
34. Boyle p.45
35. Fraser p.134.
36. Longfield p.57.
37. Earnshaw pp.37-8.
38. Risley p.103.
39. Meredith p.25.
40. Cole p.663.
41. *Needlecraft* no.31, p.3.
42. Bowles p.101.
43. Lindsey p.11.
44. Cole p.662.
45. Bowles p.101.
46. Brenan pp.431-2.
47. Lindsey p.31.
48. Verhaegen p.190.
49. Bowles p.102.
50. Boyle p.48.
51. Cole p.667.
52. Longfield p.57.
53. Royal Dublin Society.
54. Boyle p.118.
55. Catalogue of the 1910 Exhib p.39.
56. Real Lace, article no.2.
57. Boyle p.58.
58. *Daily Mail* catalogue pp.40, 53, 55-6, 58.
59. Jones p.11.
60. Dublin Exhibition Catalogue, 1853.
61. Lindsey p.11.
62. Boyle p.46.
63. Caulfield, *Irish Lace*, p.27.
64. *The Illustrated Exhibitor*, 1851, p.330.
65. Lindsey p.11.
66. Riaño p.276.
67. Dunlevy p.53.
68. *The Illustrated Exhibitor*, 1851, p.330.
69. *Lace and Embroidery Review*, July 1911, p.16.
70. *McCalls* p.434.
71. *Lace and Embroidery Review*, 1911, p.73.
72. Maguire, W.A. pp.31-2.
73. *Lace and Embroidery Review*, May 1912, p.22.
74. Ibid, April 1912, p.3.
75. *Weldons* no.213, p.3.
76. *Lace and Embroidery Review*, August 1912, p.67.
77. Anderson p.56.
78. *Needlewomen and Needlecraft* pp.17, 25.
79. Wyatt p.485.
80. Treadwin p.31.
81. Wardle p.230.
82. Wyatt vol.II,
83. Earnshaw p.83.
84. Jones pp.9-12.
85. *Needlecraft* no.31, p.3.
86. Pethebridge p.11.
87. Dillmont p.616.
88. Martin p.93.
89. Hénon p.49.
90. Caulfield, *Tambour work*, p.470.
91. *Needlecraft* no.31, p.3.
92. *The Illustrated Exhibitor*, 1852, p.46.
93. Swain, *Tamboured Muslin*, p.21.
94. *Weldons* no.213, p.12.
95. Lindsey p.102.
96. *Lace and Embroidery Review*, Sept 1912, p.71.
97. Boyle p.49.
98. *Encyclopaedia Edinensis*, vol.II, p.458, quoted by Swain, *Tamboured Muslin*.
99. *McCalls*, Feb 1903, p.434.
100. *Needlecraft* no.31, p.3.
101. *Lace and Emb Review*, Nov 1912, p.60.
102. Ibid, Sept.1912, p.67.
103. Caulfield, vol.I, *Appliqué on net*, p.11.

104. *The Art Journal*, 1851, p.208.
105. Brenan p.427.
106. Lindsey p.11.
107. *Needlecraft* no.31, p.3.
108. *Needlecraft* no.28 p.4.
111. *Irish Dept of Agriculture and Tech Instructions Jnl.*, vol XIV, no.1, Oct 1913, p.203; no.2, Jan 1914, p.420; no.4, July 1914, p.804.
112. Dunlevy p.53.
113. Felkin p.426.
114. Earnshaw p.74.
115. *Englishwoman's Dom Mag*, May 1868, p.272.
116. Ibid, June 1869, p.300.
117. Felkin p.426.
118. *Lace and Emb Review*, Nov 1912, pp.40, 74.
119. Felkin p.424.
120. Ibid.
121. Fouriscot p.20.
122. *Dans un Quartier de Nancy*
123. Brazzi p.25.
124. Felkin pp.403, 414.
125. Ferguson p.72.
126. Felkin p.406.
127. Hénon p.41.
128. Felkin p.425.
129. Felkin pp.425-6.
130. *The Art Journal* p.IV**.
131. Hénon pp.54, 89.
132. Hénon p.55.
133. Hénon p.84.
134. Felkin p.426.
135. Felkin p.425.
136. Ibid p.424.
137. Morgan p.75.
138. Felkin p.425.
139. Ibid pp.425-6.
140. Sorber pp.3-4.
141. Verhaegen p.171.
142. Ibid p.175.
143. Ibid p.173.
144. Ibid p.172.
145. Ibid p.171.
146. Ierssel-Ruygers p.43.
147. Van den Eijnden, English text, p.3.
148. Ierssel-Ruygers p.40.
149. *Young Ladies' Jnl*, vol.v-VIII, pp.232, 234.
150. Ierssel-Ruygers p.46.
151. Charles and Pagès, Pt 1, p.148.
152. Sorber pp.52-3.
153. Verhaegen p.173.
154. *Needlecraft* no.59, pp.3, 5.
155. van den Eijnden, English text, p.3.
156. Wardle, *Practical and Needful*, pp.102-3.
157. Hannover p.108.
158. Slettebo p.4.
159. Andersen, p.25.
160. Busche p.80.
161. Felkin p.426.
162. Ilké p.3.
163. *The Illus Ex*, 1852, Pt 2, p.197.
164. Felkin p.424.
165. Iklé p.4.
166. Iklé p.5-6.
167. Maguire, W.A. pp.31-2.
168. *Lace and Embroidery Rev*, Dec 1912, p.32.
169. Felkin p.414;
170. Davanzo-Poli
171. Felkin p.403.
172. Ibid p.426.
173. Ibid.
174. Niedner p.36.
175. *Seligenstader* pp.11? 14-18.
176. Dormoy pl. 39.
177. Czarnecka pp.102, 104, 107, 109-111, 165.
178. Lochhead p.13.
179. *The Illus Ex*, 1851, p.331.
180. Ibid p.332.
181. Swain, *Tamboured Muslin* p.20.
182. Lochhead pp.16-17.
183. Henson pp.429-30, 436-7.
184. Wyatt, text relating to Pl.30.
185. Caulfield, *Tambour work*, p.470.
186. Felkin p.224.
187. Ibid p.225.
188. *Isle of Wight*
189. Palliser p.372.
190. Felkin p.226.
191. *Illustrated Lond News*, May 1956.
192. Moore p.52.
193. Surridge p.42-3.
194. *The Illus Ex*, 1851, p.830.
195. Page, p.486.
196. Beaumont p.195.
197. Page p.487.
198. Page p.484.
199. The Story of Coggeshall, pp.6-7.
200. Page p.487.
201. *Daily Telegraph*, The Englishwomen's Ex at the Central Hall, Westminster, c1930.
202. *Essex Chronicle* Coggeshall Tambour Lace, June 1, 1923.
203. *Daily Graphic* June 8, 1926.
204. Surridge p.42.
205. Page p.486.
206. Swan p.119.
207. Weissman p.5.
208. Swan p.119.
209. Vanderpoel p.107.
210. *Young Ladies' Jnl*, p.232.
211. *The Delineator* p.638.
212. *Lace and Emb Rev*, Nov 1911, pp.43, 76.
213. D'Oyley pp.57-63.
214. Ibid p.60.
215. Ibid p.8.
216. Ibid p.7.
217. Earnshaw p.110.
218. Ibid p.174.
219. *Isaacs p.99.*
220. Dufour p.87.
221. Ibid p.88.
222. Ibid p.89.

READING LIST

Andersen, Ellen and Balsløv, Gugge, Ed. *Berlingske Handarbejdsleksikon: Almuebroderi and Tyllsbroderi.* Copenhagen, 1943-4.

Anderson, Miss. Notes on the Irish Lace and Crochet Industry. *Dept of Agric and Tech Instructions Jnl.*, vol.XIV, no.1, pp.53-8; Oct 1913.

Arneberg, Halfdan. *Norsk Folkekunst Kvinnearbeid.* Oslo, 1949.

The Art Journal Illustrated Catalogue. The Industry of all Nations, 1851. London, 1851.

Ballard, Linda May. The Lilley Collection. *Ulster Folk Life*, vol.35, pp.8-14. 1989.

Beaumont, George Frederick. *A History of Coggeshall.* London, 1890.

Beyer, Otto. *Lehrbuch der Weiblichen Handarbeiten: Tüllstickerei*, pp.292-299. Leipzig.

Bowles, Daly, J. *Glimpses of Irish Industries.* pp.95-112. Ward and Downey, 1889.

Boyle, Elizabeth. *The Irish Flowerers.* Ulster Folk Museum, 1971.

Brazzi, Countess Slocomb di. A Guide to Old and New Lace in Italy. *Chicago Exhibition*, 1893.

Brenan, James. The Modern Irish Lace Industry. *Ireland Industrial and Agricultural*, pp.420-432. 1902.

Busch, Ebba. *Kniplingsimitationer i tyllstraeksyning, Arv og Eje.* Danske Museet, XVI. 1965.

Caplin, Jessie. *The Lace Book.* New York, 1932.

Catalogue of the 1910 Exhibition of the Arts and Crafts Society of Ireland.

Caulfield, S.F.A. and Saward, Blanche. *Encyclopaedia of Victorian Needlework.* Dover reprint of 1887 ed.

Charles, Marguerite and Pagès, Laurent. *Les Broderies et les Dentelles.* 2 vols; Paris, 1906.

Cole, Alan Summerby. Lacemaking in Ireland. *The English Illustrated Magazine*, pp.655-668. June 1890.

Cousine Claire. *Broderies et Dentelles.* Paris, n.d.

Czarnecka, Irena ed. *Polska Sztuka Ludowa.* Warsaw 1958.

Daily Mail Exhibition of British and Irish Lace, catalogue, 1908.

Dans un Quartier de Nancy. *La Brodeuse Perleuse de Lunéville.*

Davanzo-Poli, Doretta, ed. *Cinque secoli de merletti europei: I capolavori.* Burano, 1984.

The Delineator. Carrickmacross Lace. New York, Oct. 1905.

Dillmont, Thérèse de. *Encyclopaedia of Needlework.* Birmingham, 1920s ed.

Donner, Mizi and Schnebel, Carl. *Ich Kann Handarbeiten: Die Tullspitze*, pp.350-357. Berlin and Vienna, 1913.

Dormoy, Marie. *Exposition des Arts Decoratifs, 1925: Dentelles de l'Europe Centrale.* Paris, 1926.

The D'Oyley Show. An Exhibition of Women's Fancy Needlework. Sydney, October, 1979.

Dublin Exhibition Catalogue, 1853.

Dudding, Jean. *Creating Coggeshall Lace.* 1988.

Dufour, Mary. Inspiration from Flora. The Tasmanian Lace Exhib of 1910. *The Australian Antique Collector*, Jan-June, 1983.

Dunlevy, Mairead. Kells Lace. *Irish Arts Review*, pp.50-54. Spring, 1985.

Earnshaw, Pat. *Lace Machines and Machine Laces.* Batsford, 1986.

The Englishwoman's Domestic Magazine. May 1868. June 1869. Ward, Lock and Tyler, London.

Felkin, William. *History of the Machine-wrought Hosiery and Lace Manufactures.* David and Charles reprint of the 1867 ed., 1967.

Fouriscot, Mick. *La Broderie de Lunéville.* Metiers d'Art, p.20. April June 1992.

Fraser, James. *Handbook for Travellers in Ireland.* Dublin, 1859.

Gore Allen, W. *John Heathcoat and his Heritage.* London, 1958.

Hannover, Emil. *Tønderske kniplinger.* Copenhagen, 2nd ed, 1974.

Hardouin, Mme. *Album de Broderie sur Tulle.* Paris, c1920.

Harrods Album de Luxe of Fashions, 1913.

Hélias, Pierre Jakoz. *Coiffes et Costumes de Bretagne.* Ed. Jos. Chateaulin, n.d.

Hénon, Henri. *L'Industrie des Tulles et Dentelles Mécaniques dans le pas de Calais, 1815-1900.* Paris, 1900.

Henson, Gravenor. *History of the Framework Knitters*, 1831. David and Charles reprint, 1970.

Hijlkema, Riet. *National Costumes in Holland.* Amsterdam, 1952.

Ierssel-Ruygers, Trees van. Eine andere Seite der Spitze, Bestickter Tull aus Beveren-Wasse. *Ornamente*, June 1992.

Iklé, Leopold. *Swiss National Exhibition, Zurich 1883: report on Gp.5.* Orell Fussli and Co., 1884.

The Illustrated Exhibitor and Magazine of Art. 2 vols. London, 1852. Ditto, 1851.

Illustrated London News. May, 1956.

Isaacs, Jennifer. *The Gentle Arts.* Ure Smith Press, Australia, 1992

Jones, Laura. *Lace.* Ulster Folk Museum Yearbook,

1969-70.
Jouanic, G. and Hélias, V. *La Broderie en Basse Bretagne.* Ed.Jos. Chateaulin, n.d.
Kay-Shuttlework, Rachel. Laces on Net - other than Filet. *Embroidery*, vol.IV, no.2, 1960.
Koch, Alexander. *Stickereien und Spitzen.* Darmstadt, 1923-4.
Lace and Embroidery Review. New York, 1911-13.
Lindsey, Ben. *Irish Lace: its Origin and History.* Dublin, 1886.
Lochhead, Jessie. *Lace-making in Hamilton.* Hamilton, 1971.
Longfield, Ada. *Guide to the Collection of Lace in the National Museum of Ireland*: Irish Lace, pp.48-65. Dublin, 1937.
Maguire, John Francis. *The Industrial Movement in Ireland as Illustrated by the National Exhibition of 1852.* Cork, 1853.
Maguire, W.A. *Caught in Time.* Friars Bush Press, c1939.
Martin, R.M. *Ireland Before and After the Union.* London and Dublin, 1847 and 1851.
Martin, R.M. *The Manufacture of Lace and Crochet in Ireland.* Irish International Exhibition, Dublin, 1907.
McCalls Magazine. February 1903, p.434. and July, 1905.
Meredith, Susannah. *The Lacemakers.* London, 1865.
The Modern Priscilla. Brown, Nellie, C. Yoke in Limerick Darning. August 1910, p.9.
The Modern Priscilla. Embroidered Net Fichu. Sept.1911, p.19.
Moore, Muriel. Coggeshall Tambour Work. *Embroidery*, vol.VIII, no.2, 1956. pp.52-4 and 61.
Morgan, Roberta. *Alençon Lace.* (Trans of Histoire du Point d'Alençon, Mme G. Despierres, 1886). Aberdeen, 1987.
Murphy, William S. *The Modern Drapery and Allied Trades.* 4 vols. London, 1915.
Needlecraft, no. 28, 1st series. Carrickmacross Lace. Needlecraft Ltd, 1901.
Needlecraft, no.31, 1st series. Limerick Lace, Run and Tambour. Manchester School of Embroidery, c1902.
Needlecraft, no.59. Princess Appliqué Lace. 1906.
Needlewomen and Needlecraft. *Enchanting Design for Dress and Veil.* no.78, pp.17 and 25, c1959.
Niedner, Marie. Tull Durchzug. *Beyer's Handarbeitsbücher*, Band 7. Leipzig, 1922.
Nuova enciclopedia dei lavori femminili. Mani di Fata. Milan, n.d.
Nyten, Anna Maja. *Hemstojd.*
Page, Wm, ed. *A History of the County of Essex.* vol II. The Victoria History of the Counties of England, 1907.
Palliser, Mrs Bury. *History of Lace.* Sampson Low, 1910.
Peeters, Liekens M. *Lierse Kant.* Liers centrum voor textiele Kunsten. n.d.
Pethebridge, Jeanette E. *A Manual of Lace.* Cassell, 1947.
Polak, Ada Buch. Fra Proveduksamlingen. *Vesterlandshe Kunstindustrimuseum Årbok*, 1946-8.
Real Lace and its Story. Nos.1-4. *The Times*, Dec 1, 3, 6, 9, 1904.
Riaño, Juan Facundo. *Spanish Industrial Art.* London, 1875.
Risler, Christine. *Machine Embroidery.* Studio Vista, 1973.
Robertson, Ian. *Ireland.* Blue Guide, A and C Black, 1992.
Rome-Verbeylen, Greet. *Lierse Kant, Oud en Nieuw.* c1990.
Royal Dublin Society, Historical Notes. 1980s.
Sandtner, Hilda. *Tull Kompositionen.* Stuttgart, 1980.
Schöner, Friedrich. *Spitzen*: Tullspitze, pp.168-173. Leipzig, 1988.
Seligenstader Perlenstickerei, pp.11-18. Landshaftmuseum, Seligstad, 1981.
Sheppard, Harry. An Old and Once-Famous Island Industry. *Isle of Wight County Press*, vol.XVII, no.884. c1901.
Slettebo, Tøve. Høll'er-Høll'klae'r og Høll'kasten. *Kniplebrevet,* no.8. May 1987.
Sorber, Frieda. *Uit Tulle, Garen en Kralen, 160 Jaar Borduurwerk te Lier.* Antwerp, 1992.
The Story of Coggeshall 1700-1900, Lace pp.6-7. W.E.A., 1951.
Surridge, Edith M. *Souvenir Brochure, 1951.* Coggeshall Lace, pp.42-3.
Swain, Margaret H. *The Flowerers.* Chambers, 1955.
Swain, Margaret H. Tamboured Muslin in Scotland. *Embroidery*, vol 14, no.1. Spring 1963; pp.20-21.
Swan, Susan Burrows. *The Winterthur Guide to American Needlework.* Crown Pub, 1976.
Treadwin, Mrs. *Antique Point and Honiton Lace.* Ward Lock and Tyler, 1874.
Tüll-Durchzug und Stickerei. Beyer. Leipzig, n.d.
Vanderpoel, Emily Noyes. *American Lace and Lacemakers.* Yale University Press, 1924.
Van den Eijnden-van Raaij, A.J.M. *De West-Brabantse Muts.* Netherlands, 1992.
Verhaegen, Pierre. *La Dentelle et la Broderie sur Tulle.* Brussels, 1902.
Vydra, J. and Sedlackova-Serbouskova, M., ed. *Dentellière en Tchécoslovaquie.* Prague, 1930.
Wardle, Patricia. *Practical and Needful: Dutch Lace Schools 1850-1940.* Catalogue of an Exhibition at the Museum Boymans-van Beuningen, Rotterdam, 1992.
Wardle, Patricia. *Victorian Lace.* Herbert Jenkins, 1968.
Waring, John B. Masterpieces of Industrial Art at the International Ex. 1862. Vol.I. London, 1863.
Weissman, Judith R and Lavitt, Wendy. *Labors of Love, America's Textiles and Needlework, 1650-1930.* New York, 1987.
Weldon's Ladies' Journal. March 1915.
Weldon's Practical Limerick Lace, 1st series, vol.18, no.213, p.3, n.d.
Wyatt, Sir Matthew Digby. *The Industrial Arts of the Nineteenth Century at the Great Exhibition, 1851.* 2 vols. London, 1853.
Young Ladies' Journal. Embroidery on Net. vol.VIII, p.538. 1871.
Zulawsky, Elemérné. *Muskátli.* Budapest, 1931.

INDEX